AF389349

Advanced Materials in 3D/4D Printing Technology

Advanced Materials in 3D/4D Printing Technology

Editors

Houwen Matthew Pan
Eric Luis

MDPI • Basel • Beijing • Wuhan • Barcelona • Belgrade • Manchester • Tokyo • Cluj • Tianjin

Editors
Houwen Matthew Pan
School of Chemistry,
Chemical Engineering and
Biotechnology
Nanyang Technological
University
Singapore
Singapore

Eric Luis
Faculty of Medicine
Macau University of Science
and Technology
Taipa
Macau

Editorial Office
MDPI
St. Alban-Anlage 66
4052 Basel, Switzerland

This is a reprint of articles from the Special Issue published online in the open access journal *Polymers* (ISSN 2073-4360) (available at: www.mdpi.com/journal/polymers/special_issues/Adva_mater_3D_4D_print).

For citation purposes, cite each article independently as indicated on the article page online and as indicated below:

LastName, A.A.; LastName, B.B.; LastName, C.C. Article Title. *Journal Name* **Year**, *Volume Number*, Page Range.

ISBN 978-3-0365-5112-8 (Hbk)
ISBN 978-3-0365-5111-1 (PDF)

Contents

About the Editors

Houwen Matthew Pan

Dr. Pan Houwen Matthew received his PhD in Biomedical Engineering from National University of Singapore (NUS), 2016. He is currently a Senior Research Fellow at Nanyang Technological University (NTU), working with a multidisciplinary team of researchers including engineers, cell biologists, and polymer chemists on innovations in the area of additive manufacturing. His research interests include Bioprinting, Biomaterials, Bioencapsulation, and 3D and 4D Printing Technologies. As a young researcher, he has authored peer-reviewed articles in prestigious journals, book chapters, patents, and conference papers. He is a recipient of the NUS Research Scholarship (2011), DAAD Research Scholarship (2014), and DTU's H.C. COFUND (2019).

Eric Luis

Orthopedic sports medicine has tremendous potential for growth in Indonesia. One of the pioneers of this growth, who has played an active part in shaping the orthopedic climate of Indonesia, is Dr Guntur Eric Luis. He also has the experience of carrying out several trauma surgeries in the island nation.

Orthopedic trauma surgery makes use of highly specialized techniques. He has experience working at major hospitals such as the Sahirman Sahid Memorial Hospital in Jakarta. Dr Guntur is a reputed orthopedic specialist who has represented his country at numerous international conferences. He continues to be involved with every new development that takes place in the orthopedic field.

Preface to "Advanced Materials in 3D/4D Printing Technology"

Advances made in 3D/4D printing have opened up new avenues for innovation in regenerative medicine, aerospace, manufacturing, construction, and electronics. The current state of the art for 3D/4D printing technologies includes VAT photo-polymerization, material jetting, binder jetting, material extrusion and powder bed fusion. To match the development made in printing technologies, a myriad of printing materials with advanced functionalities have also been invented. The wide array of advanced 3D/4D printing materials include cytoprotective and cytocompatible bioinks, flexible and conductive printing inks, modular bioinks, stimuli-responsive polymer inks, hydrogel composite systems, adaptive soft materials, shape memory materials, and heat-curable silicone printing ink.

Combining cutting-edge 3D/4D printing technologies with advanced materials have given rise to disruptive breakthroughs in research such as complex designs in flexible electronics, the biofabrication of humans-sized organs, adaptive soft robotics, personalized prosthesis and biomedicine. However, significant challenges still lie ahead and problems relating to material selection, multimaterial printing, print scalability, material processability, structure integrity and stability still need to be solved before we can adopt 3D/4D printing technologies on a much larger scale.

The goal of this book is to showcase state-of-the-art reviews, original research articles and communications and letters from leaders in the field of 3D/4D printing. The focus will be on 3D/4D printing materials with novel and/or advanced functionalities, novel applications of 3DP material, and material synthesis and characterization techniques.

Houwen Matthew Pan and Eric Luis
Editors

polymers

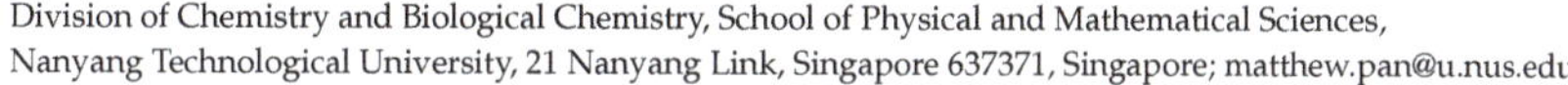

Editorial

Advanced Materials in 3D/4D Printing Technology

Houwen Matthew Pan

Division of Chemistry and Biological Chemistry, School of Physical and Mathematical Sciences, Nanyang Technological University, 21 Nanyang Link, Singapore 637371, Singapore; matthew.pan@u.nus.edu

Citation: Pan, H.M. Advanced Materials in 3D/4D Printing Technology. *Polymers* **2022**, *14*, 3255. https://doi.org/10.3390/polym14163255

Received: 26 July 2022
Accepted: 9 August 2022
Published: 10 August 2022

Publisher's Note: MDPI stays neutral with regard to jurisdictional claims in published maps and institutional affiliations.

Advances made in 3D printing have opened new avenues for innovation in dental, aerospace, soft robotics, thermal regulation, and flexible electronic devices. Current state-of-the-art for 3D printing technologies include fused deposition modelling (FDM), selective laser sintering (SLS), direct ink writing (DIW), and stereolithography (SLA). Recently, 4D printing was borne from combining 3D printing technology with stimulus-responsive printing ink. Four-dimensional-printed structures evolve as a function of time and exhibit intelligent behavior such as self-assembly and actuation. To match the rapid development of 3D and 4D printing technologies, a myriad of printing materials with advanced functionalities have also been invented.

In this Special Issue, original research articles focusing on the study and creation of advanced materials for 3D/4D printing technologies have been collected. Dynamic, bio-inspired spider silks were successfully fabricated via the 4D printing of shape memory polyurethane. The shape morphing behaviors of bio-inspired spider silks were programmable via pre-stress assemblies enabled by 4D printing. Copper complexes were synthesized as novel visible light photoinitiators for the free radical photopolymerization of acrylates at the safe wavelengths of 405 and 455 nm. The new photoinitiators were used for direct laser write experiments. A facile, water-based 3D-printable ink with sustainable nanofillers and cellulose nanocrystals (CNCs) was developed for DIW to 3D print strong PI/CNC composite aerogels for advanced thermal regulation. Dental forceps were 3D printed via fused filament fabrication (FFF) and continuous fiber reinforcement (CFR) technologies with comparable properties to conventional devices in terms of extraction force. Three layers of corrugated PVC gel artificial muscles were 3D printed using a multi-material, integrated direct writing method and displayed good actuating performance with potential applications in soft robotics and flexible electronic devices. Using FDM technology, moulds were prepared for the casting of carbon fiber components. By combining FDM with chemical smoothing, low-cost moulds with a high quality surface finish were successfully produced. Micro-/nanostructures were 3D printed on surfaces with different wetting properties, and biomolecules with different functionalities were integrated into the base polymer ink. The ability to incorporate binding tags to achieve specific interactions between relevant proteins and the fabricated micro-/nanostructures, without compromising mechanical properties, paves the way for numerous biological and sensing applications.

Additionally, in-depth research of 3D printed components for specific applications has also been included. Studies were carried out on the sound reflection behavior of four different types of 3D-printed, open-porous polylactic acid (PLA) material structures, which found that sound reflection behavior was strongly affected by the type of material structure, excitation frequency, the total volume porosity, the specimen thickness, and the air gap size. Polyphenylsulfone polymer powders were evaluated as fire-resistant materials for processing by SLS. The microstructural evaluations and the mechanical property results indicate sufficient intra- and inter-layer consolidation together with reasonable tensile property responses. Lastly, a homogeneous structural element was generated that accurately represents the behavior of FDM-processed materials, using a representative volume element (RVE). The homogenization summarizes the main mechanical characteristics of the actual 3D printed structure, enabling the accurate engineering analysis of the final 3D-printed structure.

Finally, the Editors express their appreciation to all contributors to this Special Issue and hope that it inspires and guides scientists to achieve greater progress in the area of Advanced Materials for 3D and 4D Printing.

Conflicts of Interest: The authors declare no conflict of interest.

Article

Bio-Inspired 4D Printing of Dynamic Spider Silks

Guiwei Li [1,2], Qi Tian [1], Wenzheng Wu [1], Shida Yang [1], Qian Wu [2], Yihang Zhao [1], Jiaqing Wang [1], Xueli Zhou [2], Kunyang Wang [2], Luquan Ren [1,2], Ji Zhao [1,3] and Qingping Liu [2,*]

[1] Advanced Materials Additive Manufacturing ((AM)2) Lab, School of Mechanical and Aerospace Engineering, Jilin University, Changchun 130025, China; ligw@jlu.edu.cn (G.L.); tianqi1819@mails.jlu.edu.cn (Q.T.); wzwu@jlu.edu (W.W.); yangsd9919@mails.jlu.edu.cn (S.Y.); zhaoyh9919@mails.jlu.edu.cn (Y.Z.); wjq9919@mails.jlu.edu.cn (J.W.); lqren@jlu.edu.cn (L.R.); jzhao@jlu.edu.cn (J.Z.)
[2] Key Laboratory of Bionic Engineering (Ministry of Education), Jilin University, Changchun 130022, China; wuqian21@mails.jlu.edu.cn (Q.W.); xlzhou@jlu.edu.cn (X.Z.); kywang@jlu.edu.cn (K.W.)
[3] School of Mechanical Engineering and Automation, Northeastern University, Shenyang 110004, China
* Correspondence: liuqp@jlu.edu.cn

Abstract: Spider silks exhibit excellent mechanical properties and have promising application prospects in engineering fields. Because natural spider silk fibers cannot be manufactured on a large scale, researchers have attempted to fabricate bio-inspired spider silks. However, the fabrication of bio-inspired spider silks with dynamically tunable mechanical properties and stimulation–response characteristics remains a challenge. Herein, the 4D printing of shape memory polyurethane is employed to produce dynamic bio-inspired spider silks. The bio-inspired spider silks have two types of energy-absorbing units that can be adjusted, one by means of 4D printing with predefined nodes, and the other through different stimulation methods to make the bio-inspired spider silks contract and undergo spiral deformation. The shape morphing behaviors of bio-inspired spider silks are programmed via pre-stress assemblies enabled by 4D printing. The energy-absorbing units of bio-inspired spider silks can be dynamically adjusted owing to stress release generated with the stimuli of temperature or humidity. Therefore, the mechanical properties of bio-inspired spider silks can be controlled to change dynamically. This can further help in developing applications of bio-inspired spider silks in engineering fields with dynamic changes of environment.

Keywords: 4D printing; bio-inspired spider silks; adjustable mechanical properties; shape morphing; stimulus response

Citation: Li, G.; Tian, Q.; Wu, W.; Yang, S.; Wu, Q.; Zhao, Y.; Wang, J.; Zhou, X.; Wang, K.; Ren, L.; et al. Bio-Inspired 4D Printing of Dynamic Spider Silks. *Polymers* **2022**, *14*, 2069. https://doi.org/10.3390/polym14102069

Academic Editors: Houwen Matthew Pan and Eric Luis

Received: 18 April 2022
Accepted: 17 May 2022
Published: 19 May 2022

Publisher's Note: MDPI stays neutral with regard to jurisdictional claims in published maps and institutional affiliations.

1. Introduction

The spider web has excellent mechanical properties [1–3], combines high tensile strength with high ductility, and simultaneously exhibits excellent impact resistance [4,5]. The tensile strength of silk fibers is comparable to that of steel, and the elasticity of natural spider silk is almost the same as that of rubber [6]. Spider silk has good application prospects in various fields, such as textiles [7,8], biomedicine [9,10], military affairs [11], and aerospace [12]. However, spiders are carnivores—they do not like to live in groups and even kill each other—rendering it difficult for spiders to be bred on a large scale [13,14]. Therefore, producing natural spider fibers on an industrial scale is challenging. Thus, the manufacturing and molding of bio-inspired spider silk has become a popular research topic.

Numerous studies have focused on the manufacture of bio-inspired spider webs. Vendrely et al., through genetic engineering methods, employed two biotechnological production strategies to successfully apply spider silk protein for the production of spider silk [15]. Venkatesan et al. fabricated engineered major ampullate spidroin 2 (eMaSp2) fiber by introducing foreign N- and C-terminal domains, and demonstrated that the fiber has good shape memory effects triggered by humidity factors and the ability to restore stress [16]. Inspired by the microstructure of spider silk, Chan et al. effectively utilized the localized β-sheet domains in the amorphous network considering the shortcomings of existing processing technology, and synthesized a super composite material with a spider silk-like "amorphous/β-folding" microstructure [17]. Dou et al. self-assembled homemade hydrogel fibers to create artificial spider silk. This faux spider silk fiber consists of a plastic sheath placed around an elastic core, which results in spider silk-like strength and stretchability [18]. When used in impact reduction applications, the bio-inspired spider silk fiber has a negligible rebound, allowing it to be applied to kinetic energy cushioning and shock reduction. Zou et al. successfully designed a transparent composite material with super-impact resistance using the SBHL strategy. Zou et al.'s biologically inspired spider webs are able to dissipate a significant amount of energy, as they absorb impact and seize the projectile like a natural spider web [19]. Qin et al. created spider-web mimics composed of elastic thin filaments and investigated the mechanical response of the elastic web under a variety of loading conditions [20].

The studies discussed above focused on static bio-inspired spider silk to create artificial bio-inspired fibers with mechanical properties similar to those of natural spider silk. However, natural spider silk has a hierarchical structure and multiple properties, such as humidity/water responses, light transmittance, and shape memory [21]. The actual application conditions for bio-inspired spider silk often change dynamically. For the impact resistance of spider silk it is often necessary to adapt to dynamic adjustments. We used 4D printing to create a spider silk structure. Based on 3D printing, 4D printing technology has an additional fourth dimension: the time dimension [22,23]. Thus, 4D-printed objects can change their shape/structure and mechanical properties over time after being stimulated [24–26]. Therefore, bio-inspired spider silk obtained by 4D printing can be stimulated by different conditions, such as temperature and humidity, to achieve dynamic changes. By controlling the difference between the stimulus and preset printing conditions, the mechanical properties of the bio-inspired spider silk can be controlled to change dynamically, which can widen the application prospects of bio-inspired spider silk in the engineering field.

2. Materials and Methods

2.1. Mechanical Simulation Analysis

We chose CATIA V5R21 modeling software to establish the spider silk model and used Workbench 2021 R1 software for the finite element simulation analysis of the model. In the finite element analysis, one end of the model was set as a fixed support and the other end was loaded with a tension of 1 N. For the material setting of the model, the density was set to 1250 kg/m^3, the Young's modulus was 1.4 MPa, and the ultimate tensile strength was 10 MPa. This solved the equivalent effect diagram of the model.

2.2. Printing the Bio-Inspired Spider Silk Samples

The raw material selected for printing was polyurethane (SK96B09, Kyoraku Co., Ltd., Tokyo, Japan). For the 4D printing of bio-inspired spider silk, the following process parameters were set: the bottom layer printing speed was 300 mm/min, the top layer printing speed was 7000 mm/min, the layer height was 0.2 mm, the nozzle diameter was 0.4 mm, the extrusion magnification was 1.2, the extrusion line width was 0.4 mm, the print nozzle temperature was 180 °C, and the substrate temperature was 20 °C. This printed the required bio-inspired spider silk with different nodes. At the same time, more than 10%, 30%, 50%, and 70% filling densities were employed.

2.3. Mechanical Property Characterization Experiment

For the mechanical property characterization experiments, the electronic universal testing machine (UTM6104, Shenzhen Sansi Longitudinal and Hori-zontal Technology Co., Ltd., Shenzhen, China) was used. We mainly conducted tensile property characterization experiments on bio-inspired spider silk to investigate the effects of tensile speed, stimulus time, and spider silk humidity on the tensile properties of bio-inspired spider silk. First, we investigated the tensile properties of the unstimulated sample and the samples stimulated for 1 min at 60 °C with different densities of nodes at a tensile speed of 30 mm/min. Second, we investigated the tensile properties of the unstimulated sample and the samples stimulated for 2 min at 60 °C with the same density of nodes at different tensile speeds. Finally, the tensile properties of bio-inspired spider silks with different stimulation times were investigated at a tensile speed of 30 mm/min. A universal mechanical testing machine was used to test the tensile properties of bio-inspired spider silk at a constant tensile speed until the specimens were pulled off. The stretching speed was chosen to be 30 mm/min and 30, 50, 70, 90, 150, and 210 mm/min when the effects of different stretching speeds were explored. For each set of tensile property characterization experiments, at least three mechanical samples were tested to determine the yield point and the corresponding elastic limit force, as well as the fracture point and the corresponding maximum tensile force. The elongation at the break of the bio-inspired spider was obtained by dividing the value of the deformation displacement at the break by the original length of the sample. The average value was recorded, and the standard deviation was marked on the histogram. Then, the sample test result closest to the average value was selected, according to which the force–tension displacement curve of the bio-inspired spider was drawn in the same way.

2.4. The DSC and TGA Experiments

The DSC (DSC25, TA Instruments, New Castle, DE, USA) experiments were conducted to test the difference in the degree of glass transition of bio-inspired spider samples with different moisture absorption, and the temperature rise rate was set to 5 °C/min from 0 to 200 °C. The TGA (TGA55, TA Instruments, New Castle, DE, USA) experiments were conducted to verify the moisture content absorbed by the bio-inspired spider silk sample after immersion for different times. The immersion times were chosen to be 4, 8, and 16 min at 60 °C and 30 and 60 min at 40 °C. The heating rate of the TGA experiment was 5 °C/min in the range of 30 °C to 280 °C.

2.5. Bio-Inspired Spider Web Stimulation Experiments

The webs were stimulated with 150 W infrared light from a distance of 35 mm to observe the shrinkage and deformation of both the unstimulated webs and high-temperature-stimulated webs. The infrared light (MYK020, Shanghai Yafu Lighting Electric Co., Ltd., Shanghai, China) was equipped with a Philips infrared bulb (PAR38E 150W 230V, Amsterdam, The Netherlands). Next, infrared light and humidity were used to stimulate the high-temperature-stimulated webs to compare the differences between the two stimulation methods. The high-temperature stimulation of the bio-inspired spider silk was performed using water at a constant temperature of 80 °C. Simulations were performed three times; each time, after 20 s of stimulation, the bio-inspired spider silk was quickly removed, gently straightened, and stimulated again three times. The humidity stimulation method involved soaking and humidifying the bio-inspired spider silk with constant-temperature water at 25 °C.

2.6. Impact Resistance Experiment and Invisible Interception Experiment

Application experiments were also conducted, including an impact resistance experiment. Bio-inspired spider webs 100 × 100 mm in size were spaced 12.5 mm apart, and the impact experiments were conducted using a glass ball and tennis ball. During the impact test, it was ensured that the test was conducted under windless conditions. The vertical distance between the center of the ball and the net surface was 30 mm for both balls. No additional force was attached to the ball during its fall, and the impact was taken in a free-fall manner. The diameter of the glass balls was 25 mm, and the weight of the tennis balls was 55.638 g. Photographs were taken using an industrial camera (OSG030-815UM, Shenzhen Yingshi Technology Co., Ltd., Shenzhen, China) equipped with a 2 million 1/2″ target surface industrial lens (HF-Z0412, Shenzhen Yingshi Technology Co., Ltd., Shenzhen, China) to demonstrate the feasibility of the bio-inspired spider silk applied to stealth interception. The dynamic bio-inspired spider web was used to demonstrate the feasibility of intercepting impacts.

Next, bio-inspired spider web stealth interception experiments were performed, in which the same background was chosen and spider silk with different stimulation levels was used to weave the web; the difference in the stealthiness of the bio-inspired spider web was observed under different stimulation levels, and the feasibility of its application to stealth interception was analyzed. The bio-inspired spider silk was stimulated once, three times, and five times at a constant temperature of 80 °C. The samples were removed quickly after each stimulation for 20 s and gently straightened so that the corresponding number of repetitions could obtain three types of bio-inspired spider silk with different stimulation levels. Notably, bio-inspired spider silk was removed more slowly and not straightened during the fifth stimulation to maintain the spiral effect.

3. Results and Discussion

3.1. Bio-Inspired 4D Printing of Dynamic Spider Silks

A biological prototype of spider silk is shown in Figure 1a. It is both elastic and rigid, and can be considered to be composed of two parts: a semi-amorphous domain and a nanocrystal [19]. When bio-inspired spider silk is stretched by force, the semi-amorphous domain part inside the bio-inspired spider silk becomes elastic, and the spiral part inside it is stretched by force. It can absorb energy when the bio-inspired spider silk is stressed, which provides the bio-inspired spider silk with good impact resistance. Based on this bio-inspired principle, as shown in Figure 1b, we programmed the process parameters. It is possible to print two layers at different speeds and then bury the preset nodes. Under stimulus, the linear part in the middle is spirally deformed owing to the difference in the shrinkage of the double-layer material, while the preset nodes retain their bending at the nodes so that a double preset can be achieved. The design enables the bio-inspired spider silk to achieve spiral deformation, as shown on the right side of Figure 1b, under certain stimulation conditions.

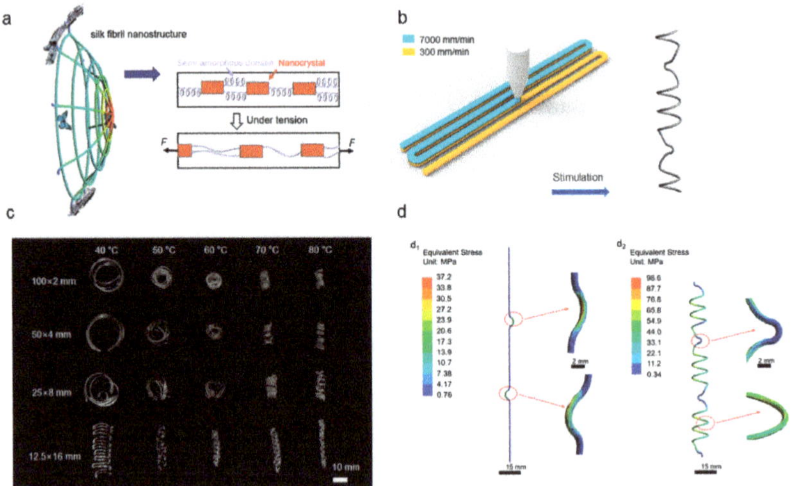

Figure 1. Schematic of bio-inspired spider webs catching prey when impacted (**a**); printing principle demonstration (**b**); bio-inspired spider silk is deformed with various nodes by stimulated shrinkage (**c**); simulation analysis of stress applied on two types of bio-inspired spider silk (**d**).

Stimulation experiments were performed on the printed spider silk with different nodes. The stimulation experiments were performed under the specific conditions mentioned in Figure 1c, where bio-inspired spider silk with different nodes was placed in constant-temperature water for the same amount of time. The time selected was 1 min, the temperature gradient was 10 °C, and the temperature range was 40 to 80 °C. The density of nodes of the bio-inspired spider silk differed; 100 × 2 indicates that the bio-inspired spider silk had a total of two segments, each with a length of 100 mm. The preset node 50 × 4 means that the bio-inspired spider silk consisted of four segments, each with a length of 50 mm and three preset nodes. Thus, under the premise that the density of the nodes of the bio-inspired spider silk differs, the total length of the selected segments is the same (200 mm), and the influence of different node densities and different temperatures on the stimulation of the bio-inspired spider silk can be observed. Next, mechanical property characterization experiments were performed. The force simulation of the bio-inspired spider silk is shown in Figure 1d; the figure shows that when the printed spider silk with nodes is stretched by force, because the equivalent force at the node is significantly greater than that at a smooth straight line, the node will lead the force expansion to offset part of the pull impact, as shown in Figure 1(d_1). Similarly, the mechanical simulation of the stimulated spider silk is shown in Figure 1(d_2); at this time, because the linear part in the middle of the original spider silk is bent spirally, the maximum equivalent force appears at the relatively uniform spiral. When the bio-inspired spider silk is stimulated and stretched by the force, the spiral part first unfolds due to the stretching force to offset part of the impact, and then the nodes are expanded by the force.

We prepared the bio-inspired spider silk with different nodes by pre-programming the process parameters. As shown in Figure 1c, the higher the temperature during stimulation, the more pronounced the spiral bending phenomenon of the bio-inspired spider silk. On the basis of the original bending of the nodes being retained, the linear part of the printed bio-inspired spider silk also undergoes spiral deformation. Additionally, when the bio-inspired spider silk is subjected to tensile deformation, the spiral deformation region generated by the linear part will be the first to be subjected to a larger value of equivalent force and unfold to resist the tensile impact. In this way, together with the bending at the original nodes, a double pre-setting is achieved to enhance the tensile impact resistance of the bio-inspired spider silk. This method uses a double-preset approach to provide a better energy-absorption effect.

3.2. Effect of Energy-Absorbing Units on the Tensile Properties of Bio-Inspired Spider Silks

The size and density of energy-absorbing units affect the stress distribution of the bio-inspired spider silk sample during stretching, which affects the dynamics of the mechanical properties of the sample under different conditions. Two types of energy absorption units exist in the samples. The density and size of the nodes are preset when printing the bio-inspired spider silk sample as the preset energy-absorbing unit. The spiral deformation of the sample after stimulation can be regarded as a large node as the additional energy absorption unit. The energy-absorbing unit is mainly controlled by temperature during stimulation, which can control the degree of spiral deformation. Thereby, the influences of the density, the size of nodes, and the stimulation temperature on the tensile properties of the bio-inspired spider silk samples were investigated.

3.2.1. Effect of Node Density

The density of the nodes significantly affected the tensile properties of the bio-inspired spider silk. Four types of spider silk with different densities of nodes were used to conduct tensile experiments under both unstimulated and stimulated conditions to investigate specific difference phenomena. Figure 2a–d show the stretching of the four unstimulated raw silk materials. As shown in Figure 1(d_1), the mechanical simulation results of the bio-inspired spider silk sample showed that when the raw bio-inspired spider silk sample was stretched by force, the equivalent force value at the node was large, and the first to be damaged. When the density of nodes was small, for example, in the raw bio-inspired spider silk sample of 100 × 2 with only one preset node, the stress was concentrated which first damaged the node, and the elastic limit force of the bio-inspired spider silk was small.

When the density of the nodes was large, for example, the raw bio-inspired spider silk of 50×4 and 25×8, which had three and seven nodes to share the corresponding tensile stress, respectively, the stress first damaged these nodes. However, when the density of the nodes was large for the 12.5×16 spider silk sample with 15 nodes, the stress concentration area was too dense, affecting the straight section; thus, the elastic tensile limit and maximum tensile force of the unstimulated sample were reduced.

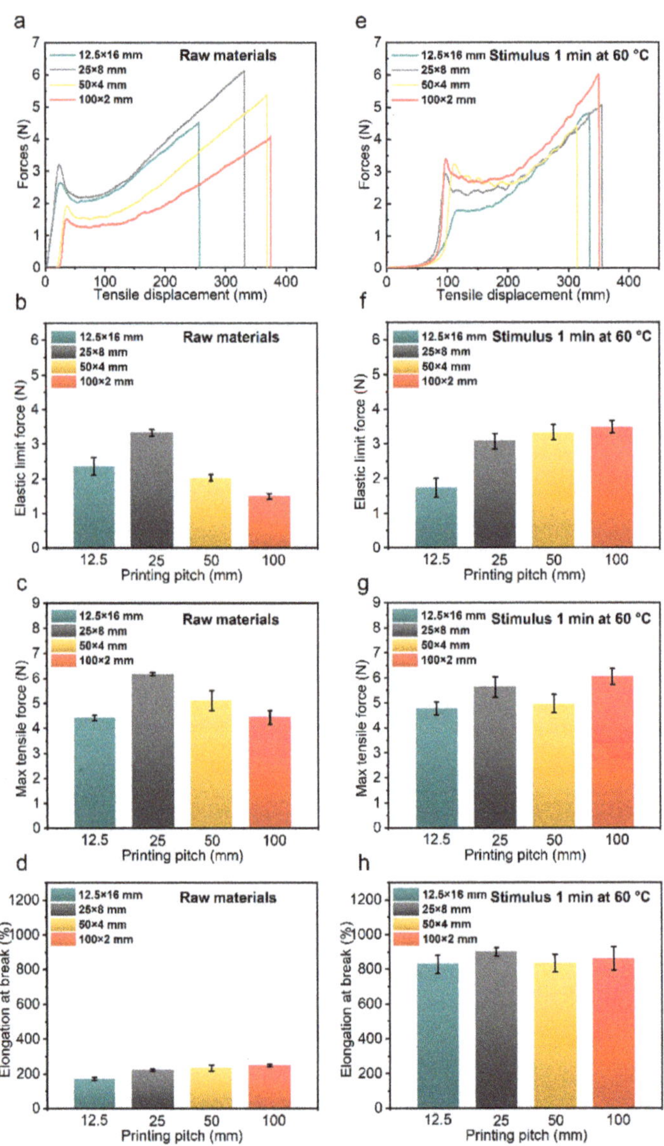

Figure 2. Force and tensile displacement curves of the unstimulated bio-inspired spider silk samples with different nodes (**a**); elastic limit force (**b**); maximum tensile force (**c**); elongation at break (**d**). Force versus tensile displacement curves of bio-inspired spider silk samples with different nodes stimulated at 60 °C for 1 min (**e**); elastic limit force (**f**); maximum tensile force (**g**); elongation at break (**h**).

The unstimulated biomimetic silk had only the first type of energy absorbing units with the preset nodes. When the sample was subjected to tension, the nodes were subjected to a large value of equivalent force. However, the stress concentration at the nodes also affected the linear part. When the linear section was short (e.g., 12.5 mm), the elastic tensile limit and the maximum tensile force of the unstimulated sample was reduced. Therefore, adopting appropriate nodal densities (e.g., 50 × 4 and 25 × 8) can improve the tensile mechanical properties of bio-inspired spider silk.

We further investigated the differences in the tensile properties of the stimulated bio-inspired spider silk, as shown in Figure 2e–h. The elongation at the break of the bio-inspired spiders with different densities of nodes after stimulation remained the same, and there was no significant difference in the maximum tensile force value; only the elastic limit force value increased with a reduction in the density of nodes of the raw bio-inspired spider material. This is because a smaller density of nodes results in a longer linear segment, as shown in Figure 1(d_2), and the equivalent maximum stress in the bio-inspired spider silk after stimulation when it is stretched by force occurs at the bend formed by the contraction of the linear segment after stimulation. In contrast, the 100 × 2 bio-inspired spider silk had the longest original straight segment and the most bending spirals formed at the same conditioned stimulation; thus, the elastic limit force of the bio-inspired spider silk with the lowest density of these node became the largest. This shows that the difference in the density of nodes causes a significant difference in the tensile properties of bio-inspired spider silks.

Two types of energy absorbing units are present in the stimulated bio-inspired spider silk sample. The spiral curved area formed by the stimulated bio-inspired spider silk sample acted as the second type of energy absorbing unit and was the first to be stretched by the force. Bio-inspired spider silks with different node densities have different numbers of the first type of energy absorption units—preset nodes—and they have different lengths of linear regions, which form the second type of energy-absorption units. The combined effect of the two types of energy-absorbing units leads to significant differences in the tensile properties of the stimulated bio-inspired spider silk samples.

3.2.2. Effect of Node Size

In the previous section, we showed that the densities of nodes significantly affected the tensile properties of bio-inspired spider silk, and the size of the nodes should also affect the tensile properties of bio-inspired spider silk. The filling density parameter during printing was set to 10%, 30%, 50%, and 70% to obtain four types of bio-inspired spider silk structures with different node sizes. Among them, the bio-inspired spider silk node was the largest at the 10% filling density and the smallest at the 70% filling density. As shown in Figure 3a–d, tensile experiments were performed on unstimulated raw bio-inspired spider silk materials with different filling densities. The nodes of the samples printed at 70% filling density were the smallest, and extremely small nodes lead to more obvious stress concentration, which causes the bio-inspired spider silk to fracture in the middle of stretching; the elongation at the breakage of this bio-inspired silk sample was the smallest. The maximum tensile force of the unstimulated biomimetic 4D-printed bio-inspired spider silk sample at this point decreased with the size of the nodes, as shown in Figure 3c, which could also be because smaller nodes lead to a more pronounced stress concentration phenomenon which, in turn, reduces the maximum tensile force. Next, the raw bio-inspired spider silk material with different filling densities was stimulated and subjected to tensile experiments, as shown in Figure 3e–h. At this point, the bio-inspired spider silk had two types of energy-absorbing units after stimulation. The analysis performed under this condition is shown in Figure 3f. The elastic limit force of the bio-inspired spider silk sample with 70% filling density was slightly smaller, which may have been caused by the small nodes and obvious stress concentration. The fracture tensile rate of the bio-inspired spider silk sample under this condition decreased with a reduction in node size; the value of elongation at the break of the sample printed at 70% filling density was also the smallest.

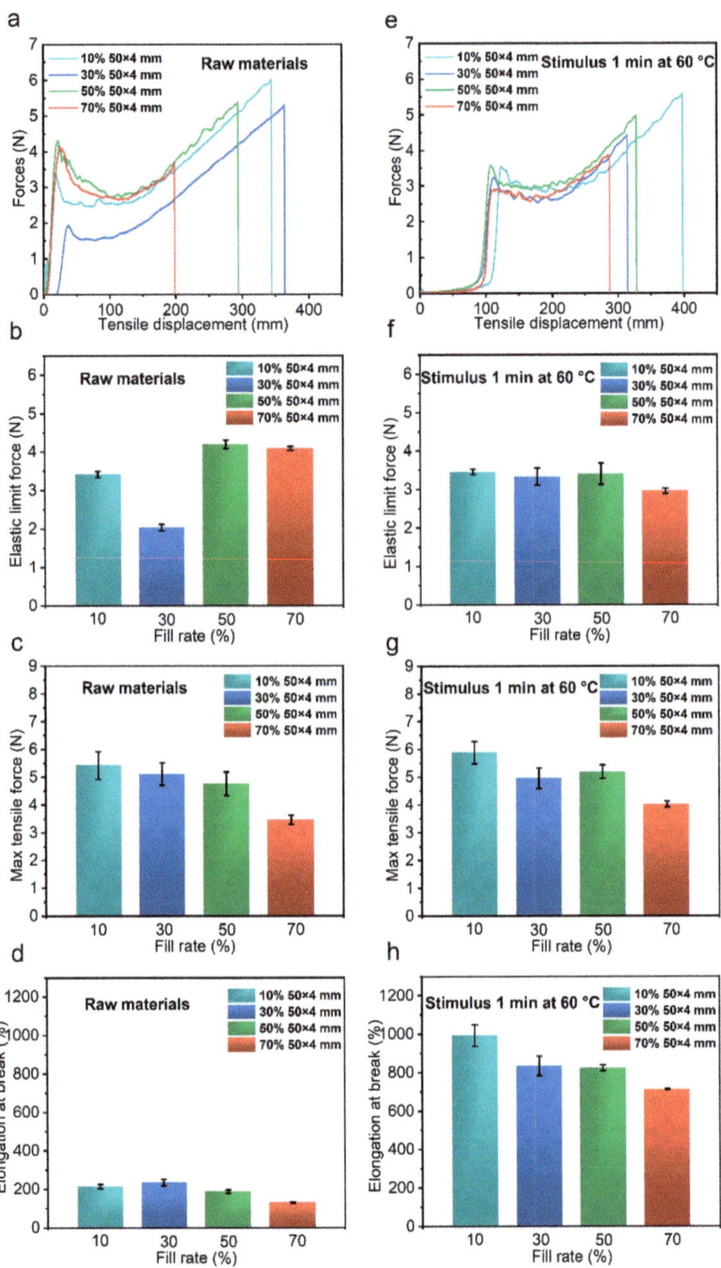

Figure 3. Force versus tensile displacement curves of the unstimulated bio-inspired spider silk samples with different filling densities (**a**); elastic limit force (**b**); maximum tensile force (**c**); elongation at break (**d**). Force versus tensile displacement curves of the bio-inspired spider silk samples with different filling densities stimulated for 1 min at 60 °C (**e**); elastic limit force (**f**); maximum tensile force (**g**); elongation at break (**h**).

The size of the nodes also significantly affects the tensile properties of the bio-inspired spider silk samples. When the nodes are small, the stress concentration is extremely pronounced and the degree of bending at these nodes is extremely dramatic. Thus, the smaller the node, the more likely it is to fracture due to excessive stress concentration. This results in relatively smaller values of maximum tensile force and fracture elongation for bio-inspired spider silk samples with smaller nodes.

3.2.3. Effect of Stimulation Temperature

Finally, the effect of stimulation temperature on the tensile properties of the bio-inspired spider silk samples was investigated, as shown in Figure 4a–d. The elastic limit force, maximum tensile force, and elongation at the breaks of the samples were not significantly different when the stimulation temperature was relatively low, e.g., at 40, 50, or 60 °C for the constant-temperature water stimulation. When the temperature was again increased to 70 °C, the elastic limit force and elongation at the break of the stimulated bio-inspired spider samples significantly reduced, while their maximum tensile force magnitude remained essentially unchanged. When the temperature was raised to 80 °C again, the elastic limit force, maximum tensile force, and elongation at the break increased compared with those at 70 °C. The elastic limit force and the elongation at the break were lower than those of the bio-inspired spider samples stimulated by constant-temperature water at 40, 50, and 60 °C, but the maximum tensile force remained the same. The influence of stimulation temperature on the tensile properties of bio-inspired spider samples cannot be ignored.

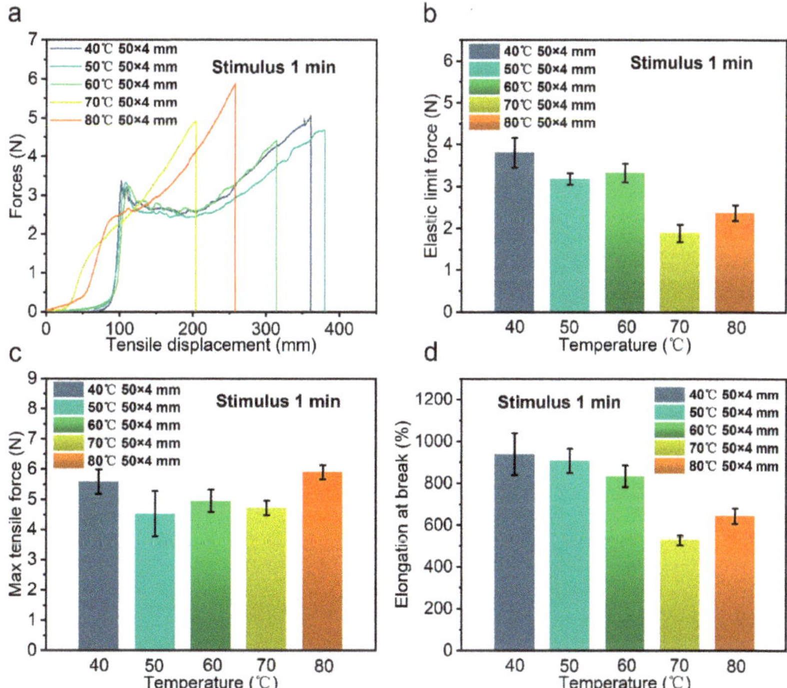

Figure 4. Force versus tensile displacement curves of the bio-inspired spider materials stimulated for 1 min at different temperatures (**a**); elastic limit force (**b**); maximum tensile force (**c**); elongation at break (**d**).

We synthesized the effects of node density and size as well as stimulation temperature on the tensile properties of bio-inspired spider silk samples. Different stimulation temperatures influenced the differences in the large nodes formed with the bio-inspired spider silk samples after helical deformation. At the nodes, the abrupt change in shape produced the stress concentration phenomenon. Differences in the type, density and size of the nodes produced differences in the stress concentration, which further influenced the bio-inspired spider silk tensile properties.

3.3. Effect of Tensile Speed and Moisture Absorption Content on the Tensile Properties of the Bio-Inspired Spider Silks

3.3.1. The DSC and TGA Results

The effects of water on the internal molecular characteristics of polyurethane were investigated via the DSC and TGA experiments. The glass transition temperature of polyurethane decreases with increases in water absorption [27]. The absorbed water can interact with the polymer, thus increasing the mobility of the polymer chains and leading to a decrease in the glass transition temperature [28].

Figure 5a–d show the DSC and TGA results. Figure 5a,c show the DSC results. Figure 5a illustrates that the glass transition temperatures of the bio-inspired spider samples soaked at 60 °C for 4, 8, and 16 min decreased with increasing soaking time. The glass transition temperature of the bio-inspired spider sample soaked for 16 min was the lowest, followed by that of the bio-inspired spider sample soaked for 8 min, and the glass transition temperature of the bio-inspired spider sample soaked for 4 min was the highest. Figure 5c illustrates the DSC experiments performed on the bio-inspired spider silk samples treated by immersion for 30 and 60 min at 40 °C. The glass transition temperature of the bio-inspired spider silk samples soaked for 30 min was higher than that of the bio-inspired spider silk samples soaked for 60 min. Figure 5b,d show the TGA results. Figure 5b shows that the moisture absorption of the bio-inspired spider silk samples soaked for 4, 8, and 16 min at 60 °C increased with increasing soaking time. The bio-inspired spider silk samples soaked for 16 min absorbed the most moisture, followed by the bio-inspired spider silk samples soaked for 8 min, and the bio-inspired spider silk samples soaked for 4 min absorbed the least moisture. Figure 5d illustrates the TGA experiments performed on the bio-inspired spider silk samples soaked at 40 °C for 30 and 60 min. The degree of moisture absorption was greater for the bio-inspired spider silk samples soaked for 60 min and lower for the bio-inspired spider silk samples soaked for 30 min. The increase in moisture absorption significantly reduced the glass transition temperature of the bio-inspired spider silk sample.

This is because the degree of moisture absorption of the bio-inspired spider silk samples can significantly affect their glass transition temperature. Therefore, we conducted experiments to investigate the effect of the degree of moisture absorption on the mechanical and tensile properties of spider silk. The magnitude of the impact of the bio-inspired spider silk samples varied by varying the stretching speed, representing the impact size of the bio-inspired spider silk and, by controlling the soaking stimulation time at 60 °C, the degree of moisture absorption of the samples could be varied.

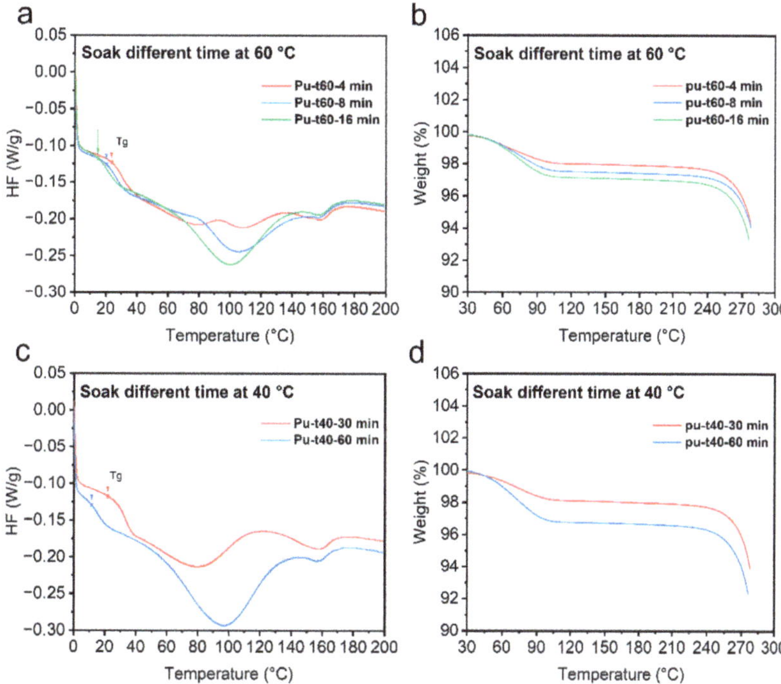

Figure 5. DCS experiments of bio-inspired spider silk samples soaked at 60 °C (**a**); TGA experiments of bio-inspired spider silk samples soaked at 60 °C (**b**); DCS experiments of bio-inspired spider silk samples soaked at 40 °C (**c**); TGA experiments of bio-inspired spider silk samples soaked at 40 °C (**d**).

3.3.2. Effect of Different Stretching Speeds

Figure 6a–d show the unstimulated raw bio-inspired spider silk samples at different stretching speeds. At lower stretching speeds, that is, 30, 60, and 90 mm/min, the stretching profiles of this photo-bio-inspired spider silk material were similar, and the tensile properties of the bio-inspired spider silk were approximately the same. At higher stretching speeds, up to 110 mm/min, a significant reduction in the elongation at the break of the original spider silk material was observed. This phenomenon was more pronounced when the stretching speed was higher (150 mm/min). Thus, when the tensile properties of the 4D-printed samples were directly characterized without stimulation, the high tensile speed indicated that the bio-inspired spider silk samples were subjected to a high impact. This led to a fracture in the bio-inspired spider silk sample without reaching the desired tensile strength limit, resulting in a significant reduction in the elongation at the break. To avoid this phenomenon, we performed tensile property characterization experiments on bio-inspired spider silk samples after deformation by stimulation under certain conditions. As shown in Figure 6e–h, the stimulation conditions were chosen to allow the bio-inspired spider silk sample to absorb moisture by submerging it in water at a constant temperature of 60 °C for 2 min. The sample did not break in the middle of the experiment, even when the tensile speed was increased to 210 mm/min. The elastic limit force, maximum tensile force, and elongation at the break of the samples remained the same for different tensile speeds, and the tensile properties of the samples did not significantly vary from 30 to 210 mm/min.

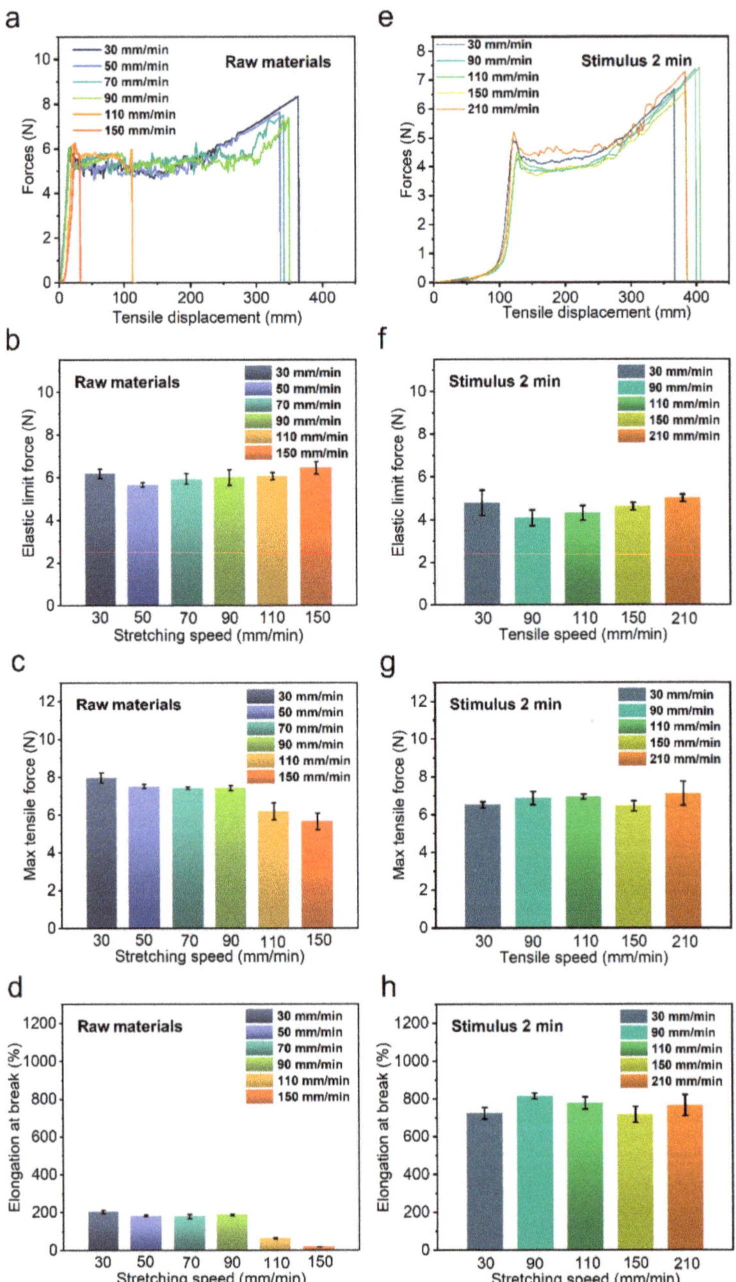

Figure 6. Force versus tensile displacement curves of unstimulated raw bio-inspired spider silk samples at different tensile speeds (**a**); elastic limit force (**b**); maximum tensile force (**c**); elongation at break (**d**). Force versus tensile displacement curves for bio-inspired spider material stimulated at 60 °C for 2 min at different tensile speeds (**e**); elastic limit force (**f**); maximum tensile force (**g**); elongation at break (**h**).

Unstimulated bio-inspired spider silk samples were subjected to larger tensile speeds, which resulted in the sudden breakage of the bio-inspired spider silk samples during stretching, causing a significant decrease in the fracture elongation of the samples at the higher tensile speeds. In order to avoid mid-rupture, the sample can be soaked to increase the relative humidity of the sample so that the tensile speed is increased from the value that causes the unstimulated spider silk sample to break, and the stimulated sample still does not break mid-rupture. Thus, by soaking and humidifying the samples, breakage in the middle of the process at a high stretching speed can be prevented.

3.3.3. Effect of the Degree of Moisture Absorption

We investigated the effect of the degree of moisture absorption on the tensile properties of the bio-inspired spider silk, as shown in Figure 7a–d, by subjecting the bio-inspired spider silk to immersion stimulation at 60 °C for 0, 1, 2, 4, 8, and 16 min, and then investigated the differences in the tensile properties of the samples under such conditions. The maximum tensile force of the samples with different degrees of moisture absorption remained the same, and there was no obvious difference; however, the curve of force versus tensile displacement of the sample with the greatest degree of moisture absorption, that is, the sample soaked for 16 min, significantly changed. The elastic limit force of the sample decreased significantly. In addition, this sample had the smallest elongation at its break compared to the previous samples that were soaked for 0, 1, 2, 4, and 8 min. The difference in soaking stimulation time significantly affected the elastic limit force and elongation at the break of the bio-inspired spider silk samples.

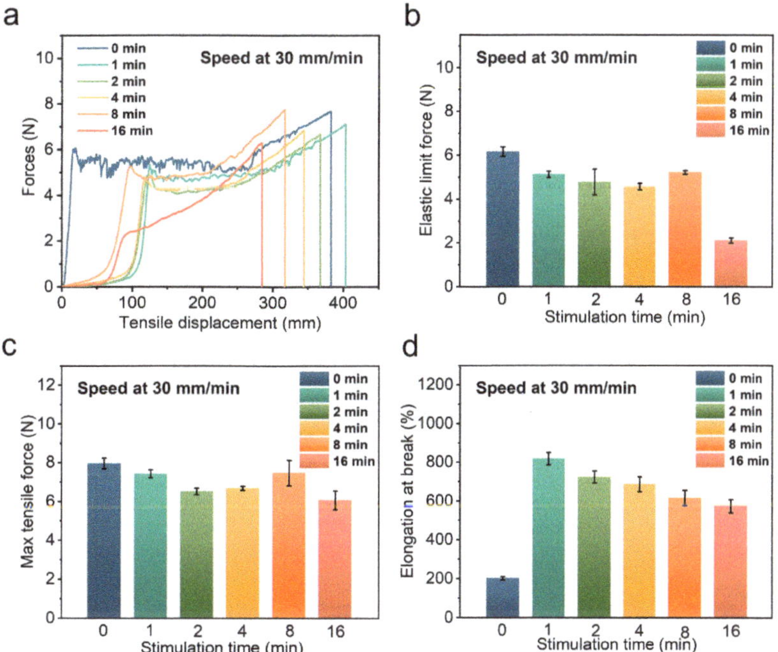

Figure 7. Force versus tensile displacement curves for bio-inspired spider material stimulated at 60 °C for different times (**a**); elastic limit force (**b**); maximum tensile force (**c**); elongation at break (**d**).

When the stimulated bio-inspired spider silk samples were immersed for longer periods of time, the elastic ultimate force and elongation at the break decreased as the degree of moisture absorption of the bio-inspired spider silk samples increased. The tensile strength of the polymer decreased with increasing water absorption. This is related to the increase in toughness of the bio-inspired spider samples after a large increase in the degree of moisture absorption of the bio-inspired spider silk.

3.4. Bio-Inspired Spider Web Stimulation Experiment

Different webs undergo relatively regular dynamic changes when stimulated in various ways. Figure 8 shows the deformation diagrams of the bio-inspired spider silk webs prepared with unstimulated and high-temperature-stimulated spider silk. Figure 8a shows the dynamic deformation process of a web woven with unstimulated spider silk. Figure 8b shows the deformation process of the web woven by the bio-inspired spider silk stimulated at high temperatures. Dynamic changes were observed in both webs under infrared light stimulation. The overall deformation is a contraction process. Thus, the response rate of the webs woven by samples stimulated at high temperatures is significantly better than that of the webs woven by unstimulated spider silk when subjected to the stimulation. This is related to the learning property of the stimulated response of the shape memory polymer [29].

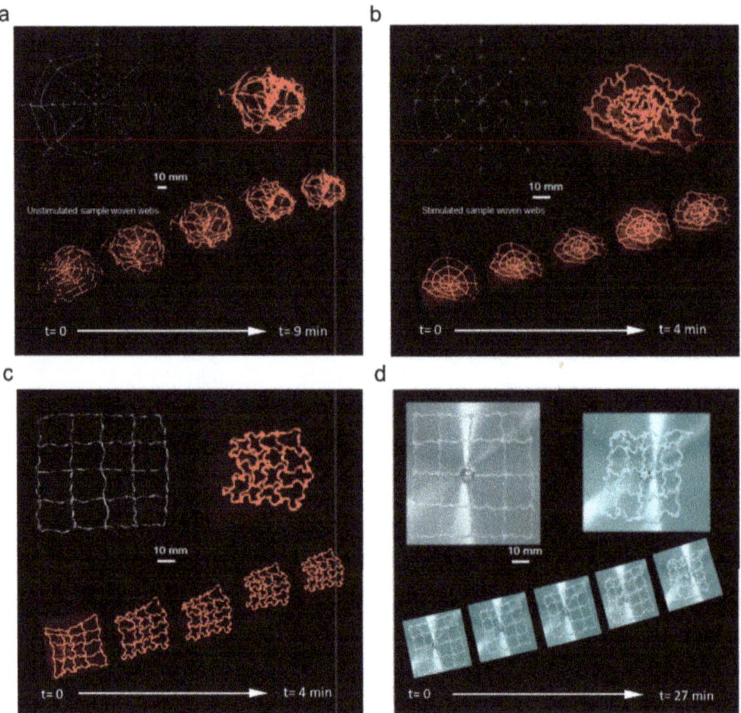

Figure 8. Stimulated deformation of the unstimulated spider webs made of bio-inspired spider silk (**a**); stimulated deformation of the webs made of bio-inspired spider silk stimulated at high temperature (**b**); stimulated deformation of square bio-inspired spider webs stimulated by infrared light (**c**); stimulated deformation of square bio-inspired spider webs stimulated by humidity (**d**).

Figure 8c,d show the results for the square-prepared bio-inspired spider silk webs under infrared and humidity stimulations, respectively. The response rate of the stimulated spider webs under infrared conditions was significantly better than that under humidity stimulation. However, the final shrinkage and deformation of the bio-inspired spider webs were similar, and it can be concluded that the different stimulation methods affected the shrinkage and deformation rates of the bio-inspired spider webs. However, the final shrinkage and deformation of the bio-inspired spider webs will not be affected if the stimulation time is sufficient, which is mainly determined by the state of the bio-inspired spider webs before this stimulation.

The woven bio-inspired spider web can shrink and deform under infrared or humidity stimulation. The degree of deformation increases with an increase in stimulation time, and the rate of shrinkage and deformation of the bio-inspired spider silk web varies under different stimulation methods. Thus, by controlling the manner, degree and timing of stimulation, bio-inspired spider silk webs with different degrees of contraction and deformation can be obtained. Furthermore, the observed dynamic changes in the shape and mechanical properties of bio-inspired spider webs can be applied in different scenarios.

3.5. Application Experiments

Figure 9a,b show the dynamic response of a bio-inspired spider web to the impact of different objects. Bio-inspired spider webs are formed by sequentially interweaving biologically inspired spider silk with a double predetermined structure, as described in the previous section. Figure 9(a_1–a_9) show the deformation of the bio-inspired spider web in response to the impact of a small glass ball; Figure 9(b_1–b_9) show the deformation of the bio-inspired spider web in response to the impact of a larger tennis ball. The entire process is shown by the forced deformation when resisting the impact, the accumulation of force to rebound, and the ball bouncing back. After each rebound, the kinetic energy of the ball is absorbed by the bio-inspired spider web, and the velocity gradually decreases.

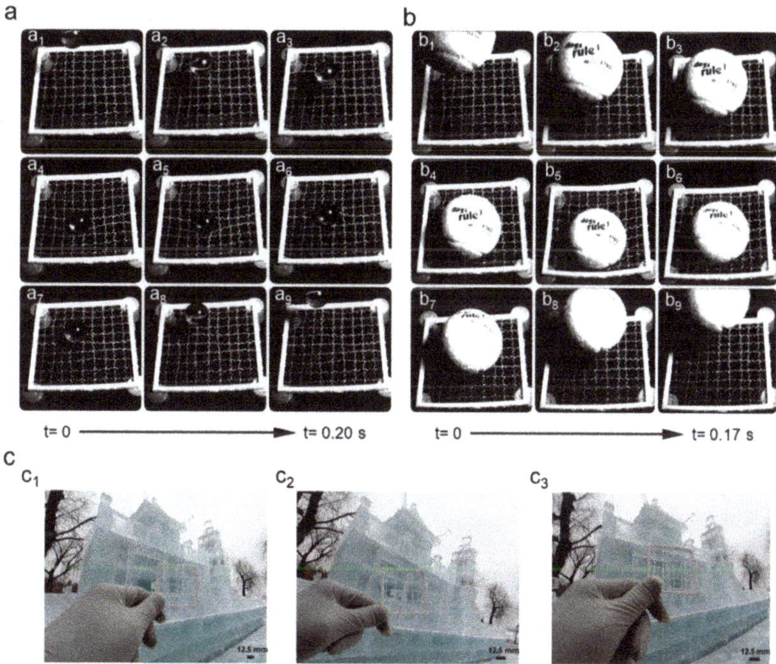

Figure 9. The glass ball impacts the bio-inspired spider silk web (**a**); tennis ball impacts the bio-inspired spider silk web (**b**); the concealment of the web woven by bio-inspired spider silk with different levels of stimulation (**c**).

Figure 9c shows the concealment of spider webs woven with bio-inspired spider silk with different levels of stimulation. The web woven by the sample that underwent only one high-temperature stimulation (Figure 9(c_1)) had better concealment under this condition and was difficult to detect. Figure 9(c_2) shows that, with three high-temperature stimulations, the bio-inspired spider silk had the second highest concealment; Figure 9(c₃) shows that, with five high-temperature stimulations, the bio-inspired spider silk had the worst concealment and could be detected more easily by observation.

As the number of stimulations increased, the contraction and spiral deformation of the bio-inspired spider silk increased, and the concealment and mechanical properties of the bio-inspired spider web dynamically changed. The concealment of the bio-inspired spider web worsened. However, the mechanical properties of the samples improved to a certain extent. Thus, a balance in practical applications must be found to achieve the desired effect. This type of bio-inspired spider web can be used for stealth interception. Generally, when the number of stimulations is lower, the concealment is better. When necessary, the number of stimulations can be rapidly increased to improve the impact resistance of the bio-inspired spider silk web for intercepting impact objects. It can be applied in underwater interception, aerospace, military bases, and many other fields with special requirements.

The bio-inspired spider web made of polyurethane can produce different dynamic deformation with the stimulation of various temperatures. When the stimulation conditions are the same, the dynamic change process can be controlled by changing the density and size of the nodes. The existing research on bio-inspired spider webs mainly focuses on generating artificial bio-inspired fibers with mechanical properties similar to those of natural spider silk, or investigating the mechanical properties exhibited by spider web structures in response to different types of impacts [30]. We took two approaches to control the dynamics of the mechanical properties of bio-inspired spider webs via pre-programming and varying the degrees of stimulation, which successfully achieved a certain degree of controlled dynamic changes in the bio-inspired spider webs.

4. Conclusions

The effects of various stimulation methods on the mechanical properties of 4D-printed bio-inspired spider silk were investigated and analyzed. The results demonstrated that the tensile properties of the bio-inspired spider silk fabricated using the 4D printing method in this study can be dynamically varied using stimulation methods such as temperature, humidity, and infrared light. The essence of this method is that by stimulating the bio-inspired spider silk material, it can be contracted and deformed to form a relatively uniform spiral structure in a localized area. By varying the stimulation method and time, the size and amount of the spiral region formed by the contraction and deformation can be controlled. This controls the shape and size of the energy-absorbing unit formed by the bio-inspired spider silk after stimulation, which further affects the mechanical properties of the bio-inspired spider silk. The energy-absorbing units can be formed not only according to the stimulation method, but also in the form of nodes that are preset into the printed bio-inspired spider silk structure. This allows the formation of a part of the energy-absorbing unit in the form of an advanced burial. Through the double preset coupling of these two types of energy-absorbing units, the proportion and size of each part of the energy-absorbing unit can be controlled so that the mechanical properties of the bio-inspired spider silk, particularly the tensile properties, can be dynamically changed. Controlling the dynamics of bio-inspired spider silk further enables dynamic changes in the impact resistance of the woven bio-inspired spider silk web. We have proposed a relatively simple way to control the dynamic changes in the mechanical properties of bio-inspired spider silk, resulting in a dynamic 4D-printed bio-inspired spider web. This can lead to the further development of application areas involving bio-inspired spider webs.

Author Contributions: Conceptualization, G.L. and Q.L.; methodology, G.L. and Q.T.; software, Q.T. and Y.Z.; investigation, G.L., Q.T., W.W., Q.W., S.Y. and J.W.; data curation, G.L., Q.T., X.Z. and K.W.; writing—original draft preparation, G.L. and Q.T.; writing—review and editing, G.L., Q.T., W.W., Q.W., S.Y., J.W., Y.Z., X.Z., K.W., J.Z. and L.R.; supervision, G.L., Q.L. and W.W.; project administration, G.L. and Q.L.; funding acquisition, G.L., K.W. and L.R. All authors have read and agreed to the published version of the manuscript.

Funding: This research is supported by National Natural Science Foundation of China (No. 52105342 and 52005209), foundation for Innovative Research Groups of the National Natural Science Foundation of China (No. 52021003).

Institutional Review Board Statement: Not applicable.

Informed Consent Statement: Not applicable.

Data Availability Statement: The data presented in this study are contained within the article.

Conflicts of Interest: The authors declare no conflict of interest.

References

1. Eisoldt, L.; Smith, A.; Scheibel, T. Decoding the secrets of spider silk. *Mater. Today* **2011**, *14*, 80–86. [CrossRef]
2. Lewis, R.V. Spider silk: Ancient ideas for new biomaterials. *Chem. Rev.* **2006**, *106*, 3762–3774. [CrossRef] [PubMed]
3. Heim, M.; Keerl, D.; Scheibel, T. Spider silk: From soluble protein to extraordinary fiber. *Angew. Chem. Int. Ed.* **2009**, *48*, 3584–3596. [CrossRef] [PubMed]
4. Du, N.; Yang, Z.; Liu, X.Y.; Li, Y.; Xu, H.Y. Structural origin of the strain-hardening of spider silk. *Adv. Funct. Mater.* **2011**, *21*, 772–778. [CrossRef]
5. Rising, A.; Johansson, J. Toward spinning artificial spider silk. *Nat. Chem. Biol.* **2015**, *11*, 309–315. [CrossRef] [PubMed]
6. Römer, L.; Scheibel, T. The elaborate structure of spider silk: Structure and function of a natural high performance fiber. *Prion* **2008**, *2*, 154–161. [CrossRef]
7. Pant, H.R.; Bajgai, M.P.; Nam, K.T.; Seo, Y.A.; Pandeya, D.R.; Hong, S.T.; Kim, H.Y. Electrospun nylon-6 spider-net like nanofiber mat containing TiO_2 nanoparticles: A multifunctional nanocomposite textile material. *J. Hazard. Mater.* **2011**, *185*, 124–130. [CrossRef]
8. Gu, Y.; Yu, L.; Mou, J.; Wu, D.; Zhou, P.; Xu, M. Mechanical properties and application analysis of spider silk bionic material. *e-Polymers* **2020**, *20*, 443–457. [CrossRef]
9. Kluge, J.A.; Rabotyagova, O.; Leisk, G.G.; Kaplan, D.L. Spider silks and their applications. *Trends Biotechnol.* **2008**, *26*, 244–251. [CrossRef]
10. Liu, Y.; Huang, W.; Meng, M.; Chen, M.; Cao, C. Progress in the application of spider silk protein in medicine. *J. Biomater. Appl.* **2021**, *36*, 859–871. [CrossRef]
11. Steffens, F.; Gralha, S.E.; Ferreira, I.L.S.; Oliveira, F.R. Military Textiles—An Overview of New Developments. *Key Eng. Mater.* **2019**, *812*, 120–126. [CrossRef]
12. Bardenhagen, A.; Sethi, V.; Gudwani, H. Spider-silk composite material for aerospace application. *Acta Astronaut.* **2021**, *193*, 704–709.
13. Edlund, A.M.; Jones, J.; Lewis, R.; Quinn, J.C. Economic feasibility and environmental impact of synthetic spider silk production from escherichia coli. *New Biotechnol.* **2018**, *42*, 12–18. [CrossRef] [PubMed]
14. Xu, J.; Dong, Q.; Yu, Y.; Niu, B.; Ji, D.; Li, M.; Huang, Y.; Chen, X.; Tan, A. Mass spider silk production through targeted gene replacement in Bombyx mori. *Proc. Natl. Acad. Sci. USA* **2018**, *115*, 8757–8762. [CrossRef]
15. Vendrely, C.; Scheibel, T. Biotechnological production of spider-silk proteins enables new applications. *Macromol. Biosci.* **2007**, *7*, 401–409. [CrossRef]
16. Venkatesan, H.; Chen, J.; Liu, H.; Kim, Y.; Na, S.; Liu, W.; Hu, J. Artificial spider silk is smart like natural one: Having humidity-sensitive shape memory with superior recovery stress. *Mater. Chem. Front.* **2019**, *3*, 2472–2482. [CrossRef]
17. Chan, N.J.-A.; Gu, D.; Tan, S.; Fu, Q.; Pattison, T.G.; O'Connor, A.J.; Qiao, G.G. Spider-silk inspired polymeric networks by harnessing the mechanical potential of β-sheets through network guided assembly. *Nat. Commun.* **2020**, *11*, 1630. [CrossRef]
18. Dou, Y.; Wang, Z.-P.; He, W.; Jia, T.; Liu, Z.; Sun, P.; Wen, K.; Gao, E.; Zhou, X.; Hu, X. Artificial spider silk from ion-doped and twisted core-sheath hydrogel fibres. *Nat. Commun.* **2019**, *10*, 5293. [CrossRef]
19. Zou, S.; Therriault, D.; Gosselin, F.P. Spiderweb-Inspired, Transparent, Impact-Absorbing Composite. *Cell Rep. Phys. Sci.* **2020**, *1*, 100240. [CrossRef]
20. Qin, Z.; Compton, B.G.; Lewis, J.A.; Buehler, M.J. Structural optimization of 3D-printed synthetic spider webs for high strength. *Nat. Commun.* **2015**, *6*, 7038. [CrossRef]
21. Li, J.; Li, S.; Huang, J.; Khan, A.Q.; An, B.; Zhou, X.; Liu, Z.; Zhu, M. Spider Silk-Inspired Artificial Fibers. *Adv. Sci.* **2021**, *9*, 2103965. [CrossRef] [PubMed]
22. Momeni, F.; Liu, X.; Ni, J. A review of 4D printing. *Mater. Des.* **2017**, *122*, 42–79. [CrossRef]
23. Kuang, X.; Roach, D.J.; Wu, J.; Hamel, C.M.; Ding, Z.; Wang, T.; Dunn, M.L.; Qi, H.J. Advances in 4D printing: Materials and applications. *Adv. Funct. Mater.* **2019**, *29*, 1805290. [CrossRef]
24. Su, J.-W.; Gao, W.; Trinh, K.; Kenderes, S.M.; Pulatsu, E.T.; Zhang, C.; Whittington, A.; Lin, M.; Lin, J. 4D printing of polyurethane paint-based composites. *Int. J. Smart Nano Mater.* **2019**, *10*, 237–248. [CrossRef]
25. Kanu, N.J.; Gupta, E.; Vates, U.K.; Singh, G.K. An insight into biomimetic 4D printing. *RSC Adv.* **2019**, *9*, 38209–38226. [CrossRef] [PubMed]

26. Mitchell, A.; Lafont, U.; Hołyńska, M.; Semprimoschnig, C. Additive manufacturing—A review of 4D printing and future applications. *Addit. Manuf.* **2018**, *24*, 606–626. [CrossRef]
27. Yang, B.; Huang, W.; Li, C.; Lee, C.; Li, L. On the effects of moisture in a polyurethane shape memory polymer. *Smart Mater. Struct.* **2003**, *13*, 191. [CrossRef]
28. Yang, B.; Huang, W.M.; Li, C.; Chor, J.H. Effects of moisture on the glass transition temperature of polyurethane shape memory polymer filled with nano-carbon powder. *Eur. Polym. J.* **2005**, *41*, 1123–1128. [CrossRef]
29. Zeng, H.; Zhang, H.; Ikkala, O.; Priimagi, A. Associative learning by classical conditioning in liquid crystal network actuators. *Matter* **2020**, *2*, 194–206. [CrossRef]
30. Su, I.; Buehler, M.J. Mesomechanics of a three-dimensional spider web. *J. Mech. Phys. Solids* **2020**, *144*, 104096. [CrossRef]

Article

Novel Copper Complexes as Visible Light Photoinitiators for the Synthesis of Interpenetrating Polymer Networks (IPNs)

Mahmoud Rahal [1,2,3], Guillaume Noirbent [4], Bernadette Graff [1,2], Joumana Toufaily [3], Tayssir Hamieh [3], Didier Gigmes [4], Frédéric Dumur [4,*] and Jacques Lalevée [1,2,*]

1   Université de Haute-Alsace, CNRS, IS2M UMR 7361, F-68100 Mulhouse, France; mahmoud-rahal@outlook.com (M.R.); bernadette.graff@uha.fr (B.G.)
2   Université de Strasbourg, F-67087 Strasbourg, France
3   Laboratory of Materials, Catalysis, Environment and Analytical Methods (MCEMA) and LEADDER Laboratory, Faculty of Sciences, Doctoral School of Sciences and Technology (EDST), Lebanese University, Beirut 6573-14, Lebanon; joumana.toufaily@ul.edu.lb (J.T.); tayssir.hamieh@ul.edu.lb (T.H.)
4   Aix Marseille Univ, CNRS, ICR UMR 7273, F-13397 Marseille, France; guillaume.noirbent@outlook.fr (G.N.); didier.gigmes@univ-amu.fr (D.G.)
*   Correspondence: frederic.dumur@univ-amu.fr (F.D.); jacques.lalevee@uha.fr (J.L.)

Abstract: This work is devoted to the study of two copper complexes (Cu) bearing pyridine ligands, which were synthesized, evaluated and tested as new visible light photoinitiators for the free radical photopolymerization (FRP) of acrylates functional groups in thick and thin samples upon light-emitting diodes (LED) at 405 and 455 nm irradiation. These latter wavelengths are considered to be safe to produce polymer materials. The photoinitiation abilities of these organometallic compounds were evaluated in combination with an iodonium (Iod) salt and/or amine (e.g., *N*-phenylglycine—NPG). Interestingly, high final conversions and high polymerization rates were obtained for both compounds using two and three-component photoinitiating systems (Cu1 (or Cu2)/Iodonium salt (Iod) (0.1%/1% *w/w*) and Cu1 (or Cu2)/Iod/amine (0.1%/1%/1% *w/w/w*)). The new proposed copper complexes were also used for direct laser write experiments involving a laser diode at 405 nm, and for the photocomposite synthesis with glass fibers using a UV-conveyor at 395 nm. To explain the obtained polymerization results, different methods and characterization techniques were used: steady-state photolysis, real-time Fourier transform infrared spectroscopy (RT-FTIR), emission spectroscopy and cyclic voltammetry.

Keywords: copper complex; photocomposite; LED; laser write; free radical photopolymerization

Citation: Rahal, M.; Noirbent, G.; Graff, B.; Toufaily, J.; Hamieh, T.; Gigmes, D.; Dumur, F.; Lalevée, J. Novel Copper Complexes as Visible Light Photoinitiators for the Synthesis of Interpenetrating Polymer Networks (IPNs). *Polymers* **2022**, *14*, 1998. https://doi.org/10.3390/polym14101998

Academic Editor: Houwen Matthew Pan

Received: 22 April 2022
Accepted: 9 May 2022
Published: 13 May 2022

Publisher's Note: MDPI stays neutral with regard to jurisdictional claims in published maps and institutional affiliations.

1. Introduction

The elaboration of polymers by photochemical means, such as free radical photopolymerization (FRP) and cationic photopolymerization (CP), have been mainly based on the use of metal-free organic dyes and photoinitiators at the industrial and academic levels [1–12], and these synthetic processes (FRP and CP) are widely used in different fields, e.g., dentistry [13–23], adhesives [24–28], coatings [29–33], composites [34], medicine [35–40], direct laser write, 3D and 4D printing [41–50], etc. On the other hand, organometallic compounds are not really used in industry; in other words, manufacturers avoid incorporating metallic compounds in their synthetic formulations due to their potential toxicity and price [51–64]. With their photochemical properties, such as high-absorption properties in the near-UV and visible range [65–68], long-lived excited states [69–74], suitable redox potentials [75–89], copper complexes can be used as photoinitiators (PIs)/photoredox catalysts able to produce active species, according to a catalytic cycle [90,91]. Therefore, it is very important to develop new metal-free photoinitiators or low-cost organometallic-based complexes [92–95].

In fact, copper complexes have attracted much attention and intense efforts have been devoted in recent years to the development of new copper complexes of improved photosensitivity, due to their competitive costs compared to other metal complexes. Copper complexes bearing a pyridine-based chelate ligand showed excellent photochemical properties for photocatalysis process, such as high-oxidation potential in the excited state [96–99], long-excited-state lifetime, high-emission quantum yields and high-absorption properties in the UV-visible region. Furthermore, copper complex derivatives have already been tested as PIs for FRP, CP, as well as IPN synthesis [100–104].

In this paper, two new copper complexes (Cu1–Cu2) (Figure 1) were synthesized and investigated as visible light photoinitiators upon exposure to LEDs at 405 and 455 nm for FRP, CP and the synthesis of interpenetrating polymer networks (IPNs) of acrylate/epoxy monomer blends. These compounds will be incorporated in two (Cu1 (or Cu2)/Iod (0.1%/1% *w/w*)) and three-component (Cu1 (or Cu2)/Iod/NPG (0.1%/1%/1% *w/w/w*)) photoinitiating systems (PISs) to produce polymer materials by free radical photopolymerization and the polymerization of acrylate/epoxy blend (IPNs). The photoinitiating ability of copper complexes will also be explained based on the interaction of Cu1 (or Cu2)/Iod and Cu1 (or Cu2)/Iod/NPG, which can be studied using different techniques and characterization processes, e.g., steady-state photolysis, cyclic voltammetry, fluorescence quenching and electron spin resonance spin trapping. Finally, to demonstrate the effectiveness of these new copper complex-based photoinitiators, experiments using direct laser writing (DLW), 3D printing and photocomposites synthesis were carried out in this work using different irradiation sources.

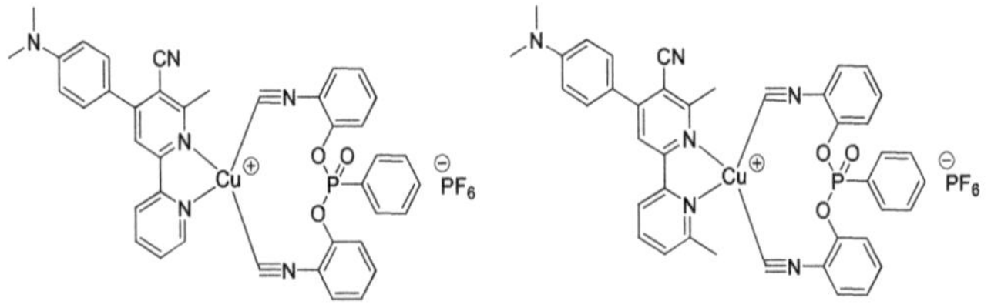

Cu1 **Cu2**

Figure 1. Copper complexes used in this work as PIs.

2. Materials and Methods

2.1. Synthesis of Chalcones, Ligands and Copper Complexes

Experimental conditions and acquisition conditions have been detailed elsewhere [13,92,100]. The two chalcones used for the design of ligands L1 and L2 were then engaged in a cyclization reaction with β-aminocrotonitrile according to a reaction reported in 1992 by Masaki Matsui [105,106]. *Bis*(2-isocyanophenyl) phenylphosphonate (binc) was synthesized by adapting a literature procedure [107,108].

Synthesis of bis(2-isocyanophenyl) phenylphosphonate (binc)

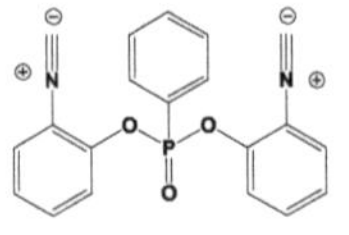

Benzoxazole (9.1 g, 76.3 mmol, 1.0 equiv.) was dissolved in dry THF (200 mL). The solution was cooled at $-78\,^{\circ}$C and *n*-BuLi (2.5 M in hexane, 32.0 mL, 80.0 mmol, 1.05 equiv.) was added. Stirring was maintained for 1.5 h at this temperature. Phenyl phosphonic dichloride (5.7 mL, 4.04 mmol, 0.53 equiv.) was added and the solution could warm to room temperature. The solution was poured in Et_2O:$NaHCO_3$ (2:1, 150 mL). The organic phase was washed with water several times, dried over magnesium sulfate and the solvent removed under reduced pressure. The residue was crystallized in pentane/ethyl acetate (4/1) to provide the ligand (55% yield) as a light brown solid. ^{1}H NMR (400 MHz, $CDCl_3$) δ(ppm): 8.23–8.11 (m, 2H), 7.69 (td, *J* = 7.4, 1.3 Hz, 1H), 7.58 (dt, *J* = 12.5, 6.3 Hz, 2H), 7.48 (d, *J* = 8.4 Hz, 2H), 7.39 (d, *J* = 7.9 Hz, 2H), 7.34 (td, *J* = 8.1, 1.6 Hz, 2H), 7.18 (t, *J* = 7.7 Hz,

2H); ^{13}C NMR (101 MHz, CDCl$_3$) δ(ppm): 169.7, 145.7, 145.7, 134.5, 132.9, 132.8, 130.7, 130.7, 129.3, 129.1, 128.3, 125.9, 124.1, 124.1, 121.7, 121.7. (isonitrile carbons not detected); HRMS (ESI MS) *m/z*: theor: 360.0664 found: 360.0666 (M$^+$. detected).

Synthesis of (E)-3-(4-(dimethylamino)phenyl)-1-(pyridin-2-yl)prop-2-en-1-one

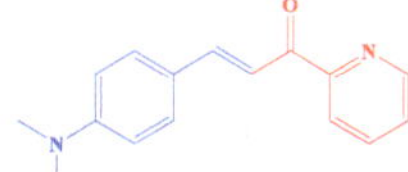

Next, 4-(Dimethylamino)benzaldehyde (1.49 g, 10.0 mmol, M = 149.19 g/mol) and 1-(pyridin-2-yl)ethan-1-one (1.21 g, 10.0 mmol, M = 121.14 g/mol) were suspended in ethanol (50 mL) and aq. KOH (40%) (10 mL) was added. After stirring overnight, the solid was filtered off, washed with ethanol and water, and dried under vacuum. The product was purified by filtration on a plug of SiO$_2$ using dichloromethane (DCM) as the eluent (2.32 g, 92% yield). ^{1}H NMR (400 MHz, CDCl$_3$) δ(ppm): 8.61–8.49 (m, 1H), 8.01 (d, *J* = 7.4 Hz, 1H), 7.92 (d, *J* = 15.8 Hz, 1H), 7.77 (d, *J* = 15.8 Hz, 1H), 7.66 (t, *J* = 7.0 Hz, 1H), 7.45 (d, *J* = 7.7 Hz, 2H), 7.31–7.22 (m, 1H), 6.49 (d, *J* = 7.8 Hz, 2H), 2.83 (s, 6H); ^{13}C NMR (101 MHz, CDCl$_3$) δ(ppm): 189.16, 155.00, 152.11, 148.69, 145.94, 136.89, 130.90, 126.40, 122.99, 122.72, 115.47, 111.74, 40.06; HRMS (ESI MS) *m/z*: theor: 253.1296 found: 253.1299 ([M + H]$^+$ detected).

Synthesis of (E)-3-(4-(dimethylamino)phenyl)-1-(6-methylpyridin-2-yl)prop-2-en-1-one

Following this, 4-(Dimethylamino)benzaldehyde (2.21 g, 14.8 mmol, M = 149.19 g/mol) and 1-(6-methylpyridin-2-yl)ethan-1-one (2.00 g, 14.8 mmol, M = 135.17 g/mol) were dissolved in ethanol (50 mL) and aq. KOH (40%) (15 mL) was added. After stirring the solution overnight, the resulting solid was filtered off. It was purified by filtration on a plug of SiO$_2$ using DCM as the eluent. For a higher purity, the solid was first dissolved in DCM and precipitated by addition of pentane (3.51 g, 89% yield). ^{1}H NMR (400 MHz, CDCl$_3$) δ(ppm): 8.02 (d, *J* = 15.8 Hz, 1H), 7.90 (d, *J* = 7.5 Hz, 1H), 7.84 (d, *J* = 15.9 Hz, 1H), 7.65 (t, *J* = 7.7 Hz, 1H), 7.58–7.53 (m, 2H), 7.23 (d, *J* = 7.4 Hz, 1H), 6.66–6.60 (m, 2H), 2.97 (s, 6H), 2.60 (s, 3H); ^{13}C NMR (101 MHz, CDCl$_3$) δ(ppm): 189.62, 157.67, 154.57, 152.04, 145.63, 136.95, 130.84, 126.02, 123.26, 119.84, 115.94, 111.77, 40.13, 24.54; HRMS (ESI MS) *m/z*: theor: 267.1453 found: 267.1451 ([M + H]$^+$ detected).

Synthesis of 4-(4-(dimethylamino)phenyl)-6-methyl-[2,2'-bipyridine]-5-carbonitrile

Potassium *tert*-butoxide (1.2 g) and β-aminocrotonitrile (4.92 g, 60 mmol, M = 82.10 g/mol) were dissolved in acetonitrile (300 mL) and the solution was heated at 35 °C for 15 min. Chalcone (E)-3-(4-(dimethylamino)phenyl)-1-(pyridin-2-yl)prop-2-en-1-one (2.52 g, 10.0 mmol, M = 252.32 g/mol) was added and stirring was maintained for three days. The solid was filtered off and washed with ethanol and water. It was purified by filtration on a plug of SiO$_2$ using DCM as the eluent (2.83 g, 90% yield). ^{1}H NMR (400 MHz, CDCl$_3$) δ(ppm): 8.70 (ddd, *J* = 4.8, 1.7, 0.9 Hz, 1H), 8.50 (dt, *J* = 8.0, 1.0 Hz, 1H), 8.40 (s, 1H), 7.85 (td, *J* = 7.8, 1.8 Hz, 1H), 7.72–7.64 (m, 2H), 7.36 (ddd, *J* = 7.5, 4.8, 1.2 Hz, 1H), 6.85–6.77 (m, 2H), 3.05 (s, 6H), 2.89 (s, 3H); ^{13}C NMR (101 MHz, CDCl$_3$) δ(ppm): 162.47, 157.15, 155.03, 154.08, 151.47, 149.38, 137.02, 129.74, 129.74, 124.47, 123.44, 122.10, 118.22, 117.68, 111.97, 111.97, 106.10, 40.19, 40.19, 24.38; HRMS (ESI MS) *m/z*: theor: 315.1565 found: 315.1564 ([M + H]$^+$ detected).

Synthesis of 4-(4-(dimethylamino)phenyl)-6,6′-methyl-[2,2′-bipyridine]-5-carbonitrile

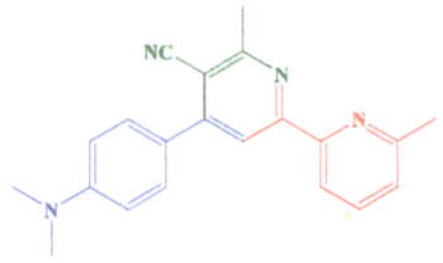

Potassium *tert*-butoxide (1.2 g) and β-aminocrotonitrile (4.92 g, 60 mmol, M = 82.10 g/mol) were dissolved in acetonitrile (300 mL) and the solution was heated at 35 °C for 15 min. Chalcone (*E*)-3-(4-(dimethylamino)phenyl)-1-(6-methylpyridin-2-yl)prop-2-en-1-one (L3) (2.66 g, 10.0 mmol, M = 266.34 g/mol) was added and stirring was maintained for three days. The solid was filtered off and washed with ethanol and water. It was purified by filtration on a plug of SiO_2 using DCM as the eluent (2.76 g, 84% yield). ^{1}H NMR (400 MHz, $CDCl_3$) δ(ppm): 8.40 (s, 1H), 8.27 (d, *J* = 7.8 Hz, 1H), 7.72 (t, *J* = 7.8 Hz, 1H), 7.69–7.63 (m, 2H), 7.21 (d, *J* = 7.6 Hz, 1H), 6.85–6.79 (m, 2H), 3.05 (s, 6H), 2.88 (s, 3H), 2.63 (s, 3H); ^{13}C NMR (101 MHz, $CDCl_3$) δ(ppm): 162.36, 158.24, 157.56, 154.37, 154.00, 151.43, 137.13, 129.70, 129.70, 124.09, 123.68, 119.11, 118.27, 117.78, 112.01, 112.01, 105.95, 40.21, 40.21, 24.61, 24.38; HRMS (ESI MS) *m/z*: theor: 329.1722 found: 329.1719 ([M + H]+ detected).

Synthesis of Cu1

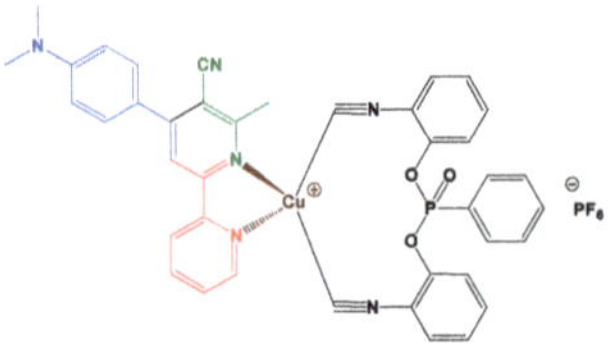

Tetrakis(acetonitrile)copper(I) hexafluorophosphate (372 mg, 1.0 mmol, M = 372.72 g/mol), *bis*(2-isocyanophenyl)phenylphosphonate (binc) (360 mg, 1.0 mmol, M = 360.31 g/mol) and 4-(4-(dimethylamino)phenyl)-6-methyl-[2,2′-bipyridine]-5-carbonitrile (314 mg, 1.0 mmol, M = 314.39 g/mol) were dissolved in DCM (100 mL) and the solution was stirred at 25 °C for 2 h. The solution was concentrated to ca. 5 mL. Diethyl ether was added, as the product was a solid (866 mg, 98% yield). ^{1}H NMR (400 MHz, $CDCl_3$) δ(ppm): 8.96 (s, 1H), 8.51 (s, 1H), 8.26 (s, 1H), 8.14 (dd, *J* = 14.2, 7.6 Hz, 2H), 7.85–7.70 (m, 4H), 7.70–7.59 (m, 4H), 7.52–7.35 (m, 5H), 7.24 (t, *J* = 8.1 Hz, 2H), 6.94 (s (br), 2H), 3.21 (s, 3H), 3.11 (s, 6H); HRMS (ESI MS) *m/z*: theor: 737.1486 found: 737.1481 (M$^+$. detected); Anal. Calc. for $C_{40}H_{31}CuF_6N_6O_3P_2$: C, 54.4; H, 3.5; O, 5.4; Found: C, 54.6; H, 3.4; O, 5.5%.

Synthesis of Cu2

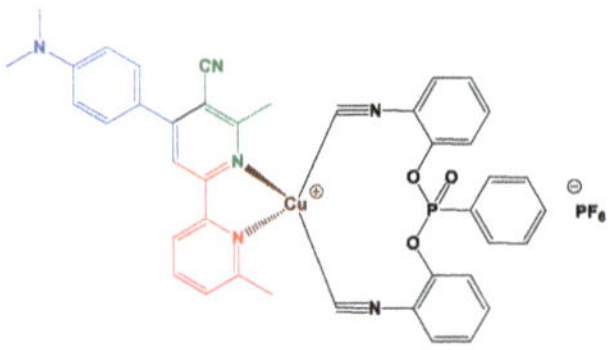

Tetrakis(acetonitrile)copper(I) hexafluorophosphate (372 mg, 1.0 mmol, M = 372.72 g/mol), *bis*(2-isocyanophenyl)phenyl phosphonate (binc) (360 mg, 1.0 mmol, M = 360.31 g/mol) and 4-(4-(dimethylamino)phenyl)-6,6′-methyl-[2,2′-bipyridine]-5-carbonitrile (328 mg, 1.0 mmol, M = 328.42 g/mol) were dissolved in DCM (100 mL) and the solution was stirred at 25 °C for 2 h. The solution was concentrated to ca. 5 mL. Addition of diethyl ether precipitated the product as a yellow solid (883 mg, 98% yield). ^{1}H NMR (400 MHz, DMSO) δ(ppm): 8.76 (s, 1H), 8.62 (s, 1H), 8.21 (s, 1H), 8.02 (dd, *J* = 13.2, 7.2 Hz, 2H), 7.79 (s, 6H), 7.67 (s, 2H), 7.56 (t, *J* = 7.9 Hz, 2H), 7.35 (dd, *J* = 21.3, 8.0 Hz, 5H), 6.92 (d, *J* = 8.8 Hz, 2H), 3.14 (s, 3H), 3.06 (s, 6H), 2.93 (s, 3H); HRMS (ESI MS) *m/z*: theor: 751.1642 found: 751.1639 (M$^+$. detected); Anal. Calc. for $C_{41}H_{33}CuF_6N_6O_3P_2$: C, 54.9; H, 3.7; O, 5.3; Found: C, 55.0; H, 3.4; O, 5.5%.

2.2. Other Chemicals

Chemical structure of the different monomers and additives are presented in Figure 2. Di-*tert*-butyl-diphenyl iodonium hexafluorophosphate (Iod) and ethyl 4-(dimethylamino)-benzoate (EDB) were obtained from Lambson Ltd. (UK). Di(trimethylolpropane) tetraacrylate (TA), trimethylolpropane triacrylate (TMPTA), (3,4-epoxycyclohexane)methyl 3,4-epoxycyclohexylcarboxylate (EPOX; Uvacure 1500), *N*-phenylglycine (NPG), *N*-vinylcarbazole (NVK) and *N,N*-dimethyl-*p*-toluidine (TMA) were obtained from Allnex or Sigma Aldrich. TA, TMPTA and EPOX were chosen as acrylic and cationic monomers for the radical and cationic polymerizations.

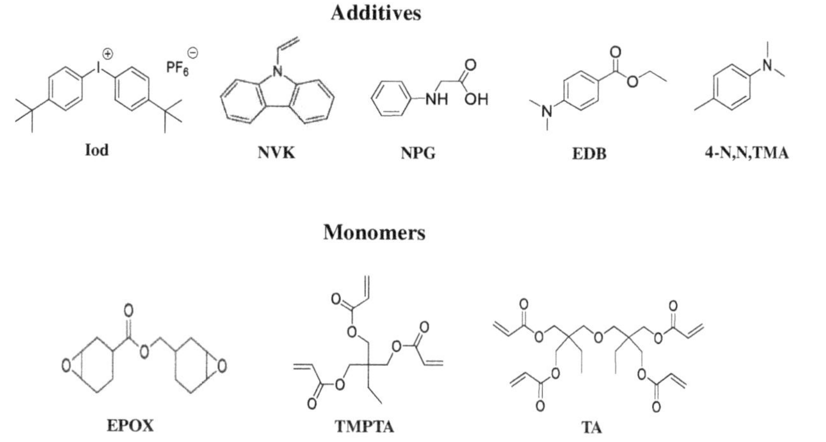

Additives

Iod   NVK   NPG   EDB   4-N,N,TMA

Monomers

EPOX   TMPTA   TA

Figure 2. Other organic compounds used.

2.3. Irradiation Sources: Light-Emitting Diodes

All the irradiation sources used during these experiments are based on light-emitting diodes (LEDs) and used as safe sources: (1) LED at 375 nm (I_0 = 75 mW·cm^{-2}) for the photolysis experiments, (2) LED at 405 nm (I_0 = 110 mW·cm^{-2}) and 455 nm (I_0 = 75 mW·cm^{-2}) for the photopolymerization experiments, (3) LED conveyor at 395 nm (I_0 = 4 W·cm^{-2}) for the photocomposite synthesis.

2.4. Photopolymerization Kinetics Determination by Real-Time Fourier Transform Infrared Spectroscopy (RT-FTIR)

In the present work, copper derivatives were used in two and three-component PISs for both FRP and CP under irradiation with LEDs at 405 and 455 nm. PISs were mainly based on two-component Cu1 (or Cu2)/Iod (0.1%-0.2%-0.5%/1% *w/w*) and three-component Cu1 (or Cu2)/Iod/amine (NPG, NVK) (0.1%/1%/1% *w/w/w*) systems. The weight percOKent of the photoinitiating (PI, co-initiator and amine) system was calculated from the global monomer content. Firstly, two different samples were studied for each photosensitive formulation in (i) thick (thickness = 1.4 mm) and (ii) thin samples (thickness = 25 μm). The epoxy and acrylate conversions were continuously followed by RT-FTIR using a JASCO FTIR 6600 (JASCO France, Lisses, France), so it was possible to determine the final conversion of reactive functions and to calculate the polymerization rate of each kinetic. Acrylate functions in thick and thin samples show peaks towards 6160 cm^{-1} and 1630 cm^{-1}, respectively, and the epoxide functions show peaks around 3600 cm^{-1} and 790 cm^{-1} for the thick and thin samples, respectively.

2.5. Redox Potentials: Electrochemical Properties

Redox potentials of copper derivatives were determined in DCM by cyclic voltammetry using tetrabutylammonium hexafluorophosphate as the supporting electrolyte (potentials vs. saturated calomel electrode (SCE)). Free energy change (ΔG_{et}) for an electron transfer reaction was calculated from Equation (1) [109], where E_{ox}, E_{red}, E^* and C represent the oxidation potential of the electron donor, the reduction potential of the electron acceptor, the excited-state energy level (determined from fluorescence experiments) and the coulombic term for the initially formed ion pair, respectively. Here, C is neglected as is usually the case for polar solvents.

$$\Delta G_{et} = E_{ox} - E_{red} - E^* + C \tag{1}$$

2.6. UV-Visible Absorption, Steady-State Photolysis and Luminescence Experiments

Acquisition conditions have been detailed elsewhere [13,92,100].

2.7. Computational Procedure

Computational conditions have been detailed elsewhere [13,92,100,110,111]. Simulated absorption spectra of copper complexes were obtained by time-dependent density functional theory at the MPW1PW91/6-31G* level of theory on the relaxed geometries calculated at the UB3LYP/6-31G* level of theory.

2.8. Photocomposite Access Using a Near-UV Conveyor

Photocomposite materials were obtained using a Dymax-UV conveyor at 395 nm. Firstly, photosensitive resins were deposited on the glass fibers (reinforcement), then, this mixture was cured using an LED conveyor @395 nm ($I = 4\ \text{W} \cdot \text{cm}^{-2}$). Distance between the belt and the LED was fixed to 15 mm, and the belt speed was fixed at 2 m/min.

2.9. Direct Laser Write (DLW) Experiment

The photosensitive formulation was deposited on a glass slide and 3D patterns were obtained under air using a computer-controlled diode laser at 405 nm (spot size = 50 μm). Analysis of the 3D patterns was carried out using a numerical optical microscope (DSX-HRSU from OLYMPUS Corporation, Rungis, France) [112].

3. Results

Light-absorption properties, initiation ability and applications (photocomposite synthesis and direct laser write) of the investigated copper complexes will be studied in this section.

3.1. Synthetic Routes to Copper Complexes Cu1 and Cu2

Copper complexes bearing a pyridine-based chelate ligand showed excellent photochemical properties for photocatalysis processes, such as a high-oxidation potential in the excited state [96–99], a long-excited-state lifetime, high-emission quantum yields and high-absorption properties in the UV-visible region. In this work, two new copper complexes have been developed, allowing, by the convenient choice of the ligands, a shift in the absorption properties in the visible range, while maintaining high efficiency.

To allow such a shift of the absorption properties towards the visible, two bipyridine ligands were synthesized starting from a chalcone. Structures of these chalcones and, therefore, of the corresponding ligands have been selected in order to induce a significant shift in the absorption spectrum of the copper complexes towards the visible range.

For the synthesis of the two chalcones, a Claisen-Schmidt condensation reaction under basic conditions between 2-acetylpyridine A1 or 2-acetyl-6-methylpyridine A2 and aldehyde A3 was carried out (See Scheme 1) [113–125].

These two chalcones engage in a cyclization reaction with β-aminocrotonitrile, allowing the ligands to be formed. This reaction and its corresponding mechanism were described in the literature in 1992 by Masaki Matsui [105,106].

As shown in Scheme 2, the mechanism proposed by Masaki Matsui involves the formation of two (L1 or L2) ligands. β-Aminocrotononitrile can exist as amino (1) and imino (2) isomers in solution. A Michael addition of thae carbanion of imino isomer 3 to C1 or C2 can give intermediate 4, which, after an intramolecular cyclization and dehydration reaction, can provide intermediate 6. At room temperature, dehydrogenation of 6 can give 7.

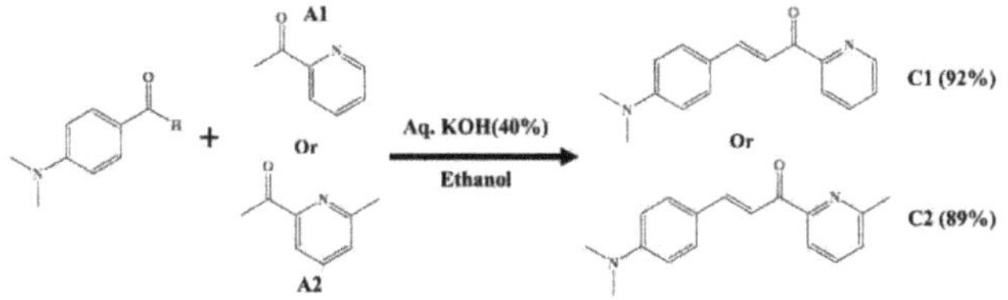

Scheme 1. Synthesis and chemical structures of 2-carbonylpyridine-based chalcones.

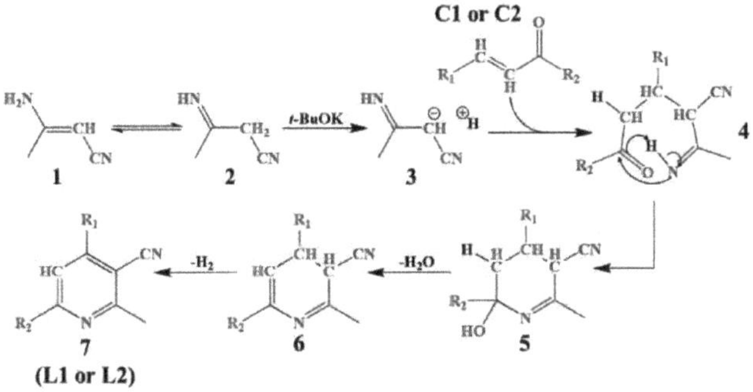

Scheme 2. Probable mechanism described by Masaki Matsui.

Starting from the conditions described by Masaki Matsui, the ligands were indeed obtained. The synthesis conditions were then optimized by using reduced quantities of solvent, base and β-aminocrotonitrile, and with a simplified reaction treatment due to the precipitation of the ligand at the end of the reaction, while maintaining a good yield. The optimized synthesis of the ligands is detailed in the synthetic procedure detailed above.

Heteroleptic copper complexes bearing a pyridine-based chelate ligand and a diphosphine ligand, such as 4,5-*bis*(diphenylphosphino)-9,9-dimethylxanthene (or Xantphos) and *bis*[2-(diphenylphosphino)phenyl]ether (or DPEphos), have previously been reported in the literature. In this work, this second ligand was replaced by a bisisonitrile ligand, namely *bis*(2-isocyanophenyl)phenylphosphonate. Indeed, as previously mentioned in a study reported by Matthias Knorn [107], copper complexes bearing this ligand showed a higher photocatalytic activity than their counterpart comprising a bisphosphine ligand. The lower activity of copper complexes comprising bisphosphine ligands can be assigned to the tendency of heteroleptic complexes to form an equilibrium with their homoleptic forms in solution [108], especially for heteroleptic copper complexes combining biphosphine and phenanthroline ligands. In contrast, in the case of the bisisonitrile ligand, studies have revealed the low ability of heteroleptic complexes to undergo ligand exchanges. The ligand was synthesized following the procedure described in the literature. Using these two ligands, the pyridine ligands and the bisisonitrile ligand in a one-step complexation reaction, the two desired copper complexes were obtained.

3.2. UV-Visible Absorption Spectra of Cu1 and Cu2

Ground-state absorption spectra of the new studied copper derivatives were determined in DCM and the results are presented in Figure 3. Extinction coefficients at different emission wavelengths used in photopolymerization experiments are reported in Table 1. New complexes are characterized by a broad absorption band, which extends between 380 nm and 650 nm, and high-extinction coefficients in the blue region, e.g., ε = 7570 M^{-1}·cm^{-1}, 7040 M^{-1}·cm^{-1} @400 nm for Cu1 and Cu2, respectively. These compounds also have high-extinction coefficients at the emission wavelengths of LEDs (at 405 nm and 455 nm) used in different experiments achieved in this work, for example, $\varepsilon_{@405nm}$ = 8460 and 7950 M^{-1}·cm^{-1} for Cu1 and Cu2, respectively. Remarkably, a bathochromic shift in the absorption spectra of Cu2 was observed compared to that of Cu1. This effect could be related to the presence of a methyl group, which is considered as an electron-donating group (inductive effect), on the pyridine ligands of Cu2 (λ_{max} = 445 nm for Cu1 and 441 nm for Cu2). This difference could also be explained by the optimized geometries, as well as the frontier orbitals (highest occupied molecular orbital—HOMO—and lowest unoccupied molecular orbital—LUMO), which are shown in Figure 4.

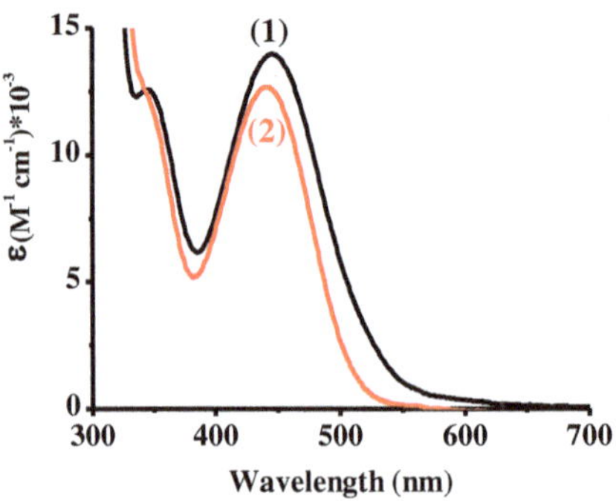

Figure 3. UV-visible absorption spectra of (**1**) Cu1 and (**2**) Cu2 in DCM.

Table 1. Maximum absorption wavelengths (λ_{max}), extinction coefficients at λ_{max}, and molar extinction coefficients for the investigated copper complexes at different emission wavelengths for different LEDs used.

	λ_{max} (nm)	ε_{max} (M^{-1}·cm^{-1})	ε_{375nm} (M^{-1}·cm^{-1})	ε_{405nm} (M^{-1}·cm^{-1})	ε_{455nm} (M^{-1}·cm^{-1})
Cu1	445	14,000	7150	8460	13,600
Cu2	441	12,700	5760	7950	11,740

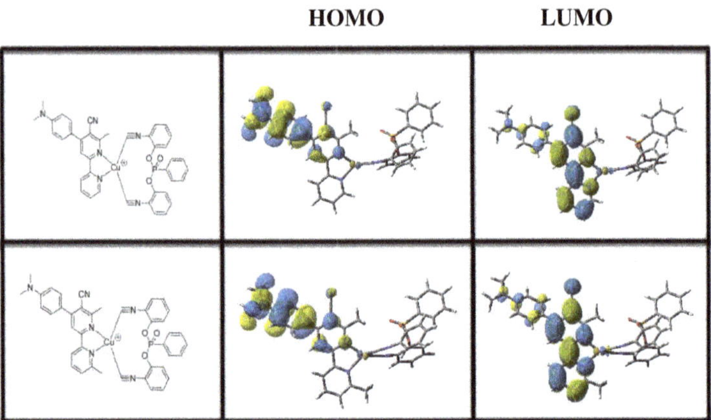

Figure 4. HOMO and LUMO for Cu1 and Cu2 at the UB3LYP/6–31G* level.

3.3. Photopolymerization Experiments

3.3.1. Free Radical Photopolymerization Using TA as a Benchmark Monomer

Due to their good light-absorption properties in the visible range, copper complexes were tested as photoinitiators for the FRP of acrylate-based monomers upon exposure to LEDs at 405 nm (I = 110 mW·cm^{-2}) and 455 nm (I = 75 mW·cm^{-2}).

In fact, photoinitiators (0.1% or 0.2%) were dissolved and mixed into the TA acrylate monomer in combination with Iod salt (1%) in order to make two-component photoinitiating systems, on the one hand, and in combination with Iod/amine (1%/1% w/w) to form three-component photoinitiating systems, on the other hand. Interestingly, these dyes exhibit excellent free radical photopolymerization conversions in thick and thin samples. The related results are gathered in Figure 5 and the data are summarized in Table 2. Remarkably, copper complexes alone, Iod and amine alone cannot polymerize the sample. Iod salt and amine are used as co-initiators in this work because they do not absorb visible light. It is important to introduce the dyes (i.e., the copper complexes) into the photosensitive formulations in order to obtain a good light absorption at 405 nm and 455 nm. The obtained results using copper derivatives in two-component PISs showed that Cu2 was more efficient than Cu1 for the FRP of TA using different PI percentages, e.g., FC ~ 64% for Cu1/Iod (0.1%/1% w/w) vs. 70% for Cu2/Iod (0.1%/1% w/w) (Figure 5A curve 1 vs. 2), and FC ~ 62% for Cu1/Iod (0.2%/1% w/w) vs. 85% for Cu2/Iod (0.2%/1% w/w) (Figure 5A curve 3 vs. 4).

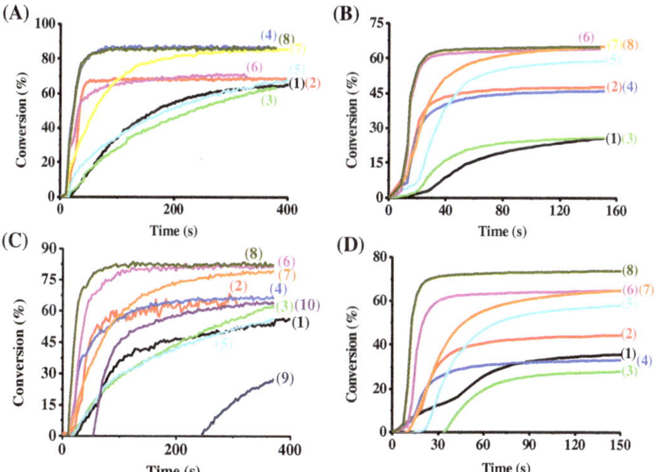

Figure 5. Free radical photopolymerization profiles of acrylate functions vs. irradiation time in: (**A**) thick samples @455 nm, (**B**) thin sample @455 nm, (**C**) thick sample @405 nm and (**D**) thin sample @405 nm: (1) Cu1/Iod (0.1%/1% w/w), (2) Cu2/Iod (0.1%/1% w/w), (3) Cu1/Iod (0.2%/1% w/w), (4) Cu2/Iod (0.2%/1% w/w), (5) Cu1/Iod/EDB (0.1%/1%/1% $w/w/w$), (6) Cu2/Iod/EDB (0.1%/1%/1% $w/w/w$), (7) Cu1/Iod/NPG (0.1%/1%/1% $w/w/w$), (8) Cu2/Iod/NPG (0.1%/1%/1% $w/w/w$), (9) Iod/EDB (1%/1% w/w) and (10) Iod/NPG (1%/1% w/w). Irradiation starts at t = 10 s.

Furthermore, Iod/NPG couple showed a weak polymerization initiation ability upon exposure to LEDs at 405 nm and 455 nm after 60 s (e.g., FC ~ 10% @405 nm). Interestingly, a greater efficiency was observed when NPG was incorporated into the formulation. Compared to their two-component system analogues, the different three-component PISs showed a better final conversion of reactive functions and a higher polymerization rate upon irradiation with LEDs at 405 nm or 455 nm (for example, an FC up to 86% is obtained with Cu1/Iod/NPG (0.1%/1%/1% $w/w/w$), and 88% using Cu2/Iod/NPG (0.1%/1%/1% $w/w/w$) with a LED @455 nm).

Table 2. Final reactive functions conversion (FC) for TA monomer using two or three-component PISs upon exposure at different wavelengths (LED at 405 and 455 nm).

	At 405 nm				At 455 nm			
	Thick Sample		Thin Sample		Thick Sample		Thin Sample	
Cu1/Iod	56% [a]	62% [b]	36% [a]	28% [b]	64% [a]	64% [b]	26% [a]	26% [b]
Cu2/Iod	64% [a]	67% [b]	45% [a]	33% [b]	70% [a]	87% [b]	48% [a]	46% [b]
Cu1/Iod/amine	57% [c]	80% [d]	58% [c]	65% [d]	69% [c]	85% [d]	59% [c]	65% [d]
Cu2/Iod/amine	82% [c]	83% [d]	65% [c]	74% [d]	71% [c]	87% [d]	64% [c]	65% [d]

[a]: Cu/Iod (0.1%/1% *w/w*), [b]: Cu/Iod (0.2%/1% *w/w*), [c]: Cu/Iod/EDB (0.1%/1%/1% *w/w/w*), [d]: Cu/Iod/NPG (0.1%/1%/1% *w/w/w*).

3.3.2. Cationic Polymerization and IPN Synthesis

Typical epoxide function conversion-time profiles for Cu1 and Cu2-based photoinitiating systems are given in Figure 6 and the data are gathered in Table 3. In fact, the cationic polymerization of the epoxide functions was carried out under air and upon irradiation at 405 nm. Indeed, the cationic polymerization is insensitive to oxygen. As expected, copper complexes alone and the additives alone were not able to initiate the CP in these irradiation conditions. The addition of Iod salt or Iod/NVK into the formulation containing the PI induced good photopolymerization profiles, i.e., the combination Cu/Iod/NVK (0.1%/2%/3% *w/w/w*) is very efficient to produce polymer materials in terms of Rp and final epoxy function conversion compared to Cu/Iod (0.1%/1% *w/w*), e.g., (FC ~ 50% for Cu1/Iod/NVK (0.1%/2%/3% *w/w/w*) vs. 27% for Cu1/Iod (0.1%/1% *w/w*)). The consumption of epoxide functions was accompanied by the formation of a polyether network (appearance of peak at ~1080 cm^{-1}), characterizing the obtained polymer.

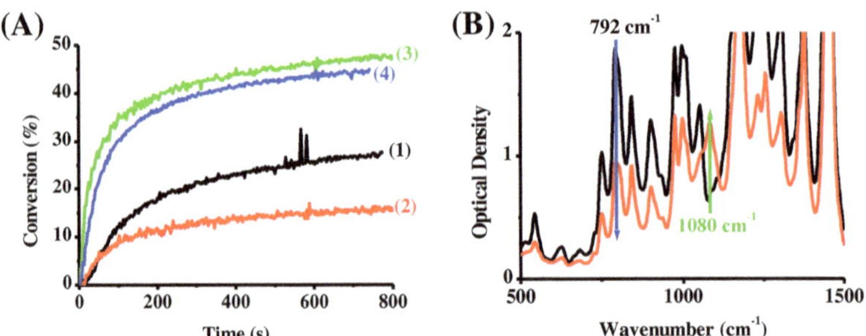

Figure 6. (**A**) Cationic polymerization (CP) profiles of epoxy functions (thin sample) vs. irradiation time upon exposure to LED at 405 nm: (1) Cu1/Iod (0.1%/1% *w/w*), (2) Cu2/Iod (0.1%/1% *w/w*), (3) Cu1/Iod/NVK (0.1%/2%/3% *w/w/w*) and (4) Cu2/Iod/NVK (0.1%/2%/3% *w/w/w*). (**B**) IR spectra recorded before and after polymerization. Irradiation starts at *t* = 10 s.

Table 3. Final reactive function conversions for EPOX monomer using two and three-component PISs upon irradiation at 405 nm during the polymerization of thin samples.

	Cu/Iod (0.1%/1% *w/w*)	**Cu/Iod/NVK (0.1%/2%/3% *w/w/w*)**
Cu1	27%	49%
Cu2	16%	45%

On the other hand, IPNs syntheses were also carried out in this work and polymerization tests were performed in thick and thin samples using LEDs at 405 nm and 455 nm (See Tables 4 and 5). Photopolymerization profiles for the IPN formation are presented in Figure 7. For example, the acrylic network formation was very fast with a high final conversion (98%) for Cu2/Iod/NPG (0.1%/1%/1% *w/w/w*) in TA/EPOX (50%/50%) upon irradiation at 455 nm, and the formation of the epoxy network was also efficient (high final conversion and Rp) using this system (FC ~ 55%).

Table 4. Final conversions of acrylate and epoxides functions for the IPN synthesis of TA/EPOX blend using Cu1 (or Cu2)/Iod/NPG (0.1%/1%/1%/ *w/w/w*) as PIS upon exposure to visible light at 405 nm.

	IPN Synthesis of TA/EPOX Blend Performed in Thick Sample at 405 nm			IPN Synthesis of TA/EPOX Blend Performed in Thin Sample at 405 nm		
	30%/70%	50%/50%	30%/70%	30%/70%	50%/50%	70%/30%
Cu1	90%/25%	90%/15%	93%/27%	88%/25%	87%/31%	84%/47%
Cu2	99%/30%	98%/20%	96%/38%	92%/22%	90%/35%	80%/32%

Table 5. Final conversions of acrylate and epoxides functions for the IPN synthesis of TA/EPOX blend using Cu1 (or Cu2)/Iod/NPG (0.1%/1%/1%/ *w/w/w*) as PIS upon exposure to visible light at 455 nm.

	IPN Synthesis of TA/EPOX Blend Performed in Thick Sample @455 nm			IPN Synthesis of TA/EPOX Blend Performed in Thin Sample @455 nm		
	30%/70%	50%/50%	30%/70%	30%/70%	50%/50%	70%/30%
Cu1	90%/30%	90%/23%	90%/41%	90%/15%	90%/25%	90%/15%
Cu2	100%/38%	98%/55%	99%/51%	98%/22%	99%/30%	98%/22%

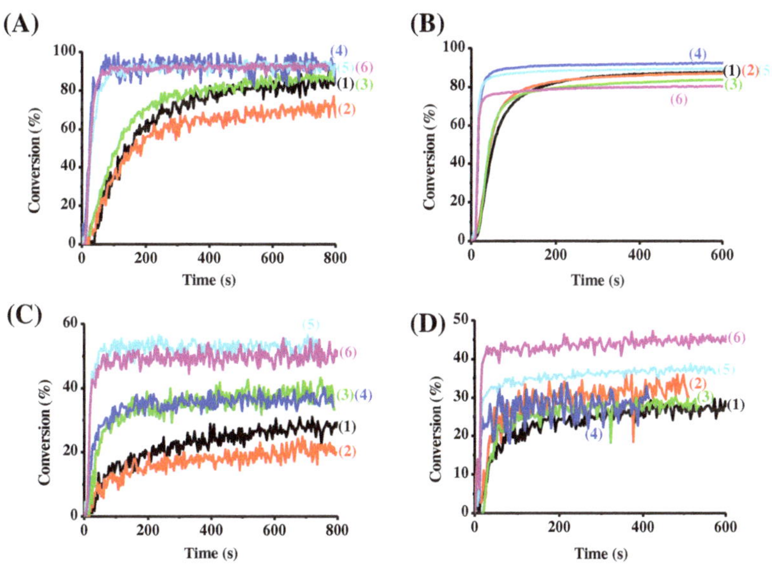

Figure 7. Photopolymerization profiles of acrylate functions (**A**) (thick sample), (**B**) (thin sample)) upon irradiation at 455 nm, and epoxide function ((**C**) (thick sample), (**D**) (thin sample)) upon exposure to LED at 405 nm of TA/EPOX using Cu1/Iod/NPG (0.1%/1%/1% *w/w/w*): (1) 30%/70%, (2) 50%/50%, (3) 70%/30%, and Cu2/Iod/NPG (0.1%/1%/1% *w/w/w*): (4) 30%/70%, (5) 50%/50%, (6) 70%/30%. The irradiation starts at t = 10 s.

3.4. Photocomposites Synthesis

Nowadays, many of our modern technologies require materials with enhanced properties. This is particularly true for materials used in aerospace, underwater and transportation applications. For example, for aeronautical applications, engineers research materials with properties of low density, rigid, solid, impact resistance, temperature and pressure resistance and obviously materials that do not easily corrode. For this purpose, composite materials have been used for different applications. By definition, a composite material is composed of a least two components that results in better properties than those of the individual components used alone: matrix (monomer blends) and reinforcement. The main advantages of composite materials are their high stiffness, strength and low density. The introduction of light for the synthesis of photocomposites will make the manufacture of these materials more ecological.

In this study, the matrix is based on acrylic monomers, such as TMPTA or TA, and the second component (reinforcement) is based on glass fibers. Firstly, the acrylic resins were deposed on the reinforcement (50%/50% w/w) and the mixtures were irradiated using an LED conveyor at 395 nm. Interestingly, a very fast polymerization on the surface and the bottom was observed with tack-free surfaces, after one pass only using one layer of glass fibers (1 mm). Increasing the reinforcement thickness by adding several layers, the polymerization on the surface is always fast and takes place after one pass, but the curing on the bottom is more complicated and will be done after several passes using Cu1 or Cu2/Iod/NPG (0.1%/1%/1% $w/w/w$) as PISs. The curing photocomposite results are depicted in Figure 8 and Table 6.

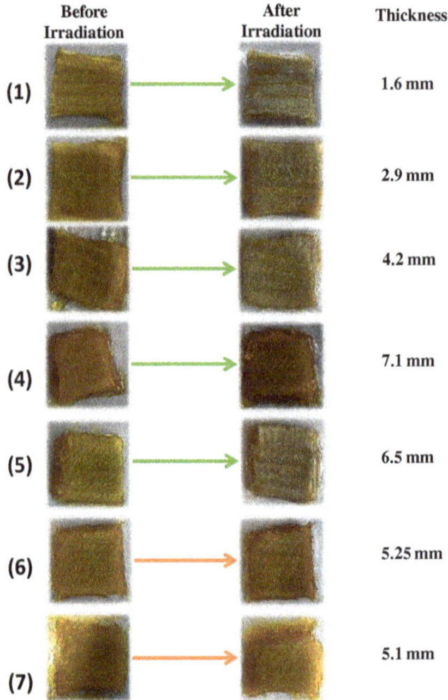

Figure 8. Free radical photopolymerization for the photocomposite synthesis upon near-UV irradiation at 395 nm (4 W·cm^{-2}) using Cu2/Iod/NPG (0.1%/1%/1% $w/w/w$) in TA [(1)–(4)], (5)) Cu2/Iod/NPG (0.1%/1%/1% $w/w/w$) in TMPTA, (6) Cu1/Iod/NPG (0.1%/1%/1% $w/w/w$) in TMPTA and (7) Cu1/Iod/NPG (0.1%/1%/1% $w/w/w$) in TA.

3.5. Direct Laser Write (DLW)

The new copper complexes were tested in some direct laser write experiments for the FRP of TMPTA or TA using a laser diode at 405 nm (spot size: 50 µm). The obtained 3D patterns were carried out under air and using different PISs based on Cu1/Iod/TMA, Cu2/Iod/TMA in TA or TMPTA (Figure 9). Due to their high ability to initiate the FRP of acrylates, these systems were able to generate high-spatial-resolution 3D patterns with a great thickness of curing (~2500 µm) in the irradiated area. As such, 3D patterns were generated with very short irradiation times (2–3 min) and they were characterized by numerical microscopy.

Table 6. Photocomposite synthesis results using TA as acrylic monomer (or TMPTA) and number of passes to reach the tack-free character on the surfaces.

	Thickness	Number of Passes to Reach Tack-Free Character on the Surface	Number of Passes to Reach Tack-Free Character on the Bottom
Cu2/Iod/NPG	1.6 mm	1	1
Cu2/Iod/NPG	2.9 mm	1	2
Cu2/Iod/NPG	4.2 mm	1	6
Cu2/Iod/NPG (TMPTA)	7.1 mm	1	25
Cu2/Iod/NPG	6.5 mm	1	30
Cu1/Iod/NPG	5.25 mm	1	40
Cu1/Iod/NPG (TMPTA)	5.1 mm	1	45

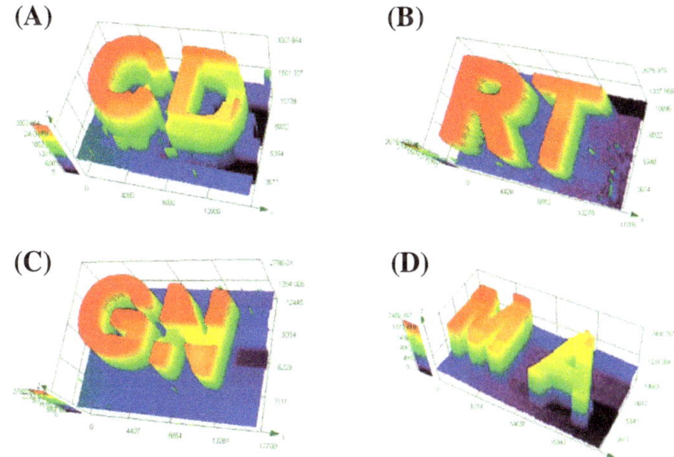

Figure 9. 3D patterns produced by free radical photopolymerization of acrylate functions (TA or TMPTA) using a laser diode at 405 nm and their characterization by numerical optical microscopy: (**A**) Cu2/Iod/TMA (0.057%/1%/0.46% *w/w/w*) in TA, (**B**) Cu1/Iod/TMA (0.058%/0.506%/0.44% *w/w/w*) in TMPTA; (**C**) Cu2/Iod/TMA (0.048%/0.65%/0.259% *w/w/w*) in TA and (**D**) Cu1/Iod/TMA (0.05%/0.5%/0.305% *w/w/w*).

3.6. Mechanical Properties: Tensile Test Measurements

The tensile strength of IPNs synthesized using different compositions of the TA/EPOX mixture are presented in Table 7. The results show that with the increase in the percentage of acrylic monomer, the tensile strength increases, which may be due to the rigid character of the acrylates (e.g., 7.2 MPa for Cu2/Iod/NPG in TA/EPOX (30%/70%) vs. 37.2 MPa for the same system in TA/EPOX (70%/30%)).

Table 7. The tensile strength of IPN materials as a result of hybrid polymerization of the TA/EPOX mixture using Cu/Iod/NPG (0.1%/1%/1% *w/w/w*) as a photoinitiating system.

Tensile Strength [MPa]	0.1%PA/Iod/NPG TA/EPOX (30% 70%) @ 395 nm	0.1% PA/Iod/NPG TA/EPOX (50% 50%) @ 395 nm	0.1% PA/Iod/NPG TA/EPOX (70% 30%) @ 395 nm
Cu1	6.5	7.3	34.3
Cu 2	7.2	26.3	37.2

4. Discussion

In order to explain the initiating ability of the organometallic complexes, their photochemical and photophysical properties were studied using different characterization techniques, allowing for the characterization of the associated chemical mechanisms.

4.1. Steady-State Photolysis of the Investigated Compounds

Photolyses of Cu1 and Cu2 dyes in DCM were investigated upon irradiation at 375 nm and 405 nm, and the related results are shown in Figure 10. First of all, no photolysis occurred for Cu1 and Cu2 alone (0% consumption) upon irradiation at 375 nm and 405 nm, but the incorporation of the iodonium salt into the photosensitive solution could promote the degradation of the dyes, so that a strong decrease in the absorbance band intensity was observed by increasing the irradiation time, e.g., consumption ~ 80% @375 nm and 82% for Cu2/Iod at 375 nm and 405 nm, respectively (Figure 11B). It is important to note that the photolysis of Cu1 in the presence of Iod salt involved the formation of a photoproduct after 60 s of irradiation in the solution, which had an absorption band more shifted in the visible (bathochromic effect) spectrum, then this photoproduct degrades under the effect of the irradiation (Figure 10A).

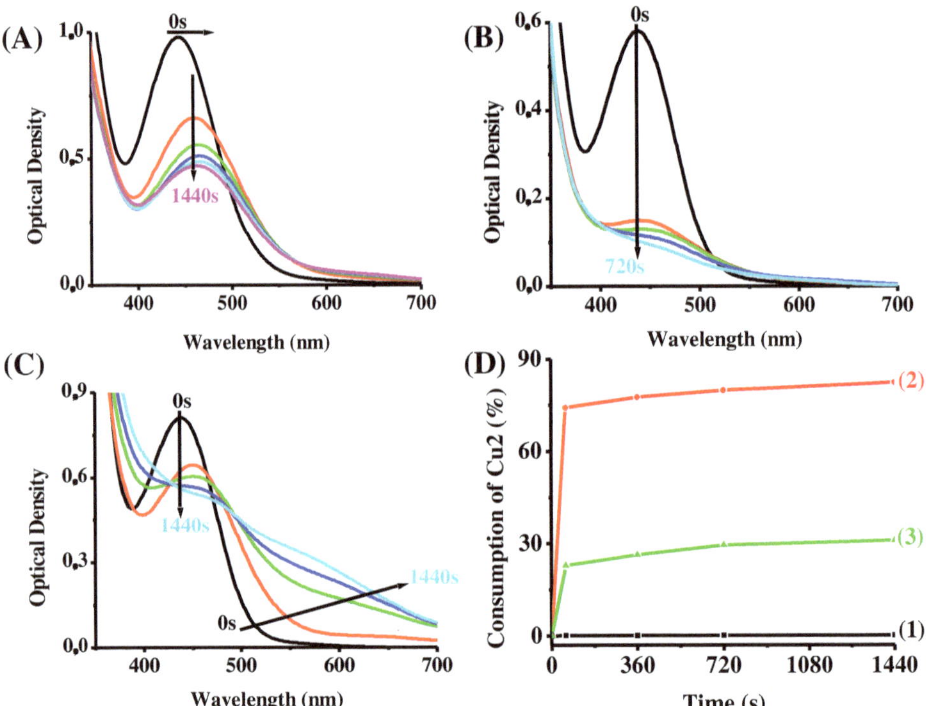

Figure 10. Photolysis experiments of (**A**) Cu1 with Iod salt (10^{-2} M), (**B**) Cu2 with Iod salt (10^{-2} M) and (**C**) Cu2/Iod/NPG (10^{-2} M) upon exposure to LED at 405 nm. (**D**) Consumption percentage of Cu2: (1) alone, (2) with Iod, (3) with Iod/NPG.

This difference between these two consumption percentages may be due to the high light-absorption ability of Cu2 at 405 nm, as well as the highest intensity of the LED at 405 nm (110 mW·cm^{-2}), compared to that at 375 nm (75 mW·cm^{-2}). Furthermore, in the case of three-component PISs, the consumption of Cu2 was lower compared to that of the two-component Cu2/Iod system. It can be confidently assigned to the regeneration of Cu2 in the three-component system due to the presence of the sacrificial amine or the formation of new photoproducts (%consumption ~ 31%).

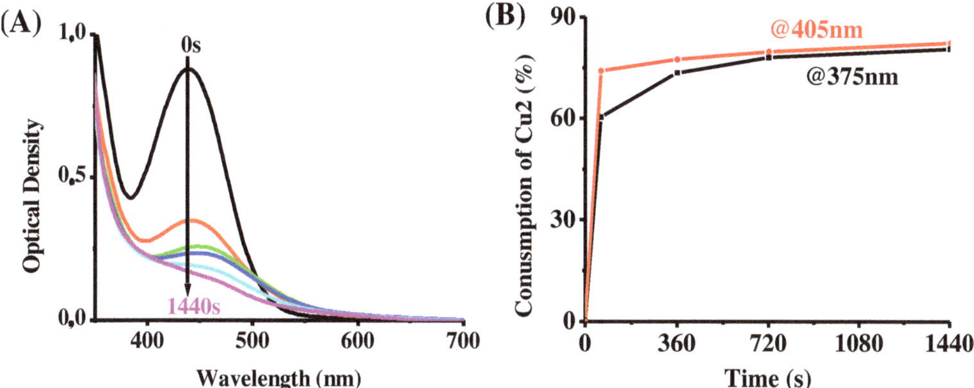

Figure 11. (**A**) Photolysis of Cu2 with Iod (10^{-2} M) upon irradiation @375 nm. (**B**) Consumption of Cu2 with irradiation @375 nm and 405 nm.

4.2. Photoluminescence and Electrochemical Properties

Fluorescence emission spectra, fluorescence quenching (measured using a JASCO FP-6200 spectrofluorimeter, Lisses, France) and the oxidation potential (measured in DCM by cyclic voltammetry, OrigaLys, Rillieux-la-Pape, France) results of the different Cu derivatives are gathered in Figure 12 and Table 8. The excited-state energy was calculated from the crossing point of the emission and absorption spectra. Using these different values, the free-energy change (ΔG) could be calculated; this parameter reflects the reactivity between Cu and Iod. In fact, a slight decrease in the fluorescence intensity was observed for Cu1/Iod, but this emission spectra showed a strong decrease for Cu2 upon addition of Iod. These behaviors explain the high reactivity of Cu2/Iod compared to Cu1/Iod e.g., ϕ = 0.55 for Cu1/Iod vs. 0.74 for Cu2/Iod. In addition, the ΔG value is negative for both complexes, so that the photo-oxidation interaction Cu1 (or Cu2)/Iod is favorable in both cases, with a superiority observed for Cu2 (ΔG = -0.74 and 0.62 eV for Cu2 and Cu1, respectively). It, therefore, explains the high-photoinitiation ability of Cu2 compared to Cu1.

Finally, the initiation ability of the new copper complexes could be explained by different characterization techniques, which allowed us to propose a chemical photoinitiation mechanism. Firstly, Cu is excited upon irradiation at 405 or 455 nm and interacts with Iod to generate aryl radical (Ar$^\bullet$) and radical Cu$^{\bullet+}$ [r1–r2]. A charge transfer complex CTC can be formed after adding NPG into the photosensitive formulation. This complex is able to produce aryl radicals as active species for the radical photopolymerization [r3–r4]. Then, ^{1,3}Cu could react with NPG and generate two radicals (NPG$_{-H}{}^\bullet$, Cu-H$^\bullet$) [r5], and the first radical can undergo a decarboxylation and produce active radicals (NPG$_{(-H,-CO2)}{}^\bullet$) [r6]. This radical can also lead to the formation of two active species after interaction with Iod salt (NPG$_{(-H,-CO2)}{}^+$, Ar$^\bullet$) [r7]. Lastly, copper complex derivatives are regenerated [r8–r9] (See Scheme 3).

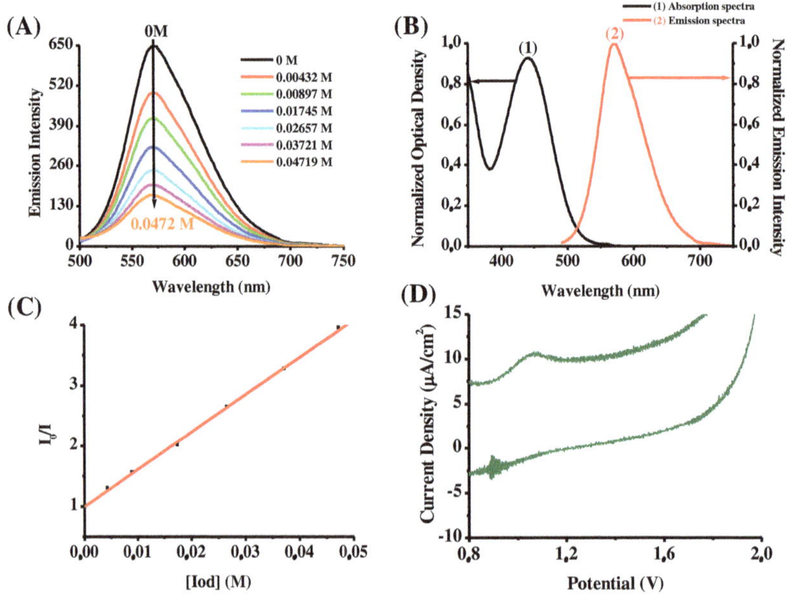

Figure 12. (**A**) Fluorescence quenching of Cu2 by Iod salt, (**B**) E_{S1} determination for Cu2, (**C**) Stern–Volmer coefficient determination of Cu2 quenched by Iod and (**D**) oxidation potential determination of Cu2.

Table 8. Parameters characterizing the chemical mechanisms between Cu1 (or Cu2) and Iod. For Iod, a reduction potential of -0.7 eV was used for the ΔGet calculations.

	E_{ox} (V)	E_{S1} (eV)	$\Delta G_{(Cu/Iod)}$ (eV)	$K_{SV(Cu/Iod)}$	$\Phi_{(Cu/Iod)}$	E_{ox} (V)
Cu1	1.07	2.34	-0.62	13.58	0.55	Cu1
Cu2	0.97	2.41	-0.74	61.83	0.74	Cu2

$Cu\ (h\nu) \rightarrow {}^{1,3}Cu$	r1
${}^{1,3}Cu + Ar_2I^+ \rightarrow Ar^\bullet + ArI + Cu^{\bullet+}$	r2
$NPG + Iod \rightarrow [NPG\text{-}Iod]_{CTC}$	r3
$[NPG\text{-}Iod]_{CTC} \rightarrow Ar^\bullet$	r4
${}^{1,3}Cu + NPG \rightarrow Cu\text{-}H^\bullet + NPG_{(-H)}^\bullet$	r5
$NPG_{(-H)}^\bullet \rightarrow NPG_{(-H;-CO2)}^\bullet + CO_2$	r6
$NPG_{(-H;-CO2)}^\bullet + Ar_2I^+ \rightarrow NPG_{(-H;-CO2)}^+ + Ar^\bullet + ArI$	r7
$Cu^{\bullet+} + NPG \rightarrow Cu + NPG^+$	r8
$Cu\text{-}H^\bullet + Ar_2I^+ \rightarrow Ar^\bullet + ArI + Cu + H^+$	r9

Scheme 3. Proposed chemical mechanisms.

5. Conclusions

In the present paper, new copper complexes were synthesized and tested as photoinitiators. These compounds have strong visible-light absorption and are able to initiate both the free radical photopolymerization and cationic polymerization. IPN synthesis through the simultaneous polymerization of acrylate/epoxy monomer blends was performed under air upon irradiation at 405 nm and 455 nm, using a very low quantity of copper complex, in two or three-component PIS. Cu2 showed a very interesting photoinitiation capacity compared to Cu1 in terms of final conversions of reactive functions and polymerization rates. The high reactivity of these compounds was demonstrated through some direct laser write experiments, where high-spatial-resolution 3D patterns were obtained. In addition, the synthesis of thick glass fiber photocomposites was possible. This work paves the way for the development of new organometallic photoinitiators.

Author Contributions: Conceptualization, F.D. and J.L.; methodology, M.R., G.N., B.G., J.T., T.H., D.G., F.D. and J.L.; software, B.G.; validation, M.R., G.N., B.G., J.T., T.H., D.G., F.D. and J.L.; formal analysis, M.R., G.N., B.G., J.T., T.H., D.G., F.D. and J.L.; investigation, M.R., G.N., B.G., J.T., T.H., D.G., F.D. and J.L.; resources, F.D. and J.L.; data curation, F.D. and J.L.; writing—original draft preparation, M.R., G.N., B.G., J.T., T.H., D.G., F.D. and J.L.; writing—review and editing, M.R., G.N., B.G., J.T., T.H., D.G., F.D. and J.L.; visualization, F.D.; supervision, M.R., G.N., B.G., J.T., T.H., D.G., F.D. and J.L.; project administration, M.R., G.N., B.G., J.T., T.H., D.G., F.D. and J.L.; funding acquisition, M.R., G.N., B.G., J.T., T.H., D.G., F.D. and J.L. All authors have read and agreed to the published version of the manuscript.

Funding: This research was funded by Aix Marseille University and the Centre National de la Recherche Scientifique.

Data Availability Statement: Not applicable.

Conflicts of Interest: The authors declare no conflict of interest.

References

1. Fouassier, J.P.; Lalevée, J. *Photoinitiators for Polymer Synthesis, Scope, Reactivity, and Efficiency*; Wiley-VCH Verlag: Weinheim, Germany, 2012.
2. Fouassier, J.P. *Photoinitiator, Photopolymerization and Photocuring: Fundamentals and Applications*; Hanser Publishers: Munich, Germany, 1995.
3. Dietliker, K.A. *Compilation of Photoinitiators Commercially Available for UV Today*; Sita Technology Ltd.: London, UK, 2002.
4. Davidson, S. *Exploring the Science, Technology and Application of UV and EB Curing*; Sita Technology Ltd.: London, UK, 1999.
5. Crivello, J.V.; Dietliker, K.; Bradley, G. *Photoinitiators for Free Radical Cationic & Anionic Photopolymerisation*; John Wiley & Sons: Chichester, UK, 1999.
6. Cunningham, A.F.; Desobry, V. *Radiation Curing in Polymer Science and Technology*; Fouassier, J.P., Rabek, J.F., Eds.; Elsevier: Barking, UK, 1993; Volume 2, p. 323.
7. Andrzejewska, E. Chapter 2—Free Radical Photopolymerization of Multifunctional Monomers. In *Three-Dimensional Microfabrication Using Two-Photon Polymerization*; Baldacchini, T., Ed.; Micro and Nano Technologies; William Andrew Publishing: Oxford, UK, 2016; pp. 62–81. ISBN 978-0-323-35321-2.
8. Luu, T.T.H.; Jia, Z.; Kanaev, A.; Museur, L. Effect of Light Intensity on the Free-Radical Photopolymerization Kinetics of 2-Hydroxyethyl Methacrylate: Experiments and Simulations. *J. Phys. Chem. B* **2020**, *124*, 6857–6866. [CrossRef] [PubMed]
9. Andrezajewska, E.; Grajek, K. Recent Advances in Photo-Induced Free-Radical Polymerization. *MOJ Polym. Sci.* **2017**, *1*, 58–60. [CrossRef]
10. Li, Z.; Zou, X.; Shi, F.; Liu, R.; Yagci, Y. Highly Efficient Dandelion-like near-Infrared Light Photoinitiator for Free Radical and Thiol-Ene Photopolymerizations. *Nat. Commun.* **2019**, *10*, 3560. [CrossRef] [PubMed]
11. Sangermano, M.; Roppolo, I.; Chiappone, A. New Horizons in Cationic Photopolymerization. *Polymers* **2018**, *10*, 136. [CrossRef]
12. Zhou, Y.; Liao, W.; Ni, X. Cationic Photopolymerization Initiated by a Photocatalytic Complex Sensitive to Visible Light at 520 Nm. *Catal. Lett.* **2021**, *151*, 1766–1775. [CrossRef]
13. Bouzrati-Zerelli, M.; Maier, M.; Dietlin, C.; Morlet-Savary, F.; Fouassier, J.P.; Klee, J.E.; Lalevée, J. A novel photoinitiating system producing germyl radicals for the polymerization of representative methacrylate resins: Camphorquinone/R3GeH/iodonium salt. *Dent. Mater.* **2016**, *32*, 1226. [CrossRef]
14. Abdul-Monem, M.M. Naturally Derived Photoinitiators for Dental and Biomaterials Applications. *Eur. Dent. Res. Biomater. J.* **2021**, *1*, 72–78. [CrossRef]
15. Alizadehgharib, S.; Östberg, A.-K.; Dahlstrand Rudin, A.; Dahlgren, U.; Christenson, K. The Effects of the Dental Methacrylates TEGDMA, Bis-GMA, and UDMA on Neutrophils in Vitro. *Clin. Exp. Dent. Res.* **2020**, *6*, 439–447. [CrossRef]
16. Arikawa, H.; Takahashi, H.; Kanie, T.; Ban, S. Effect of Various Visible Light Photoinitiators on the Polymerization and Color of Light-Activated Resins. *Dent. Mater. J.* **2009**, *28*, 454–460. [CrossRef]
17. Catarzi, D.; Cecchi, L.; Colotta, V.; Melani, F.; Filacchioni, G.; Martini, C.; Giusti, L.; Lucacchini, A. Structure-Activity Relationships of 1,2,4-Triazolo [1,5-a]Quinoxalines and Their 1-Deaza Analogs Imidazo [1,2-a]Quinoxalines at the Benzodiazepine Receptor. *J. Med. Chem.* **1994**, *37*, 2846–2850. [CrossRef]
18. Boeira, P.O.; Meereis, C.T.W.; Suárez, C.E.C.; de Almeida, S.M.; Piva, E.; da Silveira Lima, G. Coumarin-Based Iodonium Hexafluoroantimonate as an Alternative Photoinitiator for Experimental Dental Adhesives Resin. *Appl. Adhes. Sci.* **2017**, *5*, 2. [CrossRef]

19. Cadenaro, M.; Antoniolli, F.; Codan, B.; Agee, K.; Tay, F.R.; Dorigo, E.D.S.; Pashley, D.H.; Breschi, L. Influence of Different Initiators on the Degree of Conversion of Experimental Adhesive Blends in Relation to Their Hydrophilicity and Solvent Content. *Dent. Mater.* **2010**, *26*, 288–294. [CrossRef] [PubMed]
20. Dickens, S.H.; Stansbury, J.W.; Choi, K.M.; Floyd, C.J.E. Photopolymerization Kinetics of Methacrylate Dental Resins. *Macromolecules* **2003**, *36*, 6043–6053. [CrossRef]
21. Ikemura, K.; Endo, T. A Review of the Development of Radical Photopolymerization Initiators Used for Designing Light-Curing Dental Adhesives and Resin Composites. *Dent. Mater. J.* **2010**, *29*, 481–501. [CrossRef]
22. Maffezzoli, A.; Pietra, A.D.; Rengo, S.; Nicolais, L.; Valletta, G. Photopolymerization of Dental Composite Matrices. *Biomaterials* **1994**, *15*, 1221–1228. [CrossRef]
23. Neumann, M.G.; Schmitt, C.C.; Ferreira, G.C.; Corrêa, I.C. The Initiating Radical Yields and the Efficiency of Polymerization for Various Dental Photoinitiators Excited by Different Light Curing Units. *Dent. Mater.* **2006**, *22*, 576–584. [CrossRef]
24. Maruno, T.; Murata, N. Properties of a UV-Curable, Durable Precision Adhesive. *J. Adhes. Sci. Technol.* **1995**, *9*, 1343. [CrossRef]
25. Besse, V.; Derbanne, M.A.; Pham, T.-N.; Cook, W.D.; Le Pluart, L. Photopolymerization Study and Adhesive Properties of Self-Etch Adhesives Containing Bis(Acyl)Phosphine Oxide Initiator. *Dent. Mater.* **2016**, *32*, 561–569. [CrossRef]
26. Gziut, K.; Kowalczyk, A.; Schmidt, B.; Kowalczyk, K.; Weisbrodt, M. Epoxy-Based Structural Self-Adhesive Tapes Modified with Acrylic Syrups Prepared via a Free Radical Photopolymerization Process. *Polymers* **2021**, *13*, 189. [CrossRef]
27. Zhu, M.; Cao, Z.; Zhou, H.; Xie, Y.; Li, G.; Wang, N.; Liu, Y.; He, L.; Qu, X. Preparation of Environmentally Friendly Acrylic Pressure-Sensitive Adhesives by Bulk Photopolymerization and Their Performance. *RSC Adv.* **2020**, *10*, 10277–10284. [CrossRef]
28. Kilponen, L.; Uusitalo, E.; Tolvanen, M.; Varrela, J.; Vallittu, P.K. Photopolymerization of Light Curing Adhesives Used with Metal Orthodontic Brackets and Matrices. *J. Biomater. Tissue Eng.* **2016**, *6*, 659–664. [CrossRef]
29. Wu, L.; Baghdachi, J. *Functional Polymer Coatings: Principles, Methods, and Applications*; Wiley Series on Polymer Engineering and Technology: New York, NY, USA, 2015.
30. Al Rashid, A.; Ahmed, W.; Khalid, M.Y.; Koç, M. Vat Photopolymerization of Polymers and Polymer Composites: Processes and Applications. *Addit. Manuf.* **2021**, *47*, 102279. [CrossRef]
31. Pagac, M.; Hajnys, J.; Ma, Q.-P.; Jancar, L.; Jansa, J.; Stefek, P.; Mesicek, J. A Review of Vat Photopolymerization Technology: Materials, Applications, Challenges, and Future Trends of 3D Printing. *Polymers* **2021**, *13*, 598. [CrossRef] [PubMed]
32. Noè, C.; Hakkarainen, M.; Malburet, S.; Graillot, A.; Adekunle, K.; Skrifvars, M.; Sangermano, M. Frontal-Photopolymerization of Fully Biobased Epoxy Composites. *Macromol. Mater. Eng.* **2021**, 2100864. [CrossRef]
33. Dikova, T.; Maximov, J.; Todorov, V.; Georgiev, G.; Panov, V. Optimization of Photopolymerization Process of Dental Composites. *Processes* **2021**, *9*, 779. [CrossRef]
34. Mokbel, H.; Anderson, D.; Plenderleith, R.; Dietlin, C.; Morlet-Savary, F.; Dumur, F.; Gigmes, D.; Fouassier, J.P.; Lalevée, J. Copper Photoredox Catalyst "G1": A New High Performance Photoinitiator for near-UV and Visible LEDs. *Polym. Chem.* **2017**, *8*, 5580. [CrossRef]
35. Pereira, R.F.; Bártolo, P.J. Photopolymerizable hydrogels in regenerative medicine and drug delivery. In *Hot Topics in Biomaterials*; Future Science Book Series Volume 6; Future Science Ltd.: London, UK, 2014.
36. Xu, X.; Awad, A.; Robles-Martinez, P.; Gaisford, S.; Goyanes, A.; Basit, A.W. Vat Photopolymerization 3D Printing for Advanced Drug Delivery and Medical Device Applications. *J. Control. Release* **2021**, *329*, 743–757. [CrossRef]
37. Chiulan, I.; Heggset, E.B.; Voicu, Ş.I.; Chinga-Carrasco, G. Photopolymerization of Bio-Based Polymers in a Biomedical Engineering Perspective. *Biomacromolecules* **2021**, *22*, 1795–1814. [CrossRef]
38. Elisseeff, J.; Anseth, K.; Sims, D.; McIntosh, W.; Randolph, M.; Langer, R. Transdermal Photopolymerization for Minimally Invasive Implantation. *Proc. Natl. Acad. Sci. USA* **1999**, *96*, 3104–3107. [CrossRef]
39. Bao, Y.; Paunović, N.; Leroux, J.-C. Challenges and Opportunities in 3D Printing of Biodegradable Medical Devices by Emerging Photopolymerization Techniques. *Adv. Funct. Mater.* **2022**, *32*, 2109864. [CrossRef]
40. Baroli, B. Photopolymerization of Biomaterials: Issues and Potentialities in Drug Delivery, Tissue Engineering, and Cell Encapsulation Applications. *J. Chem. Technol. Biotechnol.* **2006**, *81*, 491–499. [CrossRef]
41. Lee, J.Y.; An, J.; Chua, C.K. Fundamentals and Applications of 3D Printing for Novel Materials. *Appl. Mater. Today.* **2017**, *7*, 120. [CrossRef]
42. Andreu, A.; Su, P.-C.; Kim, J.-H.; Ng, C.S.; Kim, S.; Kim, I.; Lee, J.; Noh, J.; Subramanian, A.S.; Yoon, Y.-J. 4D Printing Materials for Vat Photopolymerization. *Addit. Manuf.* **2021**, *44*, 102024. [CrossRef]
43. Bagheri, A.; Jin, J. Photopolymerization in 3D Printing. *ACS Appl. Polym. Mater.* **2019**, *1*, 593–611. [CrossRef]
44. Tang, Y.; Dai, B.; Su, B.; Shi, Y. Recent Advances of 4D Printing Technologies Toward Soft Tactile Sensors. *Front. Mater.* **2021**, *8*, 658046. [CrossRef]
45. Imrie, P.; Jin, J. Polymer 4D Printing: Advanced Shape-Change and Beyond. *J. Polym. Sci.* **2022**, *60*, 149–174. [CrossRef]
46. Zhang, Z.; Corrigan, N.; Bagheri, A.; Jin, J.; Boyer, C. A Versatile 3D and 4D Printing System through Photocontrolled RAFT Polymerization. *Angew. Chem.* **2019**, *131*, 18122–18131. [CrossRef]
47. Schwartz, J.J.; Boydston, A.J. Multimaterial Actinic Spatial Control 3D and 4D Printing. *Nat. Commun.* **2019**, *10*, 791. [CrossRef]
48. Jeong, H.Y.; Woo, B.H.; Kim, N.; Jun, Y.C. Multicolor 4D Printing of Shape-Memory Polymers for Light-Induced Selective Heating and Remote Actuation. *Sci. Rep.* **2020**, *10*, 6258. [CrossRef]

49. Shan, W.; Chen, Y.; Hu, M.; Qin, S.; Liu, P. 4D Printing of Shape Memory Polymer via Liquid Crystal Display (LCD) Stereolithographic 3D Printing. *Mater. Res. Express* **2020**, *7*, 105305. [CrossRef]
50. Choong, Y.Y.C.; Maleksaeedi, S.; Eng, H.; Yu, S.; Wei, J.; Su, P.-C. High Speed 4D Printing of Shape Memory Polymers with Nanosilica. *Appl. Mater. Today* **2020**, *18*, 100515. [CrossRef]
51. Tedla, G.; Jarabek, A.M.; Byrley, P.; Boyes, W.; Rogers, K. Human Exposure to Metals in Consumer-Focused Fused Filament Fabrication (FFF)/ 3D Printing Processes. *Sci. Total Environ.* **2022**, *814*, 152622. [CrossRef]
52. Pierau, L.; Elian, C.; Akimoto, J.; Ito, Y.; Caillol, S.; Versace, D.-L. Bio-Sourced Monomers and Cationic Photopolymerization–The Green Combination towards Eco-Friendly and Non-Toxic Materials. *Prog. Polym. Sci.* **2022**, *127*, 101517. [CrossRef]
53. Leonhardt, S.; Klare, M.; Scheer, M.; Fischer, T.; Cordes, B.; Eblenkamp, M. Biocompatibility of Photopolymers for Additive Manufacturing. *Curr. Dir. Biomed. Eng.* **2016**, *2*, 113–116. [CrossRef]
54. Klikovits, N.; Knaack, P.; Bomze, D.; Krossing, I.; Liska, R. Novel Photoacid Generators for Cationic Photopolymerization. *Polym. Chem.* **2017**, *8*, 4414–4421. [CrossRef]
55. Mueller, M.; Bandl, C.; Kern, W. Surface-Immobilized Photoinitiators for Light Induced Polymerization and Coupling Reactions. *Polymers* **2022**, *14*, 608. [CrossRef]
56. Zhou, J.; Allonas, X.; Ibrahim, A.; Liu, X. Progress in the Development of Polymeric and Multifunctional Photoinitiators. *Prog. Polym. Sci.* **2019**, *99*, 101165. [CrossRef]
57. Tomal, W.; Ortyl, J. Water-Soluble Photoinitiators in Biomedical Applications. *Polymers* **2020**, *12*, 1073. [CrossRef]
58. Photopolymerization Using Metal Oxide Semiconducting Nanoparticles for Epoxy-Based Coatings and Patterned Films. *ACS Appl. Nano Mater.* **2020**, *3*, 2875–2880. Available online: https://pubs-acs-org.lama.univ-amu.fr/doi/10.1021/acsanm.0c00147 (accessed on 24 April 2022). [CrossRef]
59. Breimer, M.A.; Yevgeny, G.; Sy, S.; Sadik, O.A. Incorporation of Metal Nanoparticles in Photopolymerized Organic Conducting Polymers: A Mechanistic Insight. *Nano Lett.* **2001**, *1*, 305–308. [CrossRef]
60. Guo, J.; Jian, J.; Wang, M.; Tomita, Y.; Cao, L.; Wang, D.; Zhang, X. Ag Nanoparticle-Enhanced Alkyl Radical Generation in Photopolymerization for Holographic Recording. *Nanophotonics* **2019**, *8*, 1795–1802. [CrossRef]
61. He, C.; Feng, Z.; Shan, S.; Wang, M.; Chen, X.; Zou, G. Highly Enantioselective Photo-Polymerization Enhanced by Chiral Nanoparticles and in Situ Photopatterning of Chirality. *Nat. Commun.* **2020**, *11*, 1188. [CrossRef] [PubMed]
62. Ibn-El-Ahrach, H.; Bachelot, R.; Lérondel, G.; Vial, A.; Grimault, A.-S.; Plain, J.; Royer, P.; Soppera, O. Controlling the Plasmon Resonance of Single Metal Nanoparticles by Near-Field Anisotropic Nanoscale Photopolymerization. *J. Microsc.* **2008**, *229*, 421–427. [CrossRef]
63. Ibn El Ahrach, H.; Bachelot, R.; Vial, A.; Lérondel, G.; Plain, J.; Royer, P.; Soppera, O. Spectral Degeneracy Breaking of the Plasmon Resonance of Single Metal Nanoparticles by Nanoscale Near-Field Photopolymerization. *Phys. Rev. Lett.* **2007**, *98*, 107402. [CrossRef] [PubMed]
64. Cai, X.; Anyaogu, K.C.; Neckers, D.C. Photopolymerization of Gold Nanoparticles: Size-Related Charge Separation and Emission. *J. Am. Chem. Soc.* **2007**, *129*, 11324–11325. [CrossRef] [PubMed]
65. Lv, J.; Lu, Y.; Wang, J.; Zhao, F.; Wang, Y.; He, H.; Wu, Y. Schiff Base-Type Copper(I) Complexes Exhibiting High Molar Extinction Coefficients: Synthesis, Characterization and DFT Studies. *J. Mol. Struct.* **2022**, *1249*, 131638. [CrossRef]
66. Kuang, D.; Ito, S.; Wenger, B.; Klein, C.; Moser, J.-E.; Humphry-Baker, R.; Zakeeruddin, S.M.; Grätzel, M. High Molar Extinction Coefficient Heteroleptic Ruthenium Complexes for Thin Film Dye-Sensitized Solar Cells. *J. Am. Chem. Soc.* **2006**, *128*, 4146–4154. [CrossRef]
67. Sciortino, G.; Maréchal, J.-D.; Fábián, I.; Lihi, N.; Garribba, E. Quantitative Prediction of Electronic Absorption Spectra of Copper(II)–Bioligand Systems: Validation and Applications. *J. Inorg. Biochem.* **2020**, *204*, 110953. [CrossRef]
68. Faggi, E.; Gavara, R.; Bolte, M.; Fajarí, L.; Juliá, L.; Rodríguez, L.; Alfonso, I. Copper(II) Complexes of Macrocyclic and Open-Chain Pseudopeptidic Ligands: Synthesis, Characterization and Interaction with Dicarboxylates. *Dalton Trans.* **2015**, *44*, 12700–12710. [CrossRef]
69. Iwamura, M.; Takeuchi, S.; Tahara, T. Ultrafast Excited-State Dynamics of Copper(I) Complexes. *Acc. Chem. Res.* **2015**, *48*, 782–791. [CrossRef]
70. Cuttell, D.G.; Kuang, S.-M.; Fanwick, P.E.; McMillin, D.R.; Walton, R.A. Simple Cu(I) Complexes with Unprecedented Excited-State Lifetimes. *J. Am. Chem. Soc.* **2002**, *124*, 6–7. [CrossRef] [PubMed]
71. Giereth, R.; Reim, I.; Frey, W.; Junge, H.; Tschierlei, S.; Karnahl, M. Remarkably Long-Lived Excited States of Copper Photosensitizers Containing an Extended π-System Based on an Anthracene Moiety. *Sustain. Energy Fuels* **2019**, *3*, 692–700. [CrossRef]
72. Egly, J.; Bissessar, D.; Achard, T.; Heinrich, B.; Steffanut, P.; Mauro, M.; Bellemin-Laponnaz, S. Copper(I) Complexes with Remotely Functionalized Phosphine Ligands: Synthesis, Structural Variety, Photophysics and Effect onto the Optical Properties. *Inorg. Chim. Acta* **2021**, *514*, 119971. [CrossRef]
73. Bergmann, L.; Hedley, G.J.; Baumann, T.; Bräse, S.; Samuel, I.D.W. Direct Observation of Intersystem Crossing in a Thermally Activated Delayed Fluorescence Copper Complex in the Solid State. *Sci. Adv.* **2016**, *2*, e1500889. [CrossRef]
74. Ochiai, H.; Miura, T.; Ikoma, T.; Minoura, M.; Nakano, H.; Matano, Y. Copper(II) Complexes of 10,20-Diaryl-5,15-Diazaporphyrin: Alternative Synthesis, Excited State Dynamics, and Substituent Effect on the $^1O_2$-Generation Efficiency. *BCSJ* **2022**, *95*, 427–432. [CrossRef]

75. Fors, B.P.; Hawker, C.J. Control of a Living Radical Polymerization of Methacrylates by light. *Angew. Chem. Int. Ed.* **2012**, *51*, 8850. [CrossRef]
76. Ohtsuki, A.; Goto, A.; Kaji, H. Visible-Light-Induced Reversible Complexation Mediated Living Radical Polymerization of Methacrylates with Organic Catalysts. *Macromolecules* **2013**, *46*, 96. [CrossRef]
77. Cope, J.D.; Valle, H.U.; Hall, R.S.; Riley, K.M.; Goel, E.; Biswas, S.; Hendrich, M.P.; Wipf, D.O.; Stokes, S.L.; Emerson, J.P. Tuning the Copper(II)/Copper(I) Redox Potential for More Robust Copper-Catalyzed C–N Bond Forming Reactions. *Eur. J. Inorg. Chem.* **2020**, *2020*, 1278–1285. [CrossRef]
78. Tano, T.; Okubo, Y.; Kunishita, A.; Kubo, M.; Sugimoto, H.; Fujieda, N.; Ogura, T.; Itoh, S. Redox Properties of a Mononuclear Copper(II)-Superoxide Complex. *Inorg. Chem.* **2013**, *52*, 10431–10437. [CrossRef]
79. Asahi, M.; Yamazaki, S.; Itoh, S.; Ioroi, T. Electrochemical Reduction of Dioxygen by Copper Complexes with Pyridylalkylamine Ligands Dissolved in Aqueous Buffer Solution: The Relationship between Activity and Redox Potential. *Dalton Trans.* **2014**, *43*, 10705–10709. [CrossRef]
80. Nishikawa, M.; Kakizoe, D.; Saito, Y.; Ohishi, T.; Tsubomura, T. Redox Properties of Copper(I) Complex Bearing 4,7-Diphenyl-2,9-Dimethyl-1,10-Phenanthroline and 1,4-Bis(Diphenylphosphino)Butane Ligands and Effects of Light in the Presence of Chloroform. *BCSJ* **2017**, *90*, 286–288. [CrossRef]
81. Araya, L.M.; Vargas, J.A.; Costamagna, J.A. Ligand Influence on the Redox Properties of Some Copper(II) Complexes with Schiff Bases Derived from Bromosalicylaldehydes and Methyl or Chloro-Substituted Anilines. *Transit. Met. Chem.* **1986**, *11*, 312–316. [CrossRef]
82. Das, A.; Ren, Y.; Hessin, C.; Murr, M.D.-E. Copper Catalysis with Redox-Active Ligands. *Beilstein J. Org. Chem.* **2020**, *16*, 858–870. [CrossRef] [PubMed]
83. Yang, K.; Yang, X.; Deng, Z.; Zhang, L.; An, J. Copper Piperazine Complex with a High Diffusion Coefficient for Dye-Sensitized Solar Cells. *ACS Appl. Energy Mater.* **2021**, *4*, 14004–14013. [CrossRef]
84. Dragonetti, C.; Magni, M.; Colombo, A.; Fagnani, F.; Roberto, D.; Melchiorre, F.; Biagini, P.; Fantacci, S. Towards Efficient Sustainable Full-Copper Dye-Sensitized Solar Cells. *Dalton Trans.* **2019**, *48*, 9703–9711. [CrossRef]
85. Giordano, M.; Volpi, G.; Bonomo, M.; Mariani, P.; Garino, C.; Viscardi, G. Methoxy-Substituted Copper Complexes as Possible Redox Mediators in Dye-Sensitized Solar Cells. *New J. Chem.* **2021**, *45*, 15303–15311. [CrossRef]
86. Conradie, J. Polypyridyl Copper Complexes as Dye Sensitizer and Redox Mediator for Dye-Sensitized Solar Cells. *Electrochem. Commun.* **2022**, *134*, 107182. [CrossRef]
87. Magni, M.; Biagini, P.; Colombo, A.; Dragonetti, C.; Roberto, D.; Valore, A. Versatile Copper Complexes as a Convenient Springboard for Both Dyes and Redox Mediators in Dye Sensitized Solar Cells. *Coord. Chem. Rev.* **2016**, *322*, 69–93. [CrossRef]
88. Michaels, H.; Benesperi, I.; Edvinsson, T.; Muñoz-Garcia, A.B.; Pavone, M.; Boschloo, G.; Freitag, M. Copper Complexes with Tetradentate Ligands for Enhanced Charge Transport in Dye-Sensitized Solar Cells. *Inorganics* **2018**, *6*, 53. [CrossRef]
89. Rui, H.; Shen, J.; Yu, Z.; Li, L.; Han, H.; Sun, L. Stable Dye-Sensitized Solar Cells Based on Copper(II/I) Redox Mediators Bearing a Pentadentate Ligand. *Angew. Chem. Int. Ed.* **2021**, *60*, 16156–16163. [CrossRef]
90. Dumur, F. Recent advances on visible light metal-based photocatalysts for polymerization under low light intensity. *Catalysts* **2019**, *9*, 736. [CrossRef]
91. Noirbent, G.; Dumur, F. Recent advances on copper complexes as visible light photoinitiators and (photo)redox initiators of polymerization. *Catalysts* **2020**, *10*, 953. [CrossRef]
92. Zivic, N.; Kuroishi, P.K.; Dumur, F.; Gigmes, D.; Dove, A.P.; Sardon, H. Recent advances and challenges in the design of organic photoacid and photobase generators for polymerizations. *Angew. Chem. Int. Ed.* **2019**, *58*, 10410–10422. [CrossRef] [PubMed]
93. Tehfe, M.-A.; Lalevée, J.; Dumur, F.; Telitel, S.; Gigmes, D.; Contal, E.; Bertin, D.; Fouassier, J.-P. Zinc-based metal complexes as new photocatalysts in polymerization initiating systems. *Eur. Polym. J.* **2013**, *49*, 1040–1049. [CrossRef]
94. Lalevée, J.; Telitel, S.; Xiao, P.; Lepeltier, M.; Dumur, F.; Morlet-Savary, F.; Gigmes, D.; Fouassier, J.-P. Metal and metal free photocatalysts: Mechanistic approach and application as photoinitiators of photopolymerization. *Beilstein J. Org. Chem.* **2014**, *10*, 863–876. [CrossRef]
95. Baralle, A.; Fensterbank, L.; Goddard, J.P.; Olivier, C. Aryl Radical Formation by Copper (I) Photocatalyzed Reduction of Diaryliodonium Salts. NMR Evidences for a Cu (II)/Cu (I) Mechanism. *Chem. Eur. J.* **2013**, *19*, 10809. [CrossRef]
96. Paria, S.; Reiser, O. Copper in Photocatalysis. *ChemCatChem* **2014**, *6*, 2477. [CrossRef]
97. Reiser, O. Shining Light on Copper: Unique Opportunities for Visible-Light-Catalyzed Atom Transfer Radical Addition Reactions and Related Processes. *Acc. Chem. Res.* **2016**, *49*, 1990. [CrossRef]
98. Hernandez-Perez, A.C.; Collins, S.K. Heteroleptic Cu-Based Sensitizers in Photoredox Catalysis. *Acc. Chem. Res.* **2016**, *49*, 1557. [CrossRef]
99. Hernandez-Perez, A.C.; Collins, S.K. A visible-light-mediated synthesis of carbazoles. *Angew. Chem. Int. Ed.* **2013**, *52*, 12696. [CrossRef]
100. Mau, A.; Dietlin, C.; Dumur, F.; Lalevée, J. Concomitant initiation of radical and cationic polymerisations using new copper complexes as photoinitiators: Synthesis and characterisation of acrylate/epoxy interpenetrated polymer networks. *Eur. Polym. J.* **2021**, *152*, 110457. [CrossRef]
101. Mau, A.; Noirbent, G.; Dietlin, C.; Graff, B.; Gigmes, D.; Dumur, F.; Lalevée, J. Panchromatic Copper Complexes for Visible Light Photopolymerization. *Photochem* **2021**, *1*, 167–189. [CrossRef]

102. Mokbel, H.; Anderson, D.; Plenderleith, R.; Dietlin, C.; Morlet-Savary, F.; Dumur, F.; Gigmes, D.; Fouassier, J.P.; Lalevée, J. Simultaneous initiation of radical and cationic polymerization reactions using the "G1" copper complex as photoredox catalyst: Applications of free radical/cationic hybrid photopolymerization in the composites and 3D printing fields. *Prog. Org. Coat.* **2019**, *132*, 50. [CrossRef]

103. AL Mousawi, A.; Kermagoret, A.; Versace, D.L.; Toufaily, J.; Hamieh, T.; Graff, B.; Dumur, F.; Gigmes, D.; Fouassier, J.P.; Lalevée, J. Copper photoredox catalysts for polymerization upon near UV or visible light: Structure/reactivity/efficiency relationships and use in LED projector 3D printing resins. *Polym. Chem.* **2017**, *8*, 568. [CrossRef]

104. Xiao, P.; Dumur, F.; Zhang, J.; Fouassier, J.-P.; Gigmes, D.; Lalevée, J. Copper complexes in radical photoinitiating systems: Applications to free radical and cationic polymerization under visible lights. *Macromolecules* **2014**, *47*, 3837–3844. [CrossRef]

105. Matsui, M.; Matsumoto, K.; Shibata, K.; Muramatsu, H. Synthesis and Characterization of 5-Cyano-6-Methyl-2,2′-Bipyridine Metal-Complex Dyes. *Dyes Pigments* **1992**, *18*, 47–55. [CrossRef]

106. Matsui, M.; Oji, A.; Hiramatsu, K.; Shibata, K.; Muramatsu, H. Synthesis and Characterization of Fluorescent 4,6-Disubstituted-3-Cyano-2-Methylpyridines. *J. Chem. Soc. Perkin Trans.* **2 1992**, *2*, 201–206. [CrossRef]

107. Knorn, M.; Rawner, T.; Czerwieniec, R.; Reiser, O. [Copper(phenanthroline)(bisisonitrile)]+-Complexes for the Visible-Light-Mediated Atom Transfer Radical Addition and Allylation Reactions. *ACS Catal.* **2015**, *5*, 5186–5193. [CrossRef]

108. Kaeser, A.; Mohankumar, M.; Mohanraj, J.; Monti, F.; Holler, M.; Cid, J.J.; Moudam, O.; Nierengarten, I.; Karmazin-Brelot, L.; Duhayon, C.; et al. Heteroleptic Copper(I) Complexes Prepared from Phenanthroline and Bis-Phosphine Ligands. *Inorg. Chem.* **2013**, *52*, 12140–12151. [CrossRef]

109. Wang, X.; Bai, X.; Su, D.; Zhang, Y.; Li, P.; Lu, S.; Gong, Y.; Zhang, W.; Tang, B. Simultaneous Fluorescence Imaging Reveals N-Methyl-d-aspartic Acid Receptor Dependent Zn^{2+}/H^+ Flux in the Brains of Mice with Depression. *Anal. Chem.* **2020**, *92*, 4101. [CrossRef]

110. Foresman, J.B.; Frisch, A. *Exploring Chemistry with Electronic Structure Methods*, 2nd ed.; Gaussian Inc.: Pittsburgh, PA, USA, 1996.

111. Frisch, M.J.; Trucks, G.W.; Schlegel, H.B.; Scuseria, G.E.; Robb, M.A.; Cheeseman, J.R.; Zakrzewski, V.G.; Montgomery, J.A.; Stratmann, J.R.E.; Burant, J.C.; et al. *Gaussian 03, Revision B-2*; Gaussian Inc.: Pittsburgh, PA, USA, 2003.

112. Al Mousawi, A.; Dumur, F.; Garra, P.; Toufaily, J.; Hamieh, T.; Goubard, F.; Bui, T.T.; Graff, B.; Gigmes, D.; Fouassier, J.P.; et al. Azahelicenes as visible light photoinitiators for cationic and radical polymerization: Preparation of photoluminescent polymers and use in high performance LED projector 3D printing resins. *J. Polym. Sci. A Polym. Chem.* **2017**, *55*, 1189. [CrossRef]

113. Dong, F.; Jian, C.; Zhenghao, F.; Kai, G.; Zuliang, L. Synthesis of Chalcones via Claisen–Schmidt Condensation Reaction Catalyzed by Acyclic Acidic Ionic Liquids. *Catal. Commun.* **2008**, *9*, 1924–1927. [CrossRef]

114. Rafiee, E.; Rahimi, F. A Green Approach to the Synthesis of Chalcones via Claisen-Schmidt Condensation Reaction Using Cesium Salts of 12-Tungstophosphoric Acid as a Reusable Nanocatalyst. *Monatsh. Chem.* **2013**, *144*, 361–367. [CrossRef]

115. Mousavi, S.R. Claisen–Schmidt Condensation: Synthesis of (1S,6R)/(1R,6S)-2-Oxo-N,4,6-Triarylcyclohex-3-Enecarboxamide Derivatives with Different Substituents in H_2O/EtOH. *Chirality* **2016**, *28*, 728–736. [CrossRef]

116. Kumar, D.; Suresh; Sandhu, J.S. An Efficient Green Protocol for the Synthesis of Chalcones by a Claisen–Schmidt Reaction Using Bismuth(III)Chloride as a Catalyst under Solvent-Free Condition. *Green Chem. Lett. Rev.* **2010**, *3*, 283–286. [CrossRef]

117. Ekanayake, U.G.M.; Weerathunga, H.; Weerasinghe, J.; Waclawik, E.R.; Sun, Z.; MacLeod, J.M.; O'Mullane, A.P.; Ostrikov, K. (Ken) Sustainable Claisen-Schmidt Chalcone Synthesis Catalysed by Plasma-Recovered MgO Nanosheets from Seawater. *Sustain. Mater. Technol.* **2022**, *32*, e00394. [CrossRef]

118. Parekh, A.K.; Desai, K.K. Synthesis and Antibacterial Activity of Chalcones and Pyrimidine-2-Ones. *E-J. Chem.* **1900**, *2*, 134830. [CrossRef]

119. Jesus, A.R.; Marques, A.P.; Rauter, A.P. An Easy Approach to Dihydrochalcones via Chalcone in Situ Hydrogenation. *Pure Appl. Chem.* **2016**, *88*, 349–361. [CrossRef]

120. Julakanti, S.R.; Patel, M.; Ponneri, V. Highly Efficient Synthesis of Chalcones from Poly Carbonyl Aromatic Compounds Using BF3–Et2O via a Regioselective Condensation Reaction. *Chem. Pharm. Bull.* **2016**, *64*, 570–576. [CrossRef]

121. Winter, C.; Caetano, J.N.; Araújo, A.B.C.; Chaves, A.R.; Ostroski, I.C.; Vaz, B.G.; Pérez, C.N.; Alonso, C.G. Activated Carbons for Chalcone Production: Claisen-Schmidt Condensation Reaction. *Chem. Eng. J.* **2016**, *303*, 604–610. [CrossRef]

122. Qian, H.; Liu, D.; Lv, C. Synthesis of Chalcones via Claisen-Schmidt Reaction Catalyzed by Sulfonic Acid-Functional Ionic Liquids. *Ind. Eng. Chem. Res.* **2011**, *50*, 1146–1149. [CrossRef]

123. Calvino, V.; Picallo, M.; López-Peinado, A.J.; Martín-Aranda, R.M.; Durán-Valle, C.J. Ultrasound Accelerated Claisen–Schmidt Condensation: A Green Route to Chalcones. *Appl. Surf. Sci.* **2006**, *252*, 6071–6074. [CrossRef]

124. Li, J.-T.; Yang, W.-Z.; Wang, S.-X.; Li, S.-H.; Li, T.-S. Improved Synthesis of Chalcones under Ultrasound Irradiation. *Ultrason. Sonochem.* **2002**, *9*, 237–239. [CrossRef]

125. Bui, T.H.; Nguyen, N.T.; Dang, P.H.; Nguyen, H.X.; Nguyen, M.T.T. Design and Synthesis of Chalcone Derivatives as Potential Non-Purine Xanthine Oxidase Inhibitors. *SpringerPlus* **2016**, *5*, 1789. [CrossRef] [PubMed]

polymers

Article

Integration of Biofunctional Molecules into 3D-Printed Polymeric Micro-/Nanostructures

Eider Berganza [1,†], Gurunath Apte [1,2,†], Srivatsan K. Vasantham [1], Thi-Huong Nguyen [2,3,*]
and Michael Hirtz [1,*]

1 Institute of Nanotechnology (INT) and Karlsruhe Nano Micro Facility (KNMFi), Karlsruhe Institute of Technology, 76131 Karlsruhe, Germany; eider.eguiarte@kit.edu (E.B.); gurunath.apte@iba-heiligenstadt.de (G.A.); srivatsan.vasantham@kit.edu (S.K.V.)
2 Institute for Bioprocessing and Analytical Measurement Techniques (iba), 37308 Heilbad Heiligenstadt, Germany
3 Faculty of Mathematics and Natural Sciences, Technische Universität Ilmenau, 98694 Ilmenau, Germany
* Correspondence: thi-huong.nguyen@iba-heiligenstadt.de (T.-H.N.); michael.hirtz@kit.edu (M.H.)
† These authors contributed equally to the work.

Abstract: Three-dimensional printing at the micro-/nanoscale represents a new challenge in research and development to achieve direct printing down to nanometre-sized objects. Here, FluidFM, a combination of microfluidics with atomic force microscopy, offers attractive options to fabricate hierarchical polymer structures at different scales. However, little is known about the effect of the substrate on the printed structures and the integration of (bio)functional groups into the polymer inks. In this study, we printed micro-/nanostructures on surfaces with different wetting properties, and integrated molecules with different functional groups (rhodamine as a fluorescent label and biotin as a binding tag for proteins) into the base polymer ink. The substrate wetting properties strongly affected the printing results, in that the lateral feature sizes increased with increasing substrate hydrophilicity. Overall, ink modification only caused minor changes in the stiffness of the printed structures. This shows the generality of the approach, as significant changes in the mechanical properties on chemical functionalization could be confounders in bioapplications. The retained functionality of the obtained structures after UV curing was demonstrated by selective binding of streptavidin to the printed structures. The ability to incorporate binding tags to achieve specific interactions between relevant proteins and the fabricated micro-/nanostructures, without compromising the mechanical properties, paves a way for numerous bio and sensing applications. Additional flexibility is obtained by tuning the substrate properties for feature size control, and the option to obtain functionalized printed structures without post-processing procedures will contribute to the development of 3D printing for biological applications, using FluidFM and similar dispensing techniques.

Keywords: FluidFM; 3D printing; microstructures; nanostructures; biofunctionalization; mechanical properties; scanning probe lithography

Citation: Berganza, E.; Apte, G.; Vasantham, S.K.; Nguyen, T.-H.; Hirtz, M. Integration of Biofunctional Molecules into 3D-Printed Polymeric Micro-/Nanostructures. *Polymers* **2022**, *14*, 1327. https://doi.org/10.3390/polym14071327

Academic Editors: Houwen Matthew Pan and Eric Luis

Received: 7 March 2022
Accepted: 22 March 2022
Published: 25 March 2022

Publisher's Note: MDPI stays neutral with regard to jurisdictional claims in published maps and institutional affiliations.

1. Introduction

Three-dimensional printing has become a versatile tool for printing biomimetic scaffolds and for other biomedical applications [1]. Sophisticated lithographically generated microstructures are, in particular, used for probing or manipulating cells at the single-cell level, for elucidating cell biology and mechanics [2,3], or influencing stem cell fate [4]. While, in particular, optical methods, such as direct laser writing (DLW) and similar techniques, have made impressive progress [5], the bioactive functionalization of such microstructures remains challenging [6].

The surrounding dynamic micro/nano environment directly influences cell behaviour. In particular, the topography, stiffness, bioactive moieties, and chemical components of the surfaces regulate the response of the cells [7–9]. We have previously found that nanostructures, together with surface modification, could effectively control the adhesion, spreading, and activation of human blood platelets [10–12]. Importantly, surfaces have been developed to mimic the structure of the extracellular matrix (ECM) [13,14]. However, these structures are almost static, and do not emulate the dynamicity and function of the ECM in vivo. To overcome this limitation, shape-memory polymers (SMPs), in the form of patterns, fibres, porous scaffolds, and microspheres, have been developed to mimic dynamic changes in the ECM structure, both in vitro and in vivo [15]. The SMPs arise from materials with unique functions that regulate cell behaviours and promote tissue growth. The adjustment of structured surfaces, as well as chemical components, can further optimize contact environments for cell response/sensing and tissue regeneration.

While methods for the fabrication of microstructures have been well established, available technology for the production of nanostructured surfaces is, to date, still limited. One recent development in the 3D printing of micro-/nanostructures is the use of nano dispensing techniques, such as FluidFM. FluidFM is based on the combination of microfluidics with atomic force microscopy (AFM), in which a hollow cantilever, with an aperture at the tip apex, can be used for precisely localized liquid dispensing [16]. Its use was quickly extended to nanolithography, e.g., for nanoparticles [17] and biomimetic membranes [18]. Depending on the size of the tip aperture, micro-/nanostructures can be printed. With this technique, hierarchical structures, at multidimensional scales, could be fabricated using the commercially available UV-curable adhesive Loctite [19,20]. This is a highly viscous ink, composed of different methacrylate esters, which are in the liquid phase and undergo cross-linking polymerization upon exposure to UV [21]. As a direct-write method, this approach is highly versatile in pattern and structure formation, and offers excellent resolution down to the tens of nanometres scale. However, little is known about the effect of substrate surface properties on the resulting printing structures, and the integration of functional groups into the printed structures to enable specific biological applications. As of now, further biofunctionalization of such structures is still lacking.

Here, we demonstrate that the substrate surface properties directly affect the geometries of the printed structures. We further show the feasibility of introducing specific protein coupling sites into the printed microstructures (Figure 1). This is conducted by admixing biotin moiety-bearing amphiphilic molecules into the base adhesive ink, which are then presented on the micro-/nanostructured surfaces after curing. As no chemical post-functionalization is needed, the approach is a facile route to introduce highly specific protein binding into the polymer structures, with the possibility of selecting proteins of interest from the extensive library of biotinylated proteins, or using sandwich approaches to add non-labelled proteins via antibody capture or DNA-directed immobilization with biotinylated oligonucleotides [22,23].

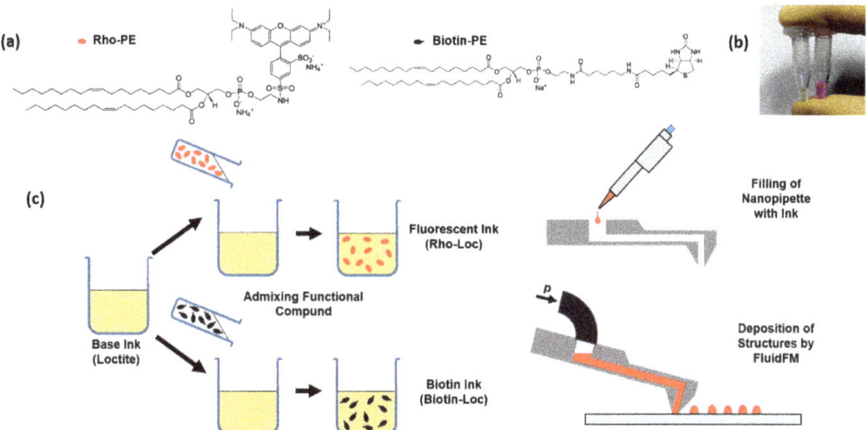

Figure 1. Ink preparation and printing. (**a**) Molecular structures of the phospholipids used for biofunctionalization. (**b**) Picture of the functionalized adhesive inks. (**c**) Scheme of functionalization and printing process.

2. Materials and Methods

2.1. Materials

The UV-curable adhesive Loctite 3491 (Henkel, Düsseldorf, Germany) was used as the base for all inks. Phospholipids used as functional admixtures were 1,2-dioleoyl-sn-glycero-3-phosphoethanolamine-*N*-(cap biotinyl) (biotin-PE) at a concentration of 10 mg/mL in chloroform, and 1,2-dioleoyl-sn-glycero-3-phosphoethanolamine-*N*-(lissamine rhodamine B sulfonyl) (Rho-PE) at a concentration of 1 mg/mL in chloroform, both from Avanti Polar Lipids, Alabaster, AL, USA. Functionalized inks were obtained by admixing either 10 vol% of the biotin-PE in chloroform solution (for the biofunctionalization ink) or 10 vol% of the Rho-PE in chloroform solution (for the fluorescent ink) with the base adhesive ink. A homogeneous mixture was obtained by vortexing the solution for 1–2 min.

2.2. FluidFM Printing

For printing of the adhesive patterns on glass or silicon substrates, the following two FluidFM systems were used: a FlexAFM (Nanosurf, Liestal, Switzerland) system and a BioAFM (JPK, Berlin, Germany) system. Experiments were performed with FluidFM nanopipettes (Cytosurge, Opfikon, Switzerland), with a nominal cantilever spring constant of 2 N/m and a 300 nm diameter nozzle/aperture at its probe end.

After the nanopipette was mounted and the reservoir was filled with 2 µL of the respective inks, 1000 mbar pressure was applied to the reservoir for 1–2 min to make the ink flow through the microchannel to the end of the probe aperture. Once the ink reached the nozzle, the ink flowed to the substrate without the need for further application of pressure. During patterning, the applied force was typically set between 10 and 20 nN. To control the feature sizes, the contact time during printing was varied between 0.5 and 5 s for nanodots whereas the writing speed was varied between 20 and 60 µm/s to print lines.

2.3. Characterization of Printed Structures with Atomic Force Microscopy (AFM)

The obtained structures were characterized by AFM, performed on a Dimension Icon system (Bruker, Berlin, Germany) in tapping mode. Tap300-G probes (Budget Sensors, Sofia, Bulgaria) with a resonance frequency of 330 kHz and nominal spring constant of 42 N/m were used. The indentation maps from which Young's modulus values were extracted were obtained on a JPK BioAFM system (Bruker, Berlin, Germany), using BL-AC40TS probes with a radius of 8 nm (Asylum Research, Santa Barbara, CA, USA), with a resonance frequency of 70 Hz and nominal force constant of 2 N/m. A total of 625 force curves were analysed from areas of 2.5 × 2.5 μm^2, with the printed feature located at the centre of the scanning area. The data obtained from the measurements were processed by fitting the force curves to the Hertz model, by selecting tip shape as pyramidical and Poisson's ratio of 0.5. The data were processed by JPKSPM data processing software and analysed using SigmaPlot (Sysstat Software GmbH, Erkrath, Germany). Measurements were also performed over a plain glass surface to obtain control values.

2.4. Protein Binding

To prevent nonspecific binding, the samples were first incubated with a 10% bovine serum albumin (BSA) solution (Sigma-Aldrich, Darmstadt Germany) for 30 min at room temperature (RT). Then, biofunctionalization was demonstrated by incubating the samples with 5 μg/mL streptavidin–FITC solution (Thermo Fisher, Waltham, MA, USA) in phosphate-buffered saline (PBS) for 30 min.

2.5. Optical Microscopy

Optical microscopy was performed on a Nikon Eclipse Ti2 inverted fluorescence microscope (Nikon, Düsseldorf, Germany). A Texas-red filter (Nikon, Germany) was used as a light filter, and a green fluorescent protein (GFP)-compatible filter was used for visualization of biotin–streptavidin bindings.

2.6. Substrate Functionalization

To assess the influence of substrate wettability on the patterning feature dimensions and spreading behavior, several glass substrates were functionalized. Prior to functionalization procedures, all coverslips were sonicated in acetone, ethanol, and DI water, subsequently. Hydrophobic substrates (S1) were prepared by exposing coverslips to oxygen plasma (200 W, 50 sccm oxygen flow, in an Atto system (Diener Electronics, Ebhausen, Germany) for 5 min, and subsequently immersing them in 7-octenyltrichlorosilane (OTS) (10 vol% in toluene) for 24 h. Medium hydrophilic surfaces (S2) were used directly after cleaning, without any surface functionalization. Another type of medium hydrophilic surfaces (S3) were induced by exposing to oxygen plasma for 5 min prior to immersing in (3-glycidyloxypropyl)- trimethoxysilan (GPTMS) (2 vol% in toluene) for 4 h. Highly hydrophilic substrates (S4) were prepared by exposure to oxygen plasma for 5 min, without any further coating.

2.7. Characterization of Substrate Wettability

The different substrates were characterized by contact angle measurements on an OCA-20 system (DataPhysics Instruments GmbH, Filderstadt, Germany). The water contact angle (WCA) for each substrate was measured by the sessile drop method. Measurements were performed at room temperature (RT), with sample droplets of 3.0 μL volume deposited at a dosing rate of 3.0 μL/s. The contact angles were determined with the onboard software. For each substrate, 3 measurements were performed at different locations.

2.8. Statistical Analysis

Statistical analyses of the data were performed using SigmaPlot (version 14.0). For the printed dot features' height measurements, 9 dots were imaged with AFM for each pulse time and on each surface. The height was then extracted in WSxM [24] from profile lines through the dots. For the printed line features' width and height measurements, averaged profiles from the printed lines AFM images were generated using the y-average tool in WSxM, and the width and height of the averaged line profiles were measured. The obtained values for width and height were averaged from 5 printed lines for each surface. The WCAs were obtained from 3 different locations for each substrate. All error ranges given in the manuscript are the standard deviation of the respective data points, unless otherwise noted.

3. Results

3.1. Printing of Pure Adhesive

Pure adhesive ink, composed of different methacrylate esters, was used to print different exemplary structures on bare glass, using a nanopipette cantilever with an aperture size of 300 nm. Different types of patterns, including dots of around 1 μm in diameter, lines, grids, and squares, can be readily printed (Figure 2 and Figure S1). Filled square patterns were obtained by drawing lines in close proximity that merged, forming structures of homogeneous thickness. It has been previously reported that, together with printing parameters such as pressure, force, and contact time, the printing direction also plays a role in the size of the printed features [19]. This effect can be observed on the grid (Figure 2b), where the line thickness varies, depending on its direction.

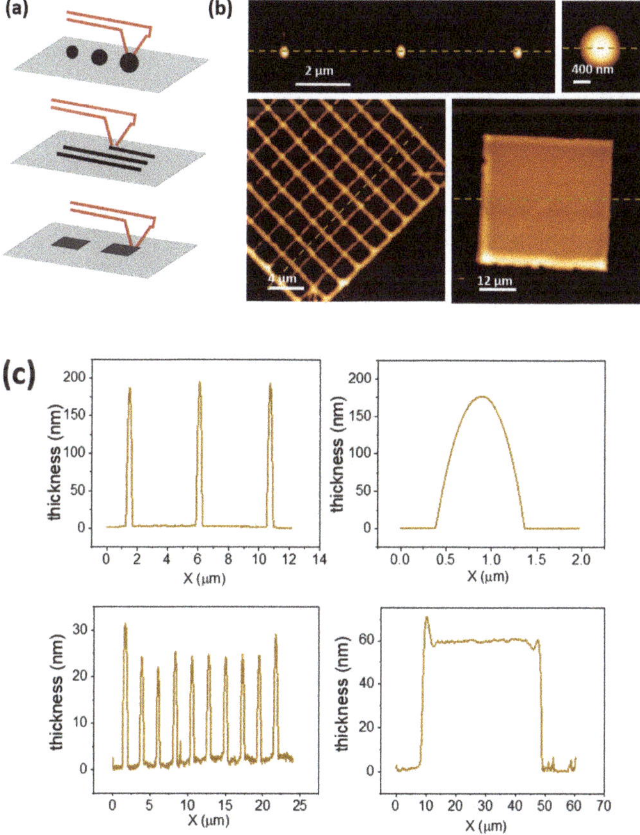

Figure 2. Basic geometrical patterns obtained with pure adhesive, Loctite. (**a**) Scheme of printing dots, lines, and squares as basic geometric patterns. (**b**) AFM topography images of exemplary printed structures and (**c**) corresponding profile sections of the structures at the yellow dashed line in (**b**).

3.2. Printing with Functionalized Adhesive

3.2.1. Preparation of Functionalized Adhesive Inks

Two types of modified inks, with different functional properties, were prepared for the experiments, by admixing functionalized phospholipids. For an easy assessment of miscibility, and to be able to observe the printed structure by fluorescence microscopy, a fluorescently labelled phospholipid (Rho-PE) was admixed. The resulting mixture turned out to be homogeneous upon visual inspection (Figure 1a), showing good compatibility of the solvents, which is extremely important, as the segregation of components could clog the FluidFM nanopipette. For the integration of biofunctional molecules into the adhesive-based patterns, a biotinylated phospholipid (biotin-PE) was admixed. The biotin–streptavidin complex [25] is widely used in biochemistry research, due to its strong binding affinity, which is commonly harnessed, e.g., in sensing applications [26].

3.2.2. Patterning on Substrates with Different Wettability

Texture [27], surface chemistry [28], and wettability [29,30] are all known to be relevant parameters in the interactions of cells, platelets, and other species with biomaterials. Since surfaces can be chemically modified to achieve the desired biological response, being able to create adhesive patterns on differently terminated functional group surfaces is of high interest.

To assess how the biofunctional inks behave when they are patterned on substrates of different surface chemistry, the biotinylated ink was used to write different features, while keeping the working parameters constant (Figure 3a,b).

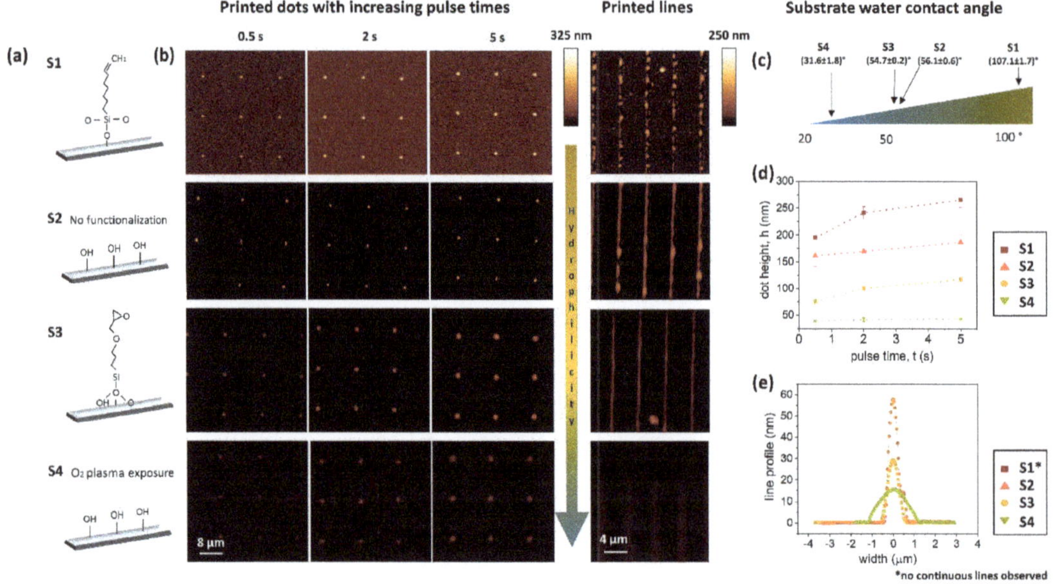

Figure 3. Influence of substrate wettability and chemistry on patterning. (**a**) Scheme of respective sample surface chemistry. (**b**) AFM images of polymer nanodots (left) and lines (right) patterned (right) on the functionalized substrates, showing increasing hydrophilicity from top to bottom (indicated by the arrow), written with the same working parameters. (**c**) Measured water contact angle for the different substrates. (**d**) Dot height on the different substrates as a function of the contact time. (**e**) Average line profiles showing different spreading behavior depending on the substrate functionalization.

For this, glass substrates were modified with different functional groups of self-assembled monolayers (SAMs). Briefly, the surfaces were functionalized by 7-octenyltrichlorosilane (S1), no treatment (S2), (3-glycidyloxypropyl)-trimethoxysilan (S3), and O_2-plasma activation (S4), inducing hydrophobic (S1), medium hydrophilic (S2,S3), and strongly hydrophilic (S4) properties, respectively. The substrates were used two days after preparation, and the water contact angle was measured immediately before patterning (Figures 3c and S2). To achieve comparable patterning, the same working conditions were employed. Nanodots were written, setting the contact time to 0.5, 2, and 5 s, subsequently, touching the substrate with 20 nN of force, and applying no pressure to the reservoir. Nanodots were successfully printed on all the substrates, and remarkably different heights were obtained, depending on the substrate treatment, while changes in the pulse time for ink dispensing had much less influence on the obtained heights (Figure 3d). The strongly hydrophobic 7-octenyltrichlorosilane-coated substrate (S1) leads to very high (over 200 nm) and confined features, while, on plasma-activated glass (S4), the dots considerably spread, causing much lower feature heights (around 40 nm). The differences between the outcomes of the patterns in S2 and S4 were surprisingly large, considering that they possessed basically the same surface chemistry, which reveals the importance of the substrate wettability (here, tuned by the oxygen plasma) in the spreading of the adhesive ink. The nanodot volume was quantified using the WSxM [24] flooding tool to prove that there were no significant differences in the amount of ink dispensed onto the different substrates (Figure S3). When considering the height change with pulse time, a clear increase in height with longer pulse time was only visible on the more hydrophobic substrates, but, even for these, the trend leveled off with longer pulse times.

The writing lines on the different SAMs turned out to be more critical, as can be inferred from Figure 3b,e. In the case of S1, the presence of the hydrophobic hydrocarbon chains compromised the stability of the features, which dewet and broke into droplets immediately after being printed. The rest of the line patterns (S2–S4), however, showed similar behaviour to that observed for the dots, where the highest lines were obtained on S2, while the spreading behaviour on the plasma-activated sample (S4) resulted in very low height features. The average values have been gathered in Table 1.

Table 1. Height of adhesive lines printed on differently functionalized substrates.

Substrate Label	S1	S2	S3	S4
Functionalization	7-octenyl trichlorosilane	None	(3-glycidyl oxypropyl)-trimethoxysilan	O_2 plasma activation
Width * (nm)	- **	300 ± 21	524 ± 26	1457 ± 65
Height (nm)	- **	68 ± 16	32 ± 3	15 ± 1
Aspect ratio ***	-	0.23 ± 0.04	0.06 ± 0.01	0.01 ± 0.01

* full width at half maximum (FWHM). ** no continuous line writing was obtained. *** calculated as height/width, error calculated by error propagation.

3.2.3. Comparison of Mechanical Properties

It has often been shown that the mechanical properties of a substrate influence cell behaviour [31]. Hence, we assessed the influence of the biofunctionalized adhesive on the mechanical properties of the printed structures. For this, nanoindentation measurements over the dot features, from both non-functionalized and functionalized adhesive, were performed (Figure 4a). The force–distance curves extracted from these measurements were then analysed with JPK software, where the Young's modulus was obtained after fitting the curves with a Hertz model (Figure 4b). The E-modulus values obtained on the functionalized and non-functionalized adhesive, together with the values on the glass surface, are presented as histograms in Figure 4c. Two peaks of distribution can be observed for all the ink samples. The measurement on the plain glass acts as a control, to differentiate the values taken on the dot features from those of the surrounding glass. The indentation maps show low E-modulus on the polymer structures (Figure 4d, dark area), while the surrounding glass exhibits a much higher value (Figure 4d, bright area). The distribution of the E-modulus values in the form of a box plot (Figure 4e) for the control glass and the inks (after subtracting the values from the glass) shows that on adding the functionality bearing phospholipids, the mechanical properties of the resulting printed features were not significantly altered. Although the addition of rhodamine or biotinylated lipids yields slightly stiffer materials, the obtained values stayed within the expected statistical variations, displayed as error bars. Additional measurements of Young's modulus for different unfunctionalized patterns are given in the Supporting Information (Figure S1).

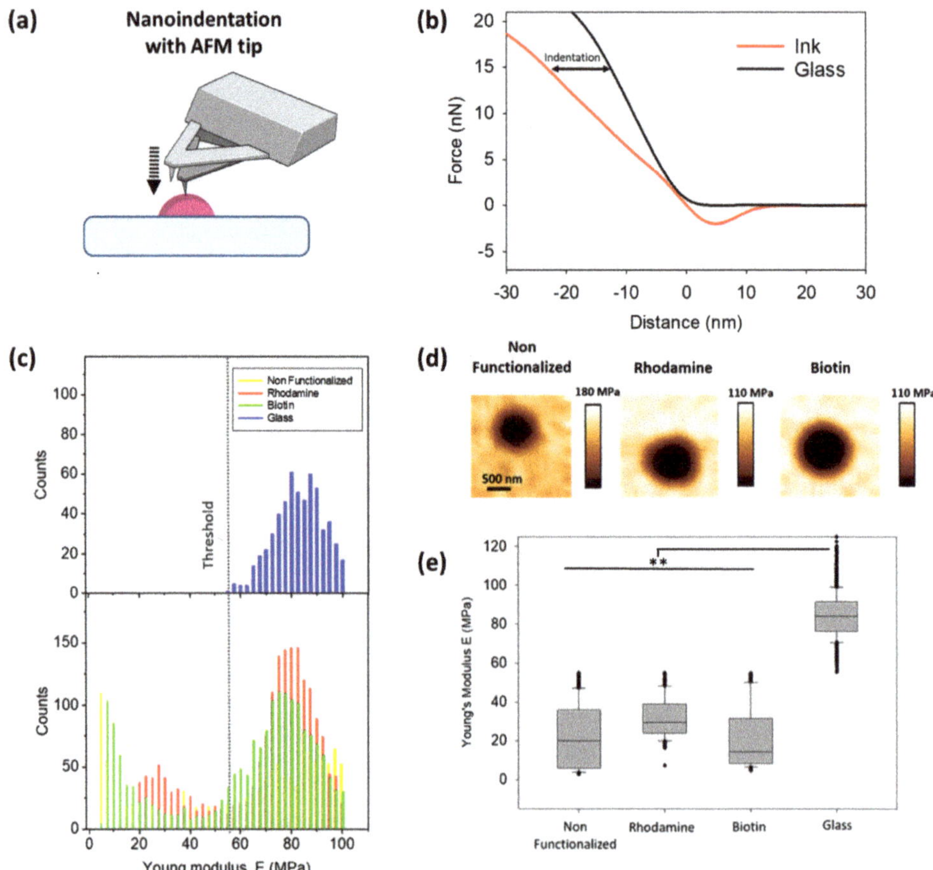

Figure 4. Influence of functional admixing on the mechanical properties of printed features determined by AFM nanoindentation. (**a**) Schematic of the nanoindentation experiments. (**b**) Typical AFM force-distance curve demonstrates the indentation on hard glass (black) to a compliant ink surface (red). (**c**) Quantification of Young's modulus of bare glass and samples with modified inks. (**d**) Indentation map for the different composition nanodots (scale bar equals 500 nm for all images). (**e**) The Young's modulus of modified nanodots shows small variation. Statistically significant difference determined by one-way ANOVA using Dunn's test ** ($p < 0.05$).

3.3. Biofunctionalization/Protein Binding

To prove the accessibility of the biotin moieties carried by the admixed phospholipids, the selective binding of a model protein was demonstrated (Figure 5a). For this purpose, new samples were prepared, where two functionalized adhesive-based inks were multiplexed onto the substrate, in the form of different geometrical micro-/nanostructures. In a preliminary stability experiment, conducted with non-cured samples, the written structures were incubated with streptavidin directly after patterning, with no further treatment, other than blocking the samples with BSA to avoid non-specific binding to the bare substrate areas. This experiment (Figure S4), showed highly specific streptavidin–biotin binding onto the functionalized structures, while the structures written with non-functionalized inks showed no sign of protein binding.

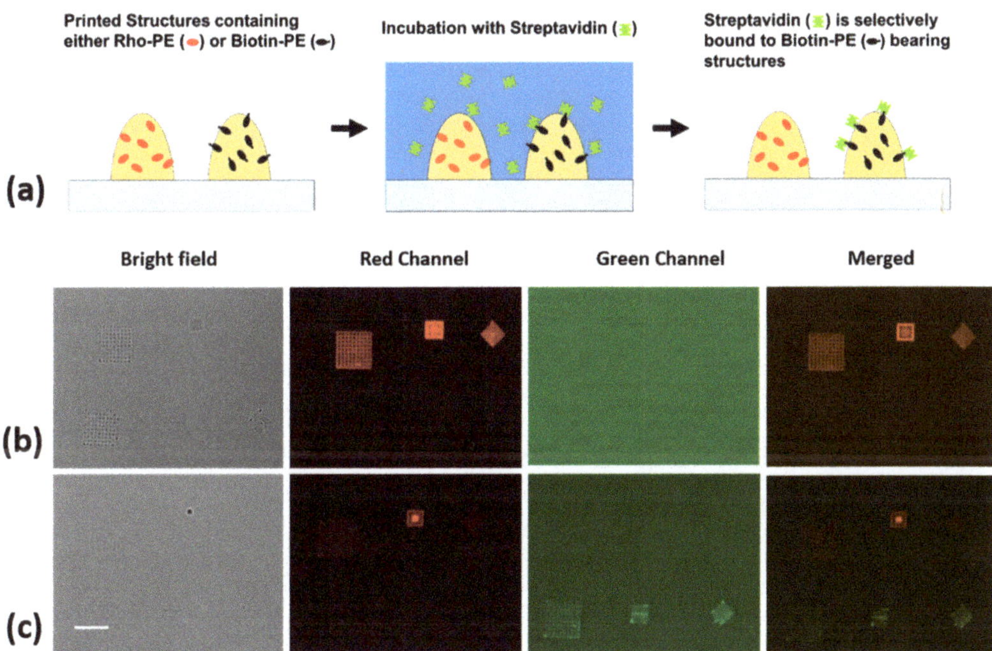

Figure 5. Biofunctionalization of adhesive-based structures by a model protein. (**a**) Scheme showing functionalization of the adhesive ink. Microscopy images of the (**b**) adhesive modified with rhodamine-PE (first row) and biotinylated adhesive structures (second row), and (**c**) the same adhesive structures after incubation with fluorescently labelled streptavidin, showing selective binding. Scale bar equals 40 µm for all images.

In a new, freshly made sample, the molecules were immobilized by curing the adhesive features under UV light for 5 min, to address the question of whether the biotin moieties stay on the surface and remain fully functional after curing. Working with fully cured samples can be critical for applications, as even when small-volume features are often already cured by exposure to ambient light, it might not be enough to fully cure thicker structures. The immersion of uncured adhesive in liquid might lead to rearrangement of the structures and the formation of droplets (Figure S4). Figure 5b,c displays the obtained fluorescent images before and after the incubation of fully cured patterns, with fluorescently labelled streptavidin. The emergence of the second row of patterns in the green channel after incubation constitutes an unequivocal sign of streptavidin binding to the biotin moieties.

4. Discussion and Conclusions

Three-dimensional nano and microprinting with UV-curable polymer inks has the potential to become a powerful and highly versatile method to produce bioactive and functional surfaces for biological research and biomedical applications. FluidFM is, in particular, suited for this kind of lithography, and widely available methacrylates, such as the commercial Loctite adhesive, can be employed [19].

We demonstrate that FluidFM allows the effective fabrication of micro-/nanostructures of different shapes. Importantly, the features of the printing structures can be adjusted by controlling the surface properties. By tuning the surface hydrophilicity with silanization or plasma cleaning, the height and width of the written structures can be modified. The highest structures were obtained on the most hydrophobic substrates, and the widest structures on the most hydrophilic substrates. This can be understood by the expected increasing contact angle of the ink with the substrate, which keeps the structure more confined, thus raising the height for the same volume of deposited ink for more hydrophobic inks. Our results can be used to purposely tune surface properties and to make informed choices for printing, when specific substrates are chosen for reasons of biological constraints.

However, it should be considered that being able to produce micro-/nanostructures in arbitrary shapes is only one aspect demanded in bioapplications. Most of the time, chemical or biochemical modifications are also key for obtaining the desired functionality. While a homogeneous chemical modification can easily be obtained in bulk by a post-functionalization step, targeted functionalization of specific micro-/nanostructures is hardly obtainable by conventional approaches. Towards this goal, we demonstrated the inherent functionalization of printed structures by admixing phospholipids to the base adhesive ink. In our approach, the resulting structures are directly functionalized without any need for further steps. Furthermore, this gives the opportunity for multiplexing, as structures can easily be written next to each other with different inks. Here, we demonstrated the concept by admixing a fluorescently labelled phospholipid or a biotinylated phospholipid, respectively. With simple conventional blocking agents, such as BSA, unspecific adhesion is blocked and streptavidin is highly selective, binding only to the structures made with the ink containing biotinylated phospholipid. This proves that the biotin moieties remain accessible at the interface of the structures to the liquid phase, probably preferentially oriented outwards of the polymer bulk by their amphiphilic natures (the hydrophilic head group is modified with the biotin moiety and hydrophobic hydrocarbon chains at the tail end). Remarkably, the biotin motif offers a wider range of possibilities for biofunctionalization, as it is one of the most widespread binding tags in biotechnology, with a myriad of compounds available with biotin modifications that could be bound over a streptavidin linker to such structures. In addition, other lipid modifications will likely work in a similar fashion, e.g., to introduce metal-chelating lipids, such as 1,2-dioleoyl-sn-glycero-3-[(*N*-(5-amino-1-carboxypentyl)iminodiacetic acid)succinyl] (nickel salt) (18:1 DGS-NTA(Ni)), to allow the binding of polyhistidine(His)-tagged proteins, further expanding the multiplexing capabilities by utilizing orthogonal binding tags. The admixing did not significantly alter the printing properties or mechanical properties of the printed structures; therefore, it is a straightforward process to include different functionalization and no interference with, e.g., changes in mechanical cues, as stiffness to the cells growing on the structures is to be expected.

Another important concern when considering bioapplications is the biocompatibility of the involved materials. While methacrylates are problematic in this regard, the toxicity of the material is mainly conveyed by residual monomers seeping out [32]. For the comparably thin structures produced in FluidFM printing processes, curing will generally be complete with not much monomer left, and, if needed, biocompatibility could be further increased by additional treatments, such as ethanol washing [33].

In summary, our results show the successful functionalization of polymer ink with model biomolecules, and provide proof of principle as to how biologically relevant species can be incorporated into the direct-patterned nanostructures, without significantly altering their global properties. Overall, these results demonstrate the potential for the direct printing of functionalized structures via FluidFM, or similar dispensing techniques, for the creation of bioactive, protein-presenting micro-/nanostructures for bioapplications.

Supplementary Materials: The following are available online at https://www.mdpi.com/article/10.3390/polym14071327/s1: Figure S1: indentation maps of different structures; Figure S2: water contact angles on the different surface functionalization; Figure S3: calculated dot volumes on substrates with different functionalization; Figure S4: additional incubation experiments.

Author Contributions: Conceptualization, T.-H.N. and M.H.; methodology, E.B., G.A. and S.K.V.; validation, E.B., G.A. and S.K.V.; formal analysis, E.B., G.A. and S.K.V.; investigation, E.B., G.A. and S.K.V.; resources, T.-H.N. and M.H.; data curation, T.-H.N. and M.H.; writing—original draft preparation, E.B., G.A., S.K.V., T.-H.N. and M.H.; writing—review and editing, E.B., G.A., S.K.V., T.-H.N. and M.H.; visualization, E.B. and G.A.; supervision, T.-H.N. and M.H.; project administration,

Polymers **2022**, *14*, 1327

T.-H.N. and M.H.; funding acquisition, E.B., T.-H.N. and M.H. All authors have read and agreed to the published version of the manuscript.

Funding: This work was partially supported by the Ministerium für Wirtschaft, Wissenschaft und Digitale Gesellschaft, TMWWDG, Freistaat Thüringen, Germany.

Data Availability Statement: The data presented in this study are available on reasonable request from the corresponding authors.

Acknowledgments: This work was partly carried out with the support of the Karlsruhe Nano Micro Facility (KNMFi, http://www.knmf.kit.edu), a Helmholtz Research Infrastructure at the Karlsruhe Institute of Technology (KIT, http://www.kit.edu). E.B. acknowledges the Karlsruhe Institute of Technology (KIT) for a postdoctoral fellowship within the Young Investigator Group Preparation Programme (YIG Prep Pro). T.-H.N. and G.A. acknowledge the support of the Freistaat Thüringen (Thüringer Ministerium für Wirtschaft, Wissenschaft und Digitale Gesellschaft, TMWWDG, Germany), and the German Research Foundation (DFG) within the project Nr. NG/133-1-2. M.H. acknowledges the additional support by the Helmholtz Association in the form of a Helmholtz ERC Recognition Award. We acknowledge support by the KIT Publication Fund of the Karlsruhe Institute of Technology.

Conflicts of Interest: The authors declare no conflict of interest. The funders had no role in the design of the study; in the collection, analyses, or interpretation of data; in the writing of the manuscript, or in the decision to publish the results.

References

1. Chung, J.J.; Im, H.; Kim, S.H.; Park, J.W.; Jung, Y. Toward Biomimetic Scaffolds for Tissue Engineering: 3D Printing Techniques in Regenerative Medicine. *Front. Bioeng. Biotechnol.* **2020**, *8*, 1–12. [CrossRef] [PubMed]
2. Hippler, M.; Lemma, E.D.; Bertels, S.; Blasco, E.; Barner-Kowollik, C.; Wegener, M.; Bastmeyer, M. 3D Scaffolds to Study Basic Cell Biology. *Adv. Mater.* **2019**, *31*, 1808110. [CrossRef] [PubMed]
3. Ouyang, X.; Zhang, K.; Wu, J.; Wong, D.S.-H.; Feng, Q.; Bian, L.; Zhang, A.P. Optical μ-Printing of Cellular-Scale Microscaffold Arrays for 3D Cell Culture. *Sci. Rep.* **2017**, *7*, 8880. [CrossRef] [PubMed]
4. Zandrini, T.; Shan, O.; Parodi, V.; Cerullo, G.; Raimondi, M.T.; Osellame, R. Multi-foci laser microfabrication of 3D polymeric scaffolds for stem cell expansion in regenerative medicine. *Sci. Rep.* **2019**, *9*, 11761. [CrossRef] [PubMed]
5. Richter, B.; Hahn, V.; Bertels, S.; Claus, T.K.; Wegener, M.; Delaittre, G.; Barner-Kowollik, C.; Bastmeyer, M. Guiding Cell Attachment in 3D Microscaffolds Selectively Functionalized with Two Distinct Adhesion Proteins. *Adv. Mater.* **2017**, *29*, 1604342. [CrossRef] [PubMed]
6. Barner-Kowollik, C.; Bastmeyer, M.; Blasco, E.; Delaittre, G.; Müller, P.; Richter, B.; Wegener, M. 3D Laser Micro- and Nanoprinting: Challenges for Chemistry. *Angew. Chem. Int. Ed.* **2017**, *56*, 15828–15845. [CrossRef] [PubMed]
7. Bettinger, C.J.; Langer, R.; Borenstein, J.T. Engineering Substrate Topography at the Micro- and Nanoscale to Control Cell Function. *Angew. Chem. Int. Ed.* **2009**, *48*, 5406–5415. [CrossRef]
8. Yang, T.; Du, Z.; Qiu, H.; Gao, P.; Zhao, X.; Wang, H.; Tu, Q.; Xiong, K.; Huang, N.; Yang, Z. From surface to bulk modification: Plasma polymerization of amine-bearing coating by synergic strategy of biomolecule grafting and nitric oxide loading. *Bioact. Mater.* **2020**, *5*, 17–25. [CrossRef]
9. Ferrari, M.; Cirisano, F.; Morán, M.C. Mammalian Cell Behavior on Hydrophobic Substrates: Influence of Surface Properties. *Colloids Interfaces* **2019**, *3*, 48. [CrossRef]
10. Apte, G.; Lindenbauer, A.; Schemberg, J.; Rothe, H.; Nguyen, T.H. Controlling Surface-Induced Platelet Activation by Agarose and Gelatin-Based Hydrogel Films. *ACS Omega* **2021**, *6*, 10963–10974. [CrossRef]
11. Bui, V.-C.; Medvedev, N.; Apte, G.; Chen, L.-Y.; Denker, C.; Greinacher, A.; Nguyen, T.-H. Response of Human Blood Platelets on Nanoscale Groove Patterns: Implications for Platelet Storage. *ACS Appl. Nano Mater.* **2020**, *3*, 6996–7004. [CrossRef]
12. Apte, G.; Börke, J.; Rothe, H.; Liefeith, K.; Nguyen, T.H. Modulation of Platelet-Surface Activation: Current State and Future Perspectives. *ACS Appl. Bio Mater.* **2020**, *3*, 5574–5589. [CrossRef] [PubMed]
13. Hauptmann, N.; Lian, Q.; Ludolph, J.; Rothe, H.; Hildebrand, G.; Liefeith, K. Biomimetic Designer Scaffolds Made of D,L-Lactide-ε-Caprolactone Polymers by 2-Photon Polymerization. *Tissue Eng. Part B Rev.* **2019**, *25*, 167–186. [CrossRef] [PubMed]
14. Kulangara, K.; Leong, K.W. Substrate topography shapes cell function. *Soft Matter* **2009**, *5*, 4072. [CrossRef]
15. Xiang, T.; Hou, J.; Xie, H.; Liu, X.; Gong, T.; Zhou, S. Biomimetic micro/nano structures for biomedical applications. *Nano Today* **2020**, *35*, 100980. [CrossRef]
16. Zambelli, T.; Aebersold, M.; Behr, P.; Han, H.; Hirt, L.; Guillaume-gentil, O.; Vörös, J. FluidFM: Development of the Instrument as well as Its Applications for 2D and 3D Lithography. In *Open-Space Microfluidics: Concepts, Implementation, and Applications*; Wiley-VCH Verlag: Weinheim, Germany, 2018; pp. 295–322. [CrossRef]

17. Grüter, R.R.; Vörös, J.; Zambelli, T. FluidFM as a lithography tool in liquid: Spatially controlled deposition of fluorescent nanoparticles. *Nanoscale* **2013**, *5*, 1097–1104. [CrossRef]
18. Berganza, E.; Hirtz, M. Direct-Write Patterning of Biomimetic Lipid Membranes In Situ with FluidFM. *ACS Appl. Mater. Interfaces* **2021**, *13*, 50774–50784. [CrossRef]
19. Ventrici de Souza, J.; Liu, Y.; Wang, S.; Dörig, P.; Kuhl, T.L.; Frommer, J.; Liu, G. Three-Dimensional Nanoprinting via Direct Delivery. *J. Phys. Chem. B* **2018**, *122*, 956–962. [CrossRef]
20. Deng, W.N.; Wang, S.; Ventrici de Souza, J.; Kuhl, T.L.; Liu, G. New Algorithm to Enable Construction and Display of 3D Structures from Scanning Probe Microscopy Images Acquired Layer-by-Layer. *J. Phys. Chem. A* **2018**, *122*, 5756–5763. [CrossRef]
21. Henkel Technical Data Sheet—Loctite®AA 3491TM. 2014. Available online: https://www.henkel-adhesives.com/de/de/produkt/uv-curing-adhesives/loctite_aa_3491.html (accessed on 24 February 2022).
22. Meyer, R.; Giselbrecht, S.; Rapp, B.E.; Hirtz, M.; Niemeyer, C.M. Advances in DNA-directed immobilization. *Curr. Opin. Chem. Biol.* **2014**, *18*, 8–15. [CrossRef]
23. Kumar, R.; Weigel, S.; Meyer, R.; Niemeyer, C.M.; Fuchs, H.; Hirtz, M. Multi-color polymer pen lithography for oligonucleotide arrays. *Chem. Commun.* **2016**, *52*, 12310–12313. [CrossRef] [PubMed]
24. Horcas, I.; Fernández, R.; Gómez-Rodríguez, J.M.; Colchero, J.; Gómez-Herrero, J.; Baro, A.M. WSXM: A software for scanning probe microscopy and a tool for nanotechnology. *Rev. Sci. Instrum.* **2007**, *78*, 013705. [CrossRef] [PubMed]
25. Wilchek, M.; Bayer, E.A. Introduction to avidin-biotin technology. In *Methods in Enzymology—Vol. 184: Avidin-Biotin Technology*; Academic Press: Cambridge, MA, USA, 1990; pp. 5–13. [CrossRef]
26. Liu, H.-Y.; Kumar, R.; Zhong, C.; Gorji, S.; Paniushkina, L.; Masood, R.; Wittel, U.A.; Fuchs, H.; Nazarenko, I.; Hirtz, M. Rapid Capture of Cancer Extracellular Vesicles by Lipid Patch Microarrays. *Adv. Mater.* **2021**, *33*, 2008493. [CrossRef] [PubMed]
27. Padmanabhan, J.; Kinser, E.R.; Stalter, M.A.; Duncan-Lewis, C.; Balestrini, J.L.; Sawyer, A.J.; Schroers, J.; Kyriakides, T.R. Engineering Cellular Response Using Nanopatterned Bulk Metallic Glass. *ACS Nano* **2014**, *8*, 4366–4375. [CrossRef] [PubMed]
28. Cyster, L.A.; Parker, K.G.; Parker, T.L.; Grant, D.M. The effect of surface chemistry and nanotopography of titanium nitride (TiN) films on primary hippocampal neurones. *Biomaterials* **2004**, *25*, 97–107. [CrossRef]
29. Salloum, D.S.; Olenych, S.G.; Keller, T.C.S.; Schlenoff, J.B. Vascular Smooth Muscle Cells on Polyelectrolyte Multilayers: Hydrophobicity-Directed Adhesion and Growth. *Biomacromolecules* **2005**, *6*, 161–167. [CrossRef]
30. Tan, G.; Ouyang, K.; Wang, H.; Zhou, L.; Wang, X.; Liu, Y.; Zhang, L.; Ning, C. Effect of Amino-, Methyl- and Epoxy-Silane Coupling as a Molecular Bridge for Formatting a Biomimetic Hydroxyapatite Coating on Titanium by Electrochemical Deposition. *J. Mater. Sci. Technol.* **2016**, *32*, 956–965. [CrossRef]
31. Breuls, R.G.M.; Jiya, T.U.; Smit, T.H. Scaffold Stiffness Influences Cell Behavior: Opportunities for Skeletal Tissue Engineering. *Open Orthop. J.* **2008**, *2*, 103–109. [CrossRef]
32. Gautam, R.; Singh, R.D.; Sharma, V.P.; Siddhartha, R.; Chand, P.; Kumar, R. Biocompatibility of polymethylmethacrylate resins used in dentistry. *J. Biomed. Mater. Res. Part B Appl. Biomater.* **2012**, *100B*, 1444–1450. [CrossRef]
33. Alifui-Segbaya, F.; George, R. Biocompatibility of 3d-printed methacrylate for hearing devices. *Inventions* **2018**, *3*, 52. [CrossRef]

 *polymers*

Article

Effect of 3D-Printed PLA Structure on Sound Reflection Properties

Katarina Monkova [1,2,*], Martin Vasina [2,3,*], Peter Pavol Monka [1,2], Jan Vanca [1] and Dražan Kozak [4]

[1] Faculty of Manufacturing Technologies, Technical University in Kosice, 080 01 Presov, Slovakia; peter.pavol.monka@tuke.sk (P.P.M.); jan.vanca@tuke.sk (J.V.)
[2] Faculty of Technology, Tomas Bata University in Zlin, Nam. T.G. Masaryka 275, 760 01 Zlin, Czech Republic
[3] Faculty of Mechanical Engineering, VŠB-Technical University of Ostrava, 17. listopadu 15/2172, 708 33 Ostrava-Poruba, Czech Republic
[4] Mechanical Engineering Faculty, University of Slavonski Brod, Trg Ivane Brlic-Mazuranic 2, HR-35000 Slavonski Brod, Croatia; dkozak@unisb.hr
* Correspondence: katarina.monkova@tuke.sk (K.M.); vasina@utb.cz (M.V.); Tel.: +421-55-602-6370 (K.M.); +420-57-603-5112 (M.V.)

Abstract: 3D printing technique is currently one of the promising emerging technologies. It is used in many areas of human activity, including acoustic applications. This paper focuses on studying the sound reflection behavior of four different types of 3D-printed open-porous polylactic acid (PLA) material structures, namely cartesian, octagonal, rhomboid and starlit structures. Sound reflection properties were evaluated by means of the normal incidence sound reflection coefficient based on the transfer function method using an acoustic impedance tube. In this study, various factors affecting the sound reflection performance of the investigated PLA samples were evaluated. It can be concluded that the sound reflection behavior of the tested PLA specimens was strongly affected by different factors. It was influenced, not only by the type of 3D-printed open-porous material structure, but also by the excitation frequency, the total volume porosity, the specimen thickness, and the air gap size behind the tested specimen inside the acoustic impedance tube.

Keywords: polylactic acid; sound reflection; excitation frequency; porosity; 3D printing technique; thickness; air gap

Citation: Monkova, K.; Vasina, M.; Monka, P.P.; Vanca, J.; Kozak, D. Effect of 3D-Printed PLA Structure on Sound Reflection Properties. *Polymers* **2022**, *14*, 413. https://doi.org/10.3390/polym14030413

Academic Editor: Houwen Matthew Pan

Received: 21 December 2021
Accepted: 18 January 2022
Published: 20 January 2022

Publisher's Note: MDPI stays neutral with regard to jurisdictional claims in published maps and institutional affiliations.

1. Introduction

Architects have often focused on designing beautiful buildings intended for musical performances, especially concert halls, operas, churches, mosques, or theatres. Acoustic design with good sound distribution is crucial in the internal spaces of such buildings [1–3]. In contrast to concert halls, in which the best possible sound acoustics are needed, in the case of main roads and other places, where there is high noise pollution, the aim is to absorb or effectively reflect noise, thus preventing the further propagation of sound waves [4]. However, other rooms which are used every day, such as classrooms, auditoriums, offices, or production halls, also require well-projected sound properties. Poor acoustic design of such workspaces can cause discomfort, dizziness, and constant exposure to excessive unwanted sounds that affect physical and mental health. Therefore, it is crucial for human health and safety to use suitable sound-absorbing materials in specific rooms, offices, manufacturing halls, vehicles, or aircraft.

Noise and sound have the same specification but are perceived differently. The speed of sound can vary due to the difference in pressure between the media through which the sound propagates [5]. Sound can be considered as the propagation of a disorder mainly in liquid or solid phases. The wave motion is triggered when the element moves the nearest air particle, which then causes a pressure difference in the medium in which the wave moves [6]. Noise is perceived as a nuisance by people, and noise pollution in the environment can also cause discomfort [7].

Sound propagates through the air or other media as a longitudinal wave, in which the mechanical vibrations run in the direction of wave propagation. The property of waves and sound associated with the echo phenomenon is known as reflection [7–10].

The sound wave changes during the passage through space when it meets an obstacle. The wave may vary due to reflection from the obstacle, diffraction around the obstacle, and transmission accompanied by a refraction to the obstacle and other space. One part of the soundwave passes through the boundary of space. The second part is reflected. The reflection itself and its intensity depend on the similarity of the spaces [11].

The sound energy reflected from a boundary/obstacle can be divided into specular or diffuse. The former is when the reflection appears at the same angle at which it strikes the surface, like the reflection of light from a mirror. In contrast, diffuse reflection occurs when the sound energy is scattered in non-specular directions [4]. The direction of wave propagation is perpendicular to the front formed by all the Huygens' wavelets. In terms of shape, suitable reflectors are used for different purposes or effects. For example, a parabolic reflector focuses a parallel sound wave to a specific point, making it easier to hear a feeble sound. Such reflectors are used in parabolic microphones to pick up sound from a distant source or select the location where the sound is observed and then focus it on the microphone. On the other hand, an elliptical shape can focus sound from one point to another [8,12].

The reflection (absorption) of sound waves is affected by several parameters, such as surface shapes, obstacles and their spatial distribution, their distance and arrangement [13]. In addition to these, one of the most important factors that has a significant effect on the proper propagation of sound is the material, including its porosity, density, thickness, and the angle of sound incidence, surface shape, excitation frequency of the acoustic wave and the type of internal structure of the material [1,6]. Zulkifli [14] and Lee [15] reported that the acoustic absorption of multilayer materials is better with perforated plates backed with airspaces. In most cases, sound-absorbing materials are measured in a reverberation chamber. In a study by Cucharero et al. [16], the authors showed how angle-dependent absorption coefficients could be measured. The results were in good agreement with sound absorption coefficients and confirmed that materials have different attenuation behavior to sound waves coming from different directions.

In recent years, with the development of manufacturing technology, porous materials have presented a challenge to many researchers to study their sound absorption or reflection properties. The process of reflecting acoustic waves in porous materials in the case of oblique impact of acoustic waves was studied by Gimaltdinov et al. [17] The results showed, that with increasing angle of impact, transmission decreases. According to [18], lightweight materials, such as foam or fabric, are often too porous to reflect sound passing through, and their energy is converted into heat with a reduction in magnitude. This approach is often used in cinemas and recording studios to reduce the room's reverberation time. While effective internally, lightweight materials are not suitable externally, so concrete is still preferred [18]. A balance between speed and density of sound and its effect on the acoustic properties of a porous material was studied by Kuczmarski et al. [19], while the effect of tortuosity, porosity, and flow resistivity on reflection and transmission of acoustic waves in porous materials was the topic of [20]. Sun et al. [21] demonstrated that the main reason for the difference in acoustic properties of materials produced via selective laser sintering is the difference in the unit structures. The investigation confirmed that transmission loss has better results with mixed structures than using individual components, and can be improved with periodically arranged scatterers inside porous materials.

To study the natural behavior of the uniform infinite layers of a porous material was the aim of [22], which considered the relation between pressure and velocity. The results of this study enabled evaluation of the intrinsic properties of the material based on plane wave acoustic surface impedance. A subsequent comprehensive literature review, with detailed description of the prediction of acoustic absorption behavior, various techniques for enhancing sound absorption and the application of Wilson's relaxation-matched equivalent fluid model, was provided by Otaru [23]. The compound impedance and sound absorption in a porous material in the first resonance cavity were examined by Fan [24]. The results showed that an acoustic board, consisting of aluminum foam as the inner structure and micro-porous board as the outer structure, with polyester between them, showed good sound absorption performance in the low-frequency zone. Similarly, Talebitooti et al. [25] investigated double-walled poroelastic composite shells to achieve diffuse sound transmission in the high-frequency range. A positive effect was observed in optimizing vibroacoustic fitness.

A summary of fabrication, evolution and prediction models for sound absorption in porous materials is described in the work of Cao et al. [26], while the work of Chen [27] demonstrated the relation between geometrical parameters and sound absorption based on the establishment of static flow resistivity and derivation of the fractal acoustic model. The finite analysis method was applied to assess sound insulation performance in a sandwich plate with lattice core, the results showing that a sandwich plate with a lattice core could outperform traditional lattice cores [28].

The present study represents an original contribution to the research of sound properties of additively produced materials with regularly distributed porous structures. It sought to investigate the sound reflection behavior of four different types of 3D-printed open-porous polylactic acid (PLA) material structures, namely cartesian, octagonal, rhomboid and starlit structures, considering various factors affecting the sound reflection performance of the investigated PLA samples. The sound reflection properties were evaluated by means of the normal incidence sound reflection coefficient based on the transfer function method using an acoustic impedance tube.

It can be seen from the overview above that the concept of using lightweight materials in technical practice is not entirely new. Porous materials give the product a specific combination of properties, especially in terms of lightness (weight reduction), while maintaining sufficient rigidity, strength, and resistance to load/stress. Their advantages include, not only their mechanical properties, but also other physical characteristics, such as acoustic, thermal, or viscous properties. Reduced material consumption will also reduce the cost of producing the component and recycling it, making porous materials more environmentally friendly. Currently, the production of cellular materials is possible mainly due to the rapid development of additive technologies. With the help of 3D printing technique, it is possible to produce even complex objects that cannot be produced by other (classical) methods. All porous materials are strongly dependent on their three-dimensional (3D) microstructures, including the porosity, pore sizes, shapes, and connectivity of the porous space [29,30]. In terms of the division into porous structures, regular and irregular, the regular arrangement of cellular structures allows better control or prediction of their properties, so the authors decided to address this type of structure in the sound reflectivity research.

To the best of the author's knowledge, no relevant studies have yet been published which examine the sound reflection of the above types of structures made of PLA plastic material, considering the influence of several factors, such as their structural type, excitation frequency, the total volume porosity, the specimen thickness, and the air gap size behind the tested specimen inside the acoustic impedance tube.

2. Materials and Methods

2.1. Characteristics of Samples

Airflow resistivity is a physical parameter characteristic of porous and fibrous materials, which quantifies, per unit length, the ability to oppose resistance to the motion of air particles inside a material. As is well-known, this property is important to evaluate the acoustic behavior of these materials in various fields of their application, such as automotive noise mitigation, architectural acoustics, and building acoustics [31,32].

The samples were produced via an additive approach from polylactic acid or polylactide (PLA) material (Prusa Polymers, Prague, Czech Republic), which is a thermoplastic made from vegetable starch. The processing of this material was carried out at a temperature of approximately (180–220) °C. The manufacturer states in the material sheet the following properties of the PLA filament used in the experiments of this research: peak melt temperature (145–160) °C, glass transition temperature (55–60) °C, specific weight 1.24 g cm^{-3}, moisture absorption 0.3% at 28 °C and yield strength of the filament 57.4 MPa.

After heating this material, it is possible to smell a sweet scent. PLA material is resistant to deformation, non-toxic, brittle, and less heat resistant than ABS material. The advantage of this material is that it is not harmful to health and is biodegradable. Another advantage of this material is that it cannot be easily torn and can be used with printers that do not have a heated printing bed. Compared to other materials, PLA cooling takes longer, and efficient cooling can be ensured by fans. Components made of this material can be finely machined, e.g., by grinding. Products extruded from this plastic should not be exposed to direct sunlight. The products are not suitable for high loads. In situations of excessive humidity, these products absorb water, which also reduces their quality. PLA material is most often used to produce food aids, prototypes, toys, design models, etc. However, it can be used to advantage in suitable technical applications in engineering and construction, and the automotive or aerospace industries [33,34].

Four different types of 3D-printed open-porous structures, namely, cartesian, octagonal, rhomboid and starlit structures, were selected to investigate sound reflection properties within the presented research. The main reason for selecting the lattice structures referred to above was the completely different type of regularly repeating cells forming the structure, as well as the nature of their distribution in the cross-sectional area (rotated around the cylindrical axis or slid in rectangular axes). Virtual models of the samples were generated in software PTC Creo 7 (Parametric Technology Corporation Inc., Boston, MA, USA). The samples were cylindrical with a diameter of 29 mm so that their outer shape and dimensions matched the impedance tube for testing; the structure itself had a diameter of 25 mm, and the outer shell of continuous material had a thickness of 2 mm. For each type of structure (cartesian, octagonal, rhomboid and starlit), samples were produced in three different volume porosities P (30%, 43% and 56%) and in three different thicknesses t (10 mm, 20 mm and 30 mm), making in total 36 pieces.

The total volume porosity of a porous material is one of the essential characteristics of a structure (apart from its geometry) and is defined by Equation (1) [35]:

$$P(\%) = \frac{V_a}{V_T} \times 100 \tag{1}$$

where V_a is the air volume in body cavities, and V_T is the total body volume. The volume porosity of the specimens was regulated by the thickness of the strut in a lattice structure and by the cell size.

If the apparent density of the samples is known (in our case it was 0.5456; 0.7068 and 0.868 gcm^{-3}), the porosity can be expressed by means of Equation (2)

$$P(\%) = \left(1 - \frac{\rho_A}{\rho}\right) \times 100 \tag{2}$$

where ρ_A is the apparent density and ρ is the density of PLA material (1.24 gcm^{-3}).

A summary of the basic characteristics of lattice structures made to investigate the sound reflection properties is given in Table 1, where $x/y/z$ are the dimensions of the base cell, where the orientation "z" is the direction of sample building and it is perpendicular to the position of the platform "x,y".

Table 1. Basic characteristics of lattice structures designed to investigate the sound reflection properties.

Structure Type	Label	Volume Porosity (%)	Structure View	Strut Diameter (mm)	Cell Size x/y/z (mm)
Cartesian	C	56		1	5/5/5
		43		1.4	5/5/5
		30		1.8	5/5/5
Octagonal	O	56		1	6/7/5
		43		1.4	6/7/5
		30		1.7	5.5/7/5
Rhomboid	R	56		1	5.5/7/7
		43		1.35	5/7/7
		30		1.7	5/7/7
Starlit	S	56		1	8/9/5
		43		1.4	7.5/9/5
		30		1.8	8/9/5

The ability to produce such complex structures in the core of an object (e.g., a future sound wall with the necessary sound reflection properties) is made possible today by technologies which work on the additive principle. The proposed test specimens were printed on a Prusa i3 Mk2 3D printer (Prusa Research a.s., Prague, Czech Republic), using fused deposition modeling (FDM) technology for 3D printing. This technology is one of the most widespread additive production technologies and is included in the group of methods used to produce solid prototypes based on the principle of extrusion. As with other technologies, including FDM, the required body is created by depositing thin layers of semi-liquid material. The material is wound in the form of a filament on a spool from where it is led to a nozzle located in the application head. The movement of the nozzle is given by a 3D model of the desired body, which is sliced into thin layers in the software before the 3D printing itself. It is "controlled" by the contours of the body in the given layer and the chosen strategy of filling continuous surfaces with the material. After application, the material adheres to the previous layer and gradually cools. The building platform, on which the body is built, drops lower with increasing layer thickness and the application process is repeated until the component is completed. The quality and strength of FDM components essentially depend on the process parameters [36,37]. The material is wound on a spool attached to the extruder with its motor, which pushes the thread through the hot end. The hot end maintains a constant temperature to melt the filament very quickly into a viscous liquid which is extruded through a small nozzle, firstly onto the platform, and then in subsequent layers, continuing to apply the material until the entire prototype is formed.

The 3D printing strategy, as well as the nozzle and platform movement, were controlled by a computer on which PrusaSlicer software of the same manufacturer as the 3D printer machine (Prusa Research a.s., Prague, Czech Republic) was installed.

The technical and technological parameters used in 3D printing were selected according to the equipment manufacturer's recommendations for PLA material. However, a range of temperatures and speeds could be used. A nozzle temperature of 220 °C and a print speed of 40 mms^{-1} for the whole sample were applied to produce a preliminary sample. In this case, the 3D printer stringing effect appeared as seen in Figure 1.

After several trials to adjust the settings, the following 3D printing parameters were finally selected: nozzle temperature 215 °C for the first layer and 210 °C for the others; bed temperature (building platform) 60 °C, printing speed on the peripheral circuits 30 mm/s and printing speed of the internal structure 20 mms^{-1}. A 1.7 mm diameter filament was used for 3D printing of the samples, which was applied in layer thickness of 0.254 mm using a 0.4 mm diameter nozzle. The samples were positioned so that their cylindrical axes were perpendicular to the building platform. They were built without supports because the lattice structures were self-supporting. The quality of the samples was checked visually, and no cracks or thin strands of plastic (as the result of 3D printer stringing) were observed.

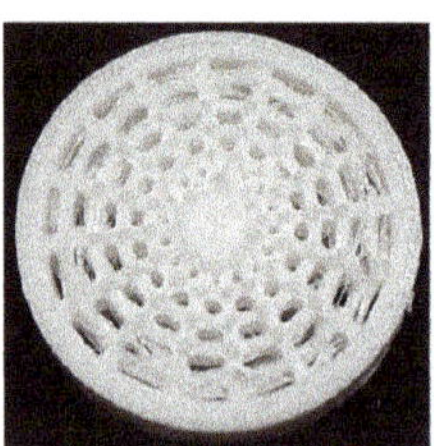

Figure 1. The 3D printer stringing effect appearance of a preliminary sample.

An example of the samples produced is shown in Figure 2. In Figure 2a, samples of thickness t = 30 mm with different types of structure are shown. In Figure 2b rhomboid samples with different volume porosities are shown, and in Figure 2c a set of samples with octagonal structure is presented.

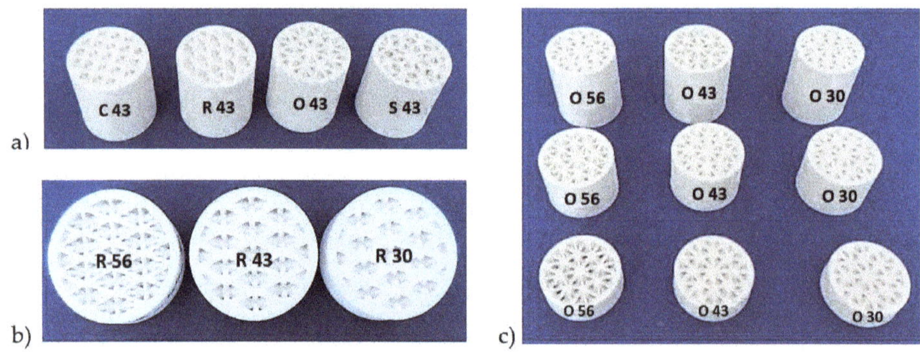

Figure 2. Examples of samples produced. (**a**) samples of thickness t = 30 mm and volume porosity P = 43% with different types of structure, (**b**) rhomboid samples with different volume porosities and (**c**) a set of samples with octagonal structure. (Note: the numbers on the sample labels in the figure indicate the total volume porosity in %).

2.2. Measurement Methodology

2.2.1. Sound Reflection Coefficient

When sound propagates from a sound source to a material's surface, the incident sound energy E_I is either reflected or absorbed by this material [38,39]. The sound reflection coefficient β is the measure of how much sound is reflected from a material [40]. It can be expressed as follows [41–44]:

$$\beta = \frac{E_R}{E_I} = \frac{E_I - E_A}{E_I} = 1 - \alpha \tag{3}$$

where E_R is the reflected sound energy, E_A is the absorbed sound energy, and α is the sound absorption coefficient.

2.2.2. Sound Reflection Properties

There are two different standardized methodologies for measuring sound reflection properties of material specimens, namely, the reverberation room and impedance tube methods [45].

Measurement of sound reflection properties using the reverberation room method [46–48] is performed in a standardized reverberation chamber and is based on the measurement of reverberation times with and without the specimen to be tested. This method is preferred by materials producers to measure the random sound reflection coefficient in a diffuse sound field. However, the reverberation room method requires a large-scale chamber of volume $V > 200$ m^3 and a large surface area (i.e., 10–12 m^2) of investigated specimens.

The impedance tube method [49,50] measures the sound reflection coefficient when sound waves propagate vertically through a test sample. Compared to the reverberation room method, the impedance tube method is a quick, low-cost method, using small-size specimens. Therefore, this method is suitable for developing new types of soundproofing materials. There are three impedance tube methods: pipe pulse, standing wave ratio, and transfer function methods [49]. The pipe pulse method is based on the separation of incident and reflection acoustic signals. Sound reflection properties are subsequently determined from the amplitude ratio of the acoustic signals. This method requires high anti-reference and capability of signal separation. The standing wave method [51] consists in measuring acoustic pressure amplitudes of the incident and reflected acoustic waves by means of a movable acoustic probe. This method is tedious, and the measuring efficiency is relatively low. The transfer function method [52] uses two precisely matched microphones and is based on a similar principle to the standing wave method. Its operational process is much simpler, and the measuring efficiency is higher than the standing wave method [49].

Experimental measurements of sound reflection properties of the investigated PLA specimens were undertaken using the transfer function method ISO 10534-2 [52] that is based on measuring sound pressures in the impedance tube by means of microphones M_1 and M_2. This method defines the normal incidence sound reflection coefficient β as follows [53–55]:

$$\beta = |R|^2 \tag{4}$$

where R is the complex reflection factor that is given by the formula:

$$R = \frac{H_{12} - H_I}{H_R - H_{12}} e^{2jk_0x_1} \tag{5}$$

where H_{12} is the pressure transfer function, H_I is the transfer function for the incident acoustic wave, H_R is the transfer function for the reflected acoustic wave, k_0 is the complex wave number, and x_1 is the distance between the microphone M_1 and the tested material sample (see Figure 3b). The above transfer functions are expressed as:

$$H_{12} = \frac{p_2}{p_1} = \frac{e^{jk_0x_2} + R\cdot e^{-jk_0x_2}}{e^{jk_0x_1} + R\cdot e^{-jk_0x_1}} \tag{6}$$

$$H_I = e^{-jk_0(x_1-x_2)} = e^{-jk_0s} \tag{7}$$

$$H_R = e^{jk_0(x_1-x_2)} = e^{jk_0s} \tag{8}$$

where p_1 and p_2 are the complex acoustic pressures measured by the microphones M_1 and M_2, x_2 is the distance between the microphone M_2 and the studied material specimen, and s is the distance between the two microphones M_1 and M_2.

The complex acoustic pressures are given by the equations:

$$p_1 = \hat{p}_I e^{jk_0x_1} + \hat{p}_R e^{-jk_0x_1} \tag{9}$$

$$p_2 = \hat{p}_I e^{jk_0x_2} + \hat{p}_R e^{-jk_0x_2} \tag{10}$$

where $\hat{p}_I$ and $\hat{p}_R$ are the amplitudes of acoustic pressures for the incident (I) and reflected (R) acoustic waves.

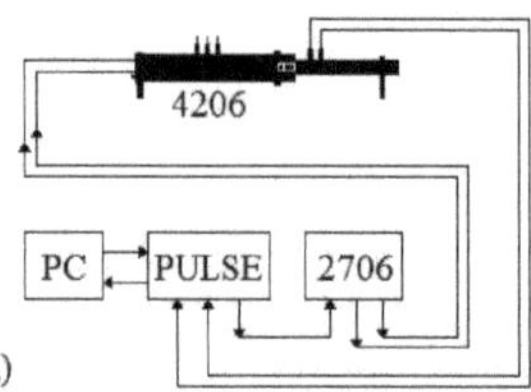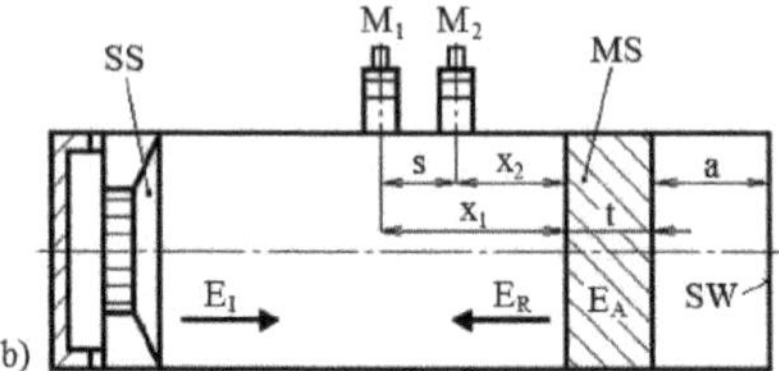

Figure 3. Circuit diagram of the experimental equipment for measuring the sound reflection coefficient (**a**) and a schematic of the acoustic impedance tube (**b**). Legend to the abbreviations: a—air gap size; E_A—absorbed sound energy; E_I—incident sound energy; E_R—reflected sound energy; M_1, M_2—measuring microphones; MS—measured specimen; s—distance between microphones M_1 and M_2; SS—sound source; SW—solid wall; t—sample thickness; x_1, x_2—microphone distances from the tested PLA sample surface.

Frequency dependencies of the normal-incidence sound reflection coefficient β of the tested 3D-printed PLA material samples were measured using a two-microphone acoustic impedance tube (BK 4206) in combination with a power amplifier (BK 2706) and a signal PULSE multi-analyzer (BK 3560-B-030) in the frequency range from 200 to 6400 Hz (Brüel & Kjær, Nærum, Denmark). A circuit diagram of the experimental equipment is depicted in Figure 3a. Normal-incidence sound reflection properties of the tested 3D-printed open-porous specimens of a given thickness t (i.e., 10, 20 and 30 mm) and a total volume porosity P (from 30 to 56%) were experimentally measured for different air gap sizes a (ranging from 0 to 100 mm) behind the studied PLA samples (see Figure 3b). Here, the solid wall SW is characterized by the perfect reflectivity (i.e., $\beta = 1$) of acoustic waves from its surface. All measurements were performed at an ambient temperature of 22 °C.

3. Results and Discussion

This section discusses various factors that influenced the sound reflection properties of the studied 3D-printed open-porous PLA structures.

3.1. Influence of Structure Type

The material structure is an important factor that has a strong influence on sound reflection performance. The effect of the structure type on sound reflection properties of the tested 3D-printed open-porous PLA materials, which was significant for thin (i.e., $t = 10$ mm) PLA material samples, is demonstrated in Figure 4. The frequency dependencies of the sound reflection coefficient for the PLA specimens of porosity $P = 30\%$ and thickness $t = 10$ mm are shown in Figure 4a. In this instance, the specimens were placed at 70 mm from the solid wall SW inside the acoustic impedance tube (see Figure 3b). Similarly, the effect of the structure type of the tested open-porous PLA materials (with $P = 56\%$, $t = 10$ mm and $a = 15$ mm) is demonstrated in Figure 4b. It is clear from Figure 4 that lower sound reflection properties were generally found for the PLA samples manufactured with the starlit structure, which were characterized by more complex pore shapes compared to the other open-porous PLA structures. For this reason, the PLA samples produced with the starlit structure exhibited higher airflow resistivity, which was accompanied by higher internal friction during the propagation of sound waves through this specimen structure, and by higher conversion of the incident acoustic energy E_I (see Figure 3b) into heat. These findings were observed mainly for low-frequency sound waves.

It can be concluded that the sound reflection performance for the investigated 3D-printed open-porous PLA materials of thickness $t = 10$ mm was significantly influenced by their volume porosity. In the case of low-porous (i.e., $P = 30\%$) PLA materials, better sound reflection behavior was observed for the specimens made with the rhomboid structure (see Figure 4a). However, the PLA materials of thickness $t = 10$ mm produced with the cartesian structure exhibited better sound reflection performance for the highest porosity value (i.e., $P = 56\%$), as shown in Figure 4b. The influence of the structure type on the sound reflection behavior of the tested 3D-printed open-porous PLA materials was negligible for the thicker (i.e., t = 20 and 30 mm) materials.

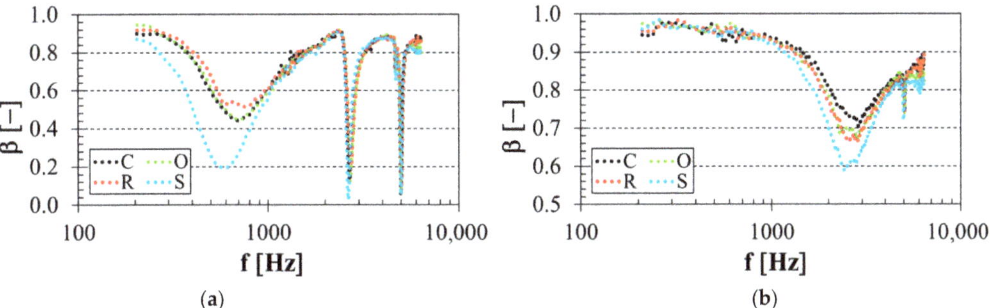

Figure 4. Influence of 3D-printed PLA structure type on the frequency dependencies of the sound reflection coefficient; (**a**) porosity P = 30%, specimen thickness t = 10 mm, air gap size a = 70 mm, (**b**) porosity P = 56%, specimen thickness t = 10 mm, air gap size a = 15 mm.

3.2. Influence of Total Volume Porosity

Sound reflection properties of the investigated four different open-porous PLA structures were significantly influenced, not only by the pore shape, but also by its size. It is known that the total volume porosity of open-porous materials is proportional to their relative density that increases with the decreasing pore size [56]. Better sound absorption properties of open-porous material structures are generally obtained at higher material density and airflow resistivity and smaller average pore sizes [57,58]. For these reasons, the increasing porosity of open-porous material structures leads to better sound reflection properties of these materials. This phenomenon was confirmed by the measured frequency dependencies of the sound reflection coefficient, as shown in Figure 5. Figure 5a represents the porosity effect on sound reflection performance of the PLA sample, which was manufactured with the rhomboid structure and thickness t = 10 mm and was placed at distance a = 40 mm in front of the solid wall SW. Similarly, the sound reflection properties of the PLA sample with the cartesian structure, thickness t = 20 mm and air gap size a = 30 mm depending on its porosity, are depicted in Figure 5b. It is clear from Figure 5 that the sound reflection coefficient generally increased with increasing sample porosity, especially at low excitation frequencies. This is because the higher porosity resulted in lower airflow resistivity during the sound wave propagation through the PLA open-porous structures, resulting in better sound reflection properties.

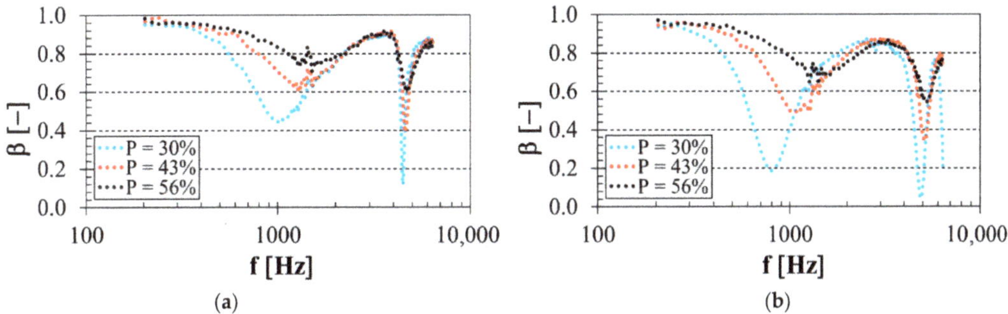

Figure 5. Effect of total volume porosity on the frequency dependencies of the sound reflection coefficient for the investigated PLA specimens: (**a**) rhomboid structure, sample thickness t = 10 mm, air gap size a = 40 mm; (**b**) cartesian structure, sample thickness t = 20 mm, air gap size a = 30 mm.

3.3. Influence of Specimen Thickness

The thickness of the 3D-printed open-porous PLA materials was also an important factor affecting their sound reflection behavior. The effect of the specimen thickness on the frequency dependencies of the sound reflection coefficient is shown in Figure 6. Figure 6a demonstrates the effect of specimen thickness on the sound reflection performance of the PLA sample with 30% porosity that was made with the starlit structure and was placed directly on the solid wall SW (i.e., a = 0 mm). Similarly, the influence of the thickness of the PLA sample with the highest porosity (i.e., P = 56%), which was produced with the octagonal structure and was placed 60 mm from the solid wall SW inside the acoustic impedance tube, is demonstrated in Figure 6b. It is clear that the sound reflection properties of the tested PLA samples decreased with increasing material thickness in a substantial part of the measured frequency range, especially in the low-frequency region. This phenomenon was caused by higher internal friction during the sound wave propagation through the thicker open-porous PLA materials, which was reflected in a greater conversion of sound wave energy into heat. Therefore, the increasing thickness of open-porous materials generally reduced their sound reflection properties. Lower sound reflection performance was found for the thicker PLA materials at the excitation frequency $f \cong 3$ kHz depending on the structure type, the volume porosity P and the air gap size a. This frequency boundary generally decreased with decreasing sample volume porosity and increasing air gap size.

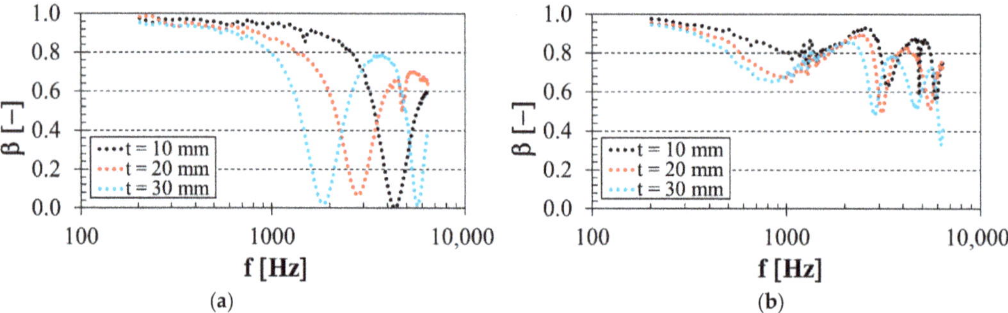

Figure 6. Effect of specimen thickness on the frequency dependencies of the sound reflection coefficient for the investigated PLA specimens: (**a**) Starlit structure, porosity P = 30%, air gap size a = 0 mm; (**b**) Octagonal structure, porosity P = 56%, air gap size a = 60 mm.

3.4. Influence of Air Gap Size

The sample distance in front of the solid wall SW inside the acoustic impedance tube was also a significant factor influencing the sound reflection properties of the investigated 3D-printed open-porous PLA materials. It is clear from Figure 7 that the frequency dependencies of the sound reflection coefficient were periodic with a certain number of minima and maxima of the sound reflection coefficient over the whole measured frequency range. This phenomenon, characteristic of open-porous materials, was demonstrated by the sound wave reflection from the wall SW inside the acoustic impedance tube (see Figure 3b), which relates to the wavelength of sound λ. The acoustic pressure will be highest at the wall, but the air particle velocity will be zero because the sound waves cannot supply enough energy to shake the solid wall [59]. If the air gap size between the solid wall and the tested sample gradually increases, the acoustic pressure decreases, but the air particle velocity increases. When the air gap is set to correspond to a quarter wavelength, the acoustic pressure is zero, and the air particle velocity is maximal. It is subsequently accompanied by higher internal friction during the propagation of sound waves through the porous specimen structure, which led to low sound reflection properties in this case. However, placing the open-porous material at a half-quarter wavelength distance from the solid wall will have a maximum reflection effect because the air particle velocity is minimum, and the acoustic pressure is maximum. Therefore, the sound reflection minima are proportional to odd multiples of quarter wavelengths and are obtained at the excitation frequencies:

$$f_{min} = \frac{c \cdot (2n+1)}{4l} = \frac{c \cdot (2n+1)}{4 \cdot \left(a + \frac{t}{2}\right)} \tag{11}$$

where c is the speed of sound, n is the integer ($n = 0, 1, 2 \ldots$), and l is the distance between the studied sample and the solid wall inside the acoustic impedance tube. Similarly, the sound reflection maxima are proportional to even multiples of quarter wavelengths and are obtained at the excitation frequencies:

$$f_{max} = \frac{c \cdot n}{2l} = \frac{c \cdot n}{2 \cdot \left(a + \frac{t}{2}\right)} \tag{12}$$

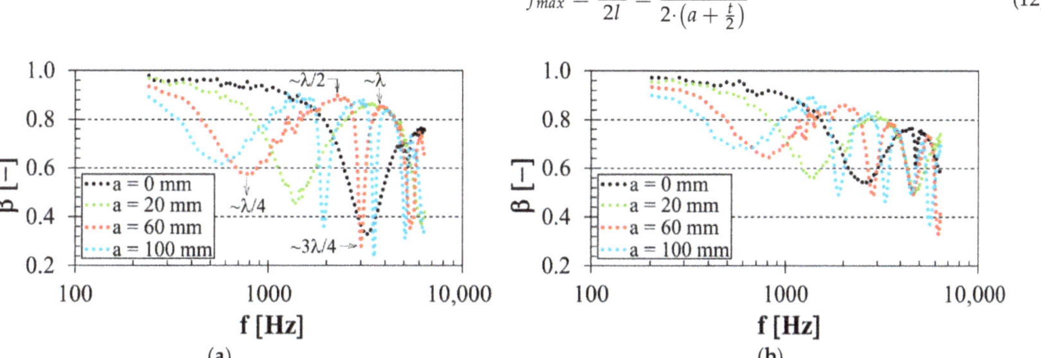

Figure 7. Influence of air gap size on the frequency dependencies of the sound reflection coefficient for the studied PLA specimens: (**a**) Starlit structure, sample thickness $t = 20$ mm, porosity $P = 43\%$; (**b**) Cartesian structure, sample thickness $t = 20$ mm, porosity $P = 56\%$.

It can be seen from Figure 7 that the excitation frequencies corresponding to sound reflection minima and maxima generally decreased with increasing air gap size, which is consistent with Equations (11) and (12) above. For this reason, the air gap size had a negative effect on sound reflection performance, especially in the low-frequency region. It is also clear that increasing air gap size led to a higher number of frequency minima and maxima over the whole measured frequency range.

Examples of the experimentally obtained values of the primary sound reflection minima and maxima, and their corresponding excitation frequencies, are shown in Tables 2 and 3. Table 2 demonstrates the measured values of the primary sound reflection minima β_{min1} (proportional to a quarter-wavelength, i.e., $\lambda/4$), the primary sound reflection maxima β_{max1} (proportional to a half-wavelength, i.e., $\lambda/2$) and the relevant frequencies f_{min1} and f_{max1} depending on the air gap size a of the most porous (i.e., $P = 56\%$) PLA samples measuring 10 mm in thickness. It is obvious that the frequencies, which correspond to the primary sound reflection minima and maxima, generally shifted to lower values of the excitation frequencies with the increasing air gap size. However, the primary sound reflection minima and maxima generally increased with increasing air gap size. It can also be seen that the lowest values of the primary sound reflection minima and maxima and their corresponding excitation frequencies were found for the PLA samples produced with the starlit structure, which is in accordance with Figure 4. Similar results were found for the low porous (i.e., $P = 30\%$) samples measuring 30 mm in thickness, as shown in Table 3. Here, the increasing air gap size led to an increase in the primary sound reflection minima and maxima and to a decrease in their corresponding excitation frequencies. However, it is also clear that the effect of the structure type of the thicker PLA samples on these excitation frequencies was practically negligible compared to the thin samples tested (see Table 2). Again, it was found that lower values of the sound reflection coefficient were observed for the specimens that were manufactured with the starlit structure. This was due to the more complex pore shapes of this 3D-printed structure than the other open-porous PLA structures, which led to multiple reflections during the propagation of sound waves through the starlit structure and consequently to greater conversion of sound energy into heat.

Table 2. Primary sound reflection minima and maxima and their corresponding excitation frequencies depending on the air gap size for the studied 3D-printed open-porous PLA material structures of thickness $t = 10$ mm and porosity $P = 56\%$.

Structure Type	a (mm)	f_{min1} (Hz)	β_{min1} (-)	f_{max1} (Hz)	β_{max1} (-)
Cartesian	25	2272	0.73	5904	0.91
	50	1392	0.75	2984	0.94
	70	1192	0.78	2296	0.95
	100	784	0.81	1424	0.97
Octagonal	25	1936	0.70	5032	0.91
	50	1336	0.74	2920	0.93
	70	1040	0.76	2256	0.93
	100	728	0.81	1360	0.97
Rhomboid	25	2024	0.71	5536	0.90
	50	1384	0.72	2960	0.91
	70	1080	0.77	2272	0.93
	100	760	0.81	1392	0.98
Starlit	25	1928	0.60	5024	0.89
	50	1320	0.64	2872	0.92
	70	960	0.70	2224	0.92
	100	680	0.75	1248	0.92

Table 3. Primary sound reflection minima and maxima and their corresponding excitation frequencies depending on the air gap size for the studied 3D-printed open-porous PLA material structures of thickness $t = 30$ mm and porosity $P = 30\%$.

Structure Type	a (mm)	f_{min1} (Hz)	β_{min1} (-)	f_{max1} (Hz)	β_{max1} (-)
Cartesian	0	1728	0.00	3608	0.79
	25	704	0.09	2056	0.86
	50	488	0.14	1760	0.87
	100	336	0.18	1288	0.91
Octagonal	0	1776	0.01	3672	0.79
	25	712	0.10	2152	0.85
	50	488	0.16	1808	0.86
	100	336	0.20	1312	0.91
Rhomboid	0	1864	0.02	3696	0.79
	25	728	0.14	2176	0.86
	50	512	0.19	1824	0.86
	100	360	0.24	1344	0.88
Starlit	0	1824	0.02	3640	0.80
	25	696	0.08	2120	0.84
	50	504	0.13	1792	0.84
	100	344	0.20	1296	0.86

Placing the porous sample at a suitable distance from the solid wall is one way to ensure higher sound reflectivity while reducing weight. Applying open-porous materials with an air gap size to reflect sound is advantageous, especially at higher excitation frequencies, because porous (e.g., fibrous and foam) materials exhibit generally good sound absorption properties at high frequencies [60–64]. Furthermore, 3D-printed open-porous PLA material structures are characterized by a relatively high stiffness/weight ratio, resulting in lower production costs.

3.5. Influence of Excitation Frequency

As stated above in Figures 4–7, the sound reflection properties of the investigated 3D-printed PLA specimens were also strongly influenced by the excitation frequency. It can be seen that good sound reflection properties were generally observed in a low-frequency range, whose upper frequency limit generally increased with increasing sample porosity and with decreasing air gap size and specimen thickness. In the case of higher excitation frequencies, better sound reflectivity can be achieved by appropriately adjusting the air gap size behind the given 3D-printed PLA sample inside the impedance tube. From Equation (12) mentioned above, the appropriate size of this air gap in the desired frequency range can be determined as follows:

$$a = \frac{c \cdot n}{2f} - \frac{t}{2} \tag{13}$$

3.6. Comparison of Sound Reflection Properties

The ability of the investigated 3D-printed open-porous PLA materials to reflect sound was compared using the mean value of the sound reflection coefficient β_m that was determined as an arithmetical average of the sound reflection coefficients over the whole measured frequency range (i.e., from 200 to 6400 Hz). Some examples of the calculated values of the mean sound reflection coefficient are shown in Table 4. Regardless of the structure type of the tested porous PLA materials, it is obvious that the mean value of the sound reflection coefficient generally increased with increasing sample porosity and with decreasing specimen thickness and air gap size behind the tested sample inside the impedance tube. It was confirmed that the PLA specimens manufactured with the starlit structure exhibited the lowest sound reflection properties because of their complex shape structure compared to the other types of studied open-porous PLA structures. It can also be seen that the sound reflection behavior of the PLA samples, which were produced with the other types (i.e., cartesian, octagonal and rhomboid) of open-porous structures, was very similar. The highest value of the mean sound reflection coefficient (i.e., $\beta_{m,max} = 0.857$), and thus the highest ability to reflect sound, was found for the PLA specimen that was produced with the cartesian structure, the highest total volume porosity (i.e., $P = 56\%$), the smallest thickness (i.e., $t = 10$ mm) and which was placed on the solid wall SW (i.e., $a = 0$ mm).

Table 4. Mean values of the sound reflection coefficient of selected 3D-printed open-porous PLA materials.

Structure Type	t (mm)	P (%)	a (mm)	β_m (-)
	10	56	0	0.857
Cartesian	20	43	50	0.702
	30	30	100	0.598
	10	56	0	0.844
Octagonal	20	43	50	0.699
	30	30	100	0.601
	10	56	0	0.842
Rhomboid	20	43	50	0.692
	30	30	100	0.610
	10	56	0	0.820
Starlit	20	43	50	0.679
	30	30	100	0.589

4. Summary

It can be concluded that the sound reflection properties of the investigated lightweight 3D-printed PLA specimens were significantly influenced by many factors, namely, their volume porosity and thickness, the structure type, the air gap size behind the tested samples and the excitation frequency. In general, good sound reflection properties were obtained using hard, rigid, and non-porous materials (e.g., concrete, stone, and metals) and low excitation frequencies. Conversely, soft, and porous materials were characterized by high soundproofing properties, mainly at high excitation frequencies. The application of thin, lightweight, open-porous 3D-printed PLA materials in the high-frequency region presents a new perspective in terms of sound reflection compared to classic rigid, heavy materials. Based on the experimental data of the investigated open-porous 3D-printed PLA specimens, better sound reflection properties can be obtained by appropriately adjusting the air gap size behind the thin specimens at a given excitation frequency. Therefore, the application of thin, lightweight 3D-printed materials is promising in terms of material weight reduction and energy savings. In future, it will be possible to develop advanced lightweight 3D-printed material structures, which cannot be produced by conventional manufacturing technologies, in order to reflect sound.

5. Conclusions

3D printing is a developing technology affecting many areas of life. 3D printing technology allows the production of lightweight materials of different shapes and structures compared to other manufacturing technologies, which leads to time and energy savings and reduces the weight of materials.

The purpose of this study was to investigate the sound reflection performance of 3D-printed open-porous PLA materials made with four different types of porous structures. In addition, the investigated PLA specimens were manufactured with different porosities and thicknesses.

It was found that the sound reflection properties of the studied PLA specimens were influenced, not only by the type of 3D-printed structure, but also by the volume porosity and thickness of the material samples and the air gap size behind the investigated PLA specimens inside the acoustic impedance tube. It can be concluded that the PLA samples manufactured with the starlit structure exhibited lower sound reflection compared to the other investigated PLA structures, especially for specimens with a thickness of 10 mm. This was because the starlit structure is created by more complex pore shapes, resulting in multiple reflections during the propagation of soundwaves through this structure and thus greater conversion of sound energy into heat. It can also be stated that a higher ability to reflect sound from the tested 3D-printed open-porous PLA materials was generally observed at low excitation frequencies and for thin, highly porous samples without the air gap size inside the impedance tube. At higher excitation frequencies, it was possible to place the porous lightweight sample at a suitable distance from the solid wall to improve the sound reflectivity while reducing its weight compared to hard, unperforated solids, which are generally characterized by good sound reflection properties. For this reason, it is possible, for these purposes, to save time, production costs and energy, which is one of the current worldwide trends.

Author Contributions: Conceptualization, K.M. and M.V.; methodology, K.M. and M.V.; software, M.V. and K.M.; validation, P.P.M.; formal analysis, P.P.M., J.V. and D.K.; investigation, M.V. and K.M.; resources, J.V. and D.K.; data curation, M.V. and K.M.; writing—original draft preparation, M.V. and K.M.; writing—review and editing, P.P.M. and D.K.; visualization, M.V. and J.V.; supervision, M.V. and K.M.; project administration, K.M. and M.V.; funding acquisition, K.M. All authors have read and agreed to the published version of the manuscript.

Funding: This work was supported by the Ministry of Education, Science, Research, and Sport of the Slovak Republic through projects APVV-19-0550 and KEGA 005TUKE-4/2021, as well as with the direct support of the European Regional Development Fund in the Research Centre of Advanced Mechatronic Systems project, project number CZ.02.1.01/0.0/0.0/16_019/0000867 within the Operational Programme Research, Development and Education.

Institutional Review Board Statement: Not applicable.

Informed Consent Statement: Not applicable.

Data Availability Statement: Not applicable.

Acknowledgments: The authors would like to thank the Ministry of Education, Science, Research, and Sport of the Slovak Republic for its direct support of this research through the projects APVV-19-0550 and KEGA 005TUKE-4/2021, as well as the European Regional Development Fund in the Research Centre of Advanced Mechatronic Systems for its financial support of this research through the project number CZ.02.1.01/0.0/0.0/16_019/0000867 within the Operational Programme Research, Development, and Education.

Conflicts of Interest: The authors declare no conflict of interest.

References

1. Brawata, K.; Kamisinski, T. Effect of Stage Curtain Legs Arrangement on Sound Distribution in Opera Houses. In Proceedings of the 2018 Joint Conference—Acoustics, Ustka, Poland, 11–14 September 2018; pp. 1–4. [CrossRef]
2. Hamdan, N.I.; Zainulabidin, M.H.; Kasron, M.Z.; Kassim, A.S.M. Effect of Perforation Size on Sound Absorption Characteristics of Membrane Absorber. *Int. J. Integr. Eng.* **2018**, *10*, 27–34. [CrossRef]
3. Othman, A.R.; Harith, C.M.; Ibrahim, N.; Ahmad, S.S. The Importance of Acoustic Design in the Mosques towards the Worshipers' Comfort. *Procd. Soc. Behv.* **2016**, *234*, 45–54. [CrossRef]
4. Rathsam, J.; Wang Lily, M. A review of Diffuse Reflections in Architectural Acoustics. *Archit. Eng.* **2006**, *14*, 1–8. [CrossRef]
5. Amares, S.; Sujatmika, E.; Hong, T.W.; Durairaj, D.; Hamid, H.S.H.B. A Review: Characteristics of Noise Absorption Material. *J. Phys. Conf. Ser.* **2017**, *908*, 012005. [CrossRef]
6. Vasina, M.; Bilek, O. Effect of Machined Surface Shape on Sound Reflection. *Manuf. Tech.* **2016**, *16*, 830–834. [CrossRef]
7. Allen, T.; Chally, A.; Moser, B.; Widenhorn, P. Sound Propagation, Reflection, and Its Relevance to Ultrasound Imaging. *Phys. Teach.* **2019**, *57*, 134–137. [CrossRef]
8. Berg, R.E. *The Physics of Sound*; Department of Physics: Meryland, MD, USA, 2021.
9. Vercammen, M. Sound Reflections from Concave Spherical Surfaces. Part I: Wave Field Approximation. *Acta Acust. Unit. Acust.* **2010**, *96*, 82–91. [CrossRef]
10. Szelag, A.; Kamisinski, T.M.; Lewinska, M.; Rubacha, J.; Ilch, A. The Characteristic of Sound Reflections from Curved Reflective Panels. *Arch. Acoust.* **2014**, *39*, 549–558. [CrossRef]
11. Mosavimehr, E.; Srikantha, P.A. Sound transmission loss characteristics of sandwich panels with a truss lattice core. *J. Acoust. Soc. Am.* **2017**, *141*, 2921–2932. [CrossRef]
12. Domenico, R.; Ruggiero, A. Choice of the optimal acoustic design of a school classroom and experimental verification. *Appl. Acoust.* **2019**, *146*, 280–287.
13. Tsukernikov, I.; Antonov, A.; Ledenev, V.; Shubin, I.; Nevenchannaya, T. Noise Calculation Method for Industrial Premises with Bulky Equipment at Mirror-diffuse Sound Reflection. *Proced. Eng.* **2017**, *176*, 218–225. [CrossRef]
14. Zulkifli, R.; Mohd Nor, M.J.; Mat Tahir, M.F.; Ismail, A.R.; Nuawi, M.Z. Acoustic Properties of Multi-Layer Coir Fibres Sound Absorption Panel. *J. Appl. Sci.* **2008**, *8*, 3709–3714. [CrossRef]
15. Lee, F.C.; Chen, W.H. Acoustic transmission analysis of multi-layer absorbers. *J. Sound. Vib.* **2001**, *248*, 621–634. [CrossRef]
16. Cucharero, J.; Hänninen, T.; Lokki, T. Angle-Dependent Absorption of Sound on Porous Materials. *Acoustics* **2020**, *2*, 753–765. [CrossRef]
17. Gimaltdinov, I.K.; Sitdikova, L.F.; Dmitriev, V.L.; Levina, T.M.; Khabeev, N.S.; Song, W.Q. Reflection of acoustic waves from a porous material at oblique incidence. *J. Eng. Phys. Thermophys.* **2017**, *90*, 1043–1052. [CrossRef]
18. Holmes, N.; Browne, A.; Montague, C. Acoustic properties of concrete panels with crumb rubber as a fine aggregate replacement. *Constr. Build. Mater.* **2014**, *73*, 195–204. [CrossRef]
19. Kuczmarski, M.A.; Johntson, J. *Acoustic Absorption in Porous Materials*; National Aeronautics and Space Administration: Pasadena, CA, USA, 2011; pp. 1–27.
20. Fellah, Z.E.A.; Fellah, M.; Depollier, C.; Ogam, E.; Mitri, F.G. Wave Propagation in Porous Materials. In *Computational and Experimental Studies of Acoustic Waves*; IntechOpen: Rijeka, Croatia, 2017; pp. 99–120.
21. Sun, X.; Jiang, F.; Wang, J. Acoustic Properties of 316L Stainless Steel Lattice Structures Fabricated via Selective Laser Melting. *Metals* **2020**, *10*, 111. [CrossRef]

22. Dragonetti, R.; Napolitano, M.; Romano, R.A. A study on the energy and the reflection angle of the sound reflected by a porous material. *J. Acoust. Soc. Am.* **2019**, *145*, 489–500. [CrossRef]

23. Otaru, A.J. Review on the Acoustical Properties and Characterization Methods of Sound Absorbing Porous Structures: A Focus on Microcellular Structures Made by a Replication Casting Method. *Met. Mater. Int.* **2020**, *26*, 915–932. [CrossRef]

24. Fan, C.; Tian, Y.; Wang, Z.Q. Structural parameter effect of porous material on sound absorption performance of double-resonance material. *IOP Conf. Ser. Mater. Sci. Eng.* **2017**, *213*, 012028. [CrossRef]

25. Talebitooti, R.; Zarastvand, M.; Darvishgohari, H. Multi-objective optimization approach on diffuse sound transmission through poroelastic composite sandwich structure. *J. Sandwich Struct. Mater.* **2019**, *23*, 1221–1252. [CrossRef]

26. Cao, L.; Fu, Q.; Si, Y.; Ding, B.; Yu, J. Porous materials for sound absorption. *Compos. Commun.* **2018**, *10*, 25–35. [CrossRef]

27. Chen, W.; Chen, T.; Wang, X.; Wu, J.; Li, S. Fractal Model for Acoustic Absorbing of Porous Fibrous Metal Materials. *Shock Vib.* **2016**, *2016*, 2890857. [CrossRef]

28. Guo, Z.K.; Hu, G.B.; Sorokin, V.; Yang, Y.; Tang, L.H. Sound transmission through sandwich plate with hourglass lattice truss core. *J. Sandwich Struct. Mater.* **2020**, *23*, 1902–1928. [CrossRef]

29. Bennett, T.D.; Coudert, F.-X.; James, S.L.; Cooper, A.I. The changing state of porous materials. *Nat. Mater.* **2021**, *20*, 1179–1187. [CrossRef]

30. Voronin, S.V.; Danilushkin, V.S.; Tregub, V.I.; Konovalov, S.V. Computer Simulation of the Process of Crack Propagation in a Brittle Porous Material. *J. Surf. Investig.* **2021**, *15*, 1212–1216. [CrossRef]

31. Chiavi, A.; Miglietta, P.; Guglielmone, C.L. Considerations on the airflow resistivity measurement of porous and fibrous materials as function of temperature. In Proceedings of the 8th European Conference on Noise Control, Euronoise, UK, 26–28 October 2009; pp. 1–11.

32. Hurrell, A.I.; Horoshenkov, K.V.; Pelegrinis, M.T. The accuracy of some models for the airflow resistivity of nonwoven materials. *Appl. Acoust.* **2018**, *130*, 230–237. [CrossRef]

33. Naghieh, S.; Karamooz Ravari, M.R.; Badrossamay, M.; Foroozmehr, E.; Kadkhodaei, M. Numerical investigation of the mechanical properties of the additive manufactured bone scaffolds fabricated by FDM: The effect of layer penetration and post-heating. *J. Mech. Behav. Biomed. Mater.* **2016**, *59*, 241–250. [CrossRef] [PubMed]

34. Pantazopoulos, G.A. A Short Review on Fracture Mechanisms of Mechanical Components Operated under Industrial Process Conditions: Fractographic Analysis and Selected Prevention Strategies. *Metals* **2019**, *9*, 148. [CrossRef]

35. Seddeq, H.S. Factors influencing acoustic performance of sound absorptive materials. Austral. *J. Basic Appl. Sci.* **2009**, *3*, 4610–4617.

36. Łukaszewski, K.; Buchwald, T.; Wichniarek, R. The FDM Technique in Processes of Prototyping Spare Parts for Servicing and Repairing Agricultural Machines: A General Outline. *Int. J. Appl. Mech. Eng.* **2021**, *26*, 145–155. [CrossRef]

37. Chawla, K.; Singh, R.; Singh, J. On recyclability of thermoplastic ABS polymer as fused filament for FDM technique of additive manufacturing. *World J. Eng.* **2021**. [CrossRef]

38. Corredor-Bedoya, A.C.; Acuna, B.; Serpa, A.L.; Mesiero, B. Effect of the excitation signal type on the absorption coefficient measurement using the impedance tube. *Appl. Acoust.* **2021**, *171*, 107659. [CrossRef]

39. Fediuk, R.; Amran, M.; Vatin, N.; Vasilev, Y.; Lesovik, V.; Ozbakkaloglu, T. Acoustic Properties of Innovative Concretes: A Review. *Materials* **2021**, *14*, 398. [CrossRef]

40. Zainulabidin, M.; Rani, M.; Nezere, N.; Tobi, A. Optimum sound absorption by materials fraction combination. *Int. J. Mech. Mechatron. Eng.* **2014**, *14*, 118–121.

41. Paul, P.; Mishra, R.; Behera, B.K. Acoustic behaviour of textile structures. *Text. Prog.* **2021**, *53*, 1–64. [CrossRef]

42. Koizumi, T.; Tsujiuchi, N.; Adachi, A. The development of sound absorbing materials using natural bamboo fibers. *High Perform. Struct. Mater.* **2002**, *4*, 157–166.

43. Bujoreanu, C.; Nedeff, F.; Benchea, M.; Agop, M. Experimental and theoretical considerations on sound absorption performance of waste materials including the effect of backing plates. *Appl. Acoust.* **2017**, *119*, 88–93. [CrossRef]

44. Wu, J.J.; Li, C.G.; Wang, D.B.; Gui, M.C. Damping and sound absorption properties of particle reinforced Al matrix composite foams. *Compos. Sci. Technol.* **2003**, *63*, 569–574. [CrossRef]

45. Mendes, C.O.B.; Nunes, M.A.D. Numerical Methodology to Obtain the Sound Absorption of Materials by Inserting the Acoustic Impedance. *Arch. Acoust.* **2021**, *46*, 649–656. [CrossRef]

46. *International Organization for Standardization ISO 10534*; Acoustics-Determination of Sound Absorption in a Reverberation Room. IOS: Geneva, Switzerland, 2003.

47. Scrosati, C.; Martellotta, F.; Pompoli, F.; Schiavi, A.; Prato, A.; D'Orazio, D.; Garai, M.; Granzotto, N.; Di Bella, A.; Scamoni, F. Towards more reliable measurements of sound absorption coefficient in reverberation rooms: An Inter-Laboratory Test. *Appl. Acoust.* **2020**, *165*, 107398. [CrossRef]

48. Shtrepi, L.; Prato, A. Towards a sustainable approach for sound absorption assessment of building materials: Validation of small-scale reverberation room measurements. *Appl. Acoust.* **2020**, *165*, 107304. [CrossRef]

49. Sun, Y.; Hua, B. System error calculation and analysis of underwater sound absorption coefficient measurement experiment. *Appl. Acoust.* **2022**, *186*, 108489. [CrossRef]

50. Amran, M.; Fediuk, R.; Murali, G.; Vatin, N.; Al-Fakih, A. Sound-Absorbing Acoustic Concretes: A Review. *Sustainability* **2021**, *13*, 10712. [CrossRef]

51. *International Organization for Standardization ISO 10534-1*; Acoustics—Determination of Sound Absorption Coefficient and Impedance in Impedance Tubes—Part 1: Method Using Standing Wave Ratio. IOS: Geneva, Switzerland, 1998.

52. *International Organization for Standardization ISO 10534-2*; Acoustics-Determination of Sound Absorption Coefficient and Impedance in Impedance Tubes-Part 2: Transfer-Function Method; ISO/TC 43/SC2 Building Acoustics. CEN, European Committee for Standardization: Brussels, Belgium, 1998; 10534–10542.

53. Doutres, O.; Salissou, Y.; Atalla, N.; Panneton, R. Evaluation of the acoustic and non-acoustic properties of sound absorbing materials using a three-microphone impedance tube. *Appl. Acoust.* **2010**, *71*, 506–509. [CrossRef]

54. Song, W.J.; Cha, D.J. Determination of an acoustic reflection coefficient at the inlet of a model gas turbine combustor for power generation. *IOP Conf. Ser. Mater. Sci. Eng.* **2017**, *164*, 012010. [CrossRef]

55. Taban, E.; Khavanin, A.; Jafari, A.J.; Faridan, M.; Tabrizi, A.K. Experimental and mathematical survey of sound absorption performance of date palm fibers. *Heliyon* **2019**, *5*, e01977. [CrossRef]

56. Jiang, B.; Wang, Z.J.; Zhao, N.Q. Effect of pore size and relative density on the mechanical properties of open cell aluminum foams. *Scr. Mater.* **2007**, *56*, 169–172. [CrossRef]

57. Zhu, W.B.; Chen, S.M.; Wang, Y.B.; Zhu, T.T.; Jiang, Y. Sound Absorption Behavior of Polyurethane Foam Composites with Different Ethylene Propylene Diene Monomer Particles. *Arch. Acoust.* **2018**, *43*, 403–411. [CrossRef]

58. Tiuc, A.E.; Vasile, O.; Usca, A.D.; Gabor, T.; Vermesan, H. The Analysis of Factors That Influence the Sound Absorption Coefficient of Porous Materials. *Rom. J. Acoust. Vib.* **2014**, *11*, 105–108.

59. Everest, F.A. Absorption of sound. In *Master Handbook of Acoustics*, 4th ed.; McGraw-Hill: New York, NY, USA, 2001; pp. 179–233, ISBN 0-07-139974-7.

60. Zhu, J.L.; Sun, J.; Tang, J.P.; Wang, J.Z.; Ao, Q.B.; Bao, T.F.; Song, W.D. Gradient-structural optimization of metal fiber porous materials for sound absorption. *Powder. Technol.* **2016**, *301*, 1235–1241. [CrossRef]

61. Liguori, C.; Ruggiero, A.; Russo, D.; Sommella, P. Estimation of the minimum measurement time interval in acoustic noise. *Appl. Acoust.* **2017**, *127*, 126–132. [CrossRef]

62. Liu, X.W.; Yu, C.L.; Xin, F.X. Gradually perforated porous materials backed with Helmholtz resonant cavity for broadband low-frequency sound absorption. *Compos. Struct.* **2021**, *263*, 113647. [CrossRef]

63. Madurelo-Sanz, R.; Acedo-Fuentes, P.; Garcia-Cobos, F.J.; Sánchez-Delgado, F.J.; Mota-López, M.I.; Meneses-Rodríguez, J.M. The recycling of surgical face masks as sound porous absorbers: Preliminary evaluation. *Sci. Total Environ.* **2021**, *786*, 147461. [CrossRef]

64. Guan, Y.-J.; Ge, Y.; Sun, H.-X.; Yuan, S.-Q.; Liu, X.-J. Low-Frequency, Open, Sound-Insulation Barrier by Two Oppositely Oriented Helmholtz Resonators. *Micromachines* **2021**, *12*, 1544. [CrossRef]

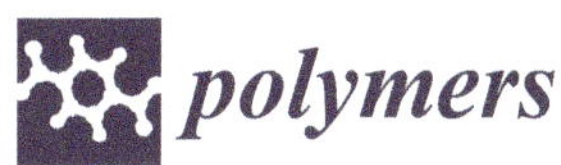 *polymers*

Article

Molds with Advanced Materials for Carbon Fiber Manufacturing with 3D Printing Technology

Patrich Ferretti *, Gian Maria Santi, Christian Leon-Cardenas *, Marco Freddi, Giampiero Donnici, Leonardo Frizziero and Alfredo Liverani

Department of Industrial Engineering, Alma Mater Studiorum University of Bologna, I-40136 Bologna, Italy; gianmaria.santi2@unibo.it (G.M.S.); marco.freddi@studio.unibo.it (M.F.); giampiero.donnici@unibo.it (G.D.); leonardo.frizziero@unibo.it (L.F.); alfredo.liverani@unibo.it (A.L.)
* Correspondence: patrich.ferretti2@unibo.it (P.F.); christian.leon2@unibo.it (C.L.-C.)

Abstract: Fused Deposition Modeling (FDM) 3D printing is the most widespread technology in additive manufacturing worldwide that thanks to its low costs, finished component applications, and the production process of other parts. The need for lighter and higher-performance components has led to an increased usage of polymeric matrix composites in many fields ranging from automotive to aerospace. The molds used to manufacture these components are made with different technologies, depending on the number of pieces to be made. Usually, they are fiberglass molds with a thin layer of gelcoat to lower the surface roughness and obtain a smooth final surface of the component. Alternatively, they are made from metal, thus making a single carbon fiber prototype very expensive due to the mold build. Making the mold using FDM technology can be a smart solution to reduce costs, but due to the layer deposition process, the roughness is quite high. The surface can be improved by reducing the layer height, but it is still not possible to reach the same degree of surface finish of metallic or gelcoat molds without the use of fillers. Thermoplastic polymers, also used in the FDM process, are generally soluble in specific solvents. This aspect can be exploited to perform chemical smoothing of the external surface of a component. The combination of FDM and chemical smoothing can be a solution to produce low-cost molds with a very good surface finish.

Keywords: FDM; chemical smoothing; vapor smoothing; PVB; carbon fiber mold

Citation: Ferretti, P.; Santi, G.M.; Leon-Cardenas, C.; Freddi, M.; Donnici, G.; Frizziero, L.; Liverani, A. Molds with Advanced Materials for Carbon Fiber Manufacturing with 3D Printing Technology. *Polymers* **2021**, *13*, 3700. https://doi.org/10.3390/polym13213700

Academic Editors: Houwen Matthew Pan and Eric Luis

Received: 29 September 2021
Accepted: 25 October 2021
Published: 27 October 2021

Publisher's Note: MDPI stays neutral with regard to jurisdictional claims in published maps and institutional affiliations.

1. Introduction

The FDM process was first patented by Stratasys in the 1990s to build a three-dimensional plastic object without the use of a mold. The parts are produced layer by layer through the extrusion of thermoplastic filaments usually wound in spools [1]. This is the most popular additive manufacturing technique nowadays as it offers a wide range of thermoplastic material choices from common PLA [2] up to engineering-grade materials such as Nylons [3]. This manufacturing process can be used to create solid components with complex shapes and geometries, as highlighted by the studies of SAVU et al. [4] and Brian et al. [5]. Even though the Additive Manufacturing (AM) processes are challenged because of their low productivity, inferior surface quality, dimensional instability, and the internal anisotropy that decreases the mechanical properties of the products [6]. This process has also shown suitable to produce end-use parts and for small series production [7,8].

1.1. 3D Printing for Supporting the Component Manufacturing Process

FDM 3D printing is the most cost-effective additive manufacturing process for thermoplastic materials on the market. This is primarily due to the relative simplicity of hardware construction compared to other technologies. For example, the closed and heated chamber is not always necessary as in the case of Selective Laser Melting (SLM). The filament production is simple, made from polymer granules, and hence there are dozens of manufacturers and a large number of thermoplastic polymers available on the market. The

FDM process includes some safety advantages as well, opposite to printing from powder technologies that always require personal protective equipment to avoid problems with the respiratory system during the powder handling and component cleaning. Subsequently, filament 3D printing showed no problems during material handling since the polymer is wound on a spool and is easy to substitute once the filament is finished. Any fumes produced during printing can be effectively removed using specific filters installed on the printer. Nevertheless, the main drawback of FDM printing is related to the surface quality of the component that is lower than that obtained with other 3D printing technologies such as SLM, Stereolithography (SLA) [9], and multi jet fusion from HP [10]. Some research findings suggest that surface noise given by spikes and peaks in the component during modelling could lead to improper print quality [11]. The challenge in correctly predicting residual stresses [12] and deformations of printed components once extruded has so far limited the use of FDM printing for structural components, thus requiring numerous trials before obtaining the finished component with the desired quality level. Overall, FDM printing can be used as a support procedure to create other components, especially in the case of very limited batches or prototypes. Other findings of Komineas et al. [13] can help to accurately calculate overall build time to industrialize elements built with AM. A noteworthy application case is the manufacturing of metal components with the lost wax casting technique [14]. In this case, the starting component on which the ceramic mold is built is no longer made of wax but is 3D printed using low melting thermoplastic polymers such as PLA or ABS [15]. These are suitably modified to reduce the creation of ashes during the polymer removal phase from the mold, which takes place in the furnace. In this case, FDM 3D printing allowed us to avoid the need to create a metal mold (made in aluminum or steel otherwise) in which the wax is casted to obtain the desired model, thus being able to lead to a significant reduction in costs. The alternative to 3D printing would be to make the wax models manually, in which case the printing process allowed us to obtain greater reproducibility and better dimensional tolerances. Additionally, it should be remembered that SLA technology is also used to produce models intended for the traditional lost-wax casting process [16], allowing us in this case to obtain finished components with surface roughness identical to that obtained with wax models. However, the SLA technique is limited to small components due to the large deformations that would occur during the resin printing process and is usually used in the goldsmith and medical sectors [17], and the cost of the resin is also usually higher than that of FDM printing filaments. Another example of how the FDM process can be used as a supporting technology is the mold creation for silicone components [18]. The advantages are similar to the previous example.

1.2. Mold to Create Carbon Fiber Components

This study was mainly focused on building a custom mold made through FDM and smoothing processes for polymer-matrix composites. This solution led to a high surface finish that can guarantee the production of continuous fiber-reinforced components such as carbon fiber composites with an epoxy matrix. The mold is necessary, as the current moldless technologies do not allow the production of carbon fiber components starting from a fabric, being constrained to a single, continuous fiber, allowing us just to make reinforcements [19].

Mold manufacturing is usually very expensive. Material options are fiberglass or metal molds. Nevertheless, a starting model is required for the construction of a fiberglass mold; it is usually made by CNC in wood or in high-density polyurethane foam. The model is then covered with a layer of release agent (Polyvinyl Alcohol) or wax. A layer of gelcoat is applied over the model to obtain a shiny and homogeneous surface on the mold, and finally the fibers are soaked in epoxy or a thermosetting resin to give rigidity and consistency to the mold. The whole process has to be done manually by experienced operators and it is difficult to mechanize. The creation of single prototypes in carbon fiber is therefore extremely expensive since it is necessary to make the mold and amortize it with a single piece. The aim of this research is to introduce a solution to this problem using 3D printing technology.

FDM 3D Printed Mold

We need to obtain cost-effective, high-quality molds to reduce the costs of prototypes or small batch production. The possibility of making molds with FDM technology is a smart solution [20,21]. The main challenge is related to the high surface roughness that would be transferred directly to the final component. A filler could be used on the mold, followed by manual sandblasting to improve the surface finish. This technique can be applied to components with a relatively simple geometry with low tolerance values, but either way it would still require an important manual intervention.

2. Materials and Methods

2.1. From Solvent Bonding to Chemical Smoothing

Thermoplastic polymers are generally soluble in some solvent compounds. The application of the solvent on the surface of the plastic component softened its surface. If two components are compressed against each other and the surfaces have been treated with the solvent, a mutual diffusion of the polymer chains is obtained, and the result is a very strong adhesion once evaporated. Solvent bonding differs from adhesive bonding since the solvent does not become permanently adhered to the adhered substrate. A further advantage is that this softening usually occurs well below the glass Transition Temperature (Tg) and therefore the overall component integrity is maintained. This process can also be used to smooth the surface of thermoplastic components [22]. This is superfluous for parts made with injection molding as little surface roughness could be achieved, but instead it could become a process to improve the surface characteristics of a component made by FDM [23]. This process is called chemical smoothing and it allows a localized reaction on the surface of the component only, keeping the main structure unchanged. This process used by research of Kuo et al. [24] can be used to ensure watertight surfaces are achieved on the mold. Once reaching the desired smoothing quality, the component must be cooled in open air or under forced ventilation to promote the evaporation of the solvent from the surface. In this research, the chosen machine for the vapor smoothing is the Polymaker Polishear (Polymaker Inc., Shanghai, China), designed specifically for Polyvinyl butyral (PVB) smoothing using Isopropyl Alcohol (IPA).

2.2. Material Choice

Polyvinyl butyral (PVB) FDM filament (Polymaker Inc., Shanghai, China) is the chosen material for this application. It is the result of a reaction between polyvinyl alcohol and butyraldehyde. This polymer is usually used in the creation of multilayer safety glass [25] in the automotive sector due to its high transparency but is not popular in FDM printing. Currently, there are only two filament manufacturers available, and the cost of this product is higher respect to the most common PLA, but considerably lower than that of Nylon and other engineering materials. Printability is excellent compared to PLA and its mechanical properties are such as the latter [26]. Having a low glass temperature (Tg), the deformations in the printing phase (warping) are limited, even on medium-sized prints, similar to what occurs with PLA. Additionally, similar to every thermoplastic material, it is soluble in a specific solvent, such as IPA alcohol. IPA is a very volatile solvent but it is not very harmful if in contact with human skin. It is sold without special regulations. PLA is also soluble in chloroform [27], but this liquid is much more dangerous and is not for sale. Other polymers such as ABS or ASA could be considered a valid alternative to PVB as they are soluble in acetone [28]. The high Tg of the latter two materials allows the thermal resistance to be greatly increased at the expense of printability. However, ASA and ABS require very high printing and bed temperatures as they require a heated chamber for medium-sized components (150 $\times$ 150 $\times$ 150 mm). Nevertheless, the onset of warping and delamination phenomena between the layers remains a serious problem. Finally, ASA and ABS contain styrene, which is a toxic component, and the production of fumes during printing could lead to various respiratory system diseases. For safety reasons it is therefore necessary to have a suitable device for filtering the fumes. Overall, PVB offers the best tradeoff between PLA and ABS [29], obtaining the good printability of the former and the solubility of the latter in readily available solvents. In Figure 1 it is possible to see the effect of smoothing in an image taken using an optical microscope at 20× magnification.

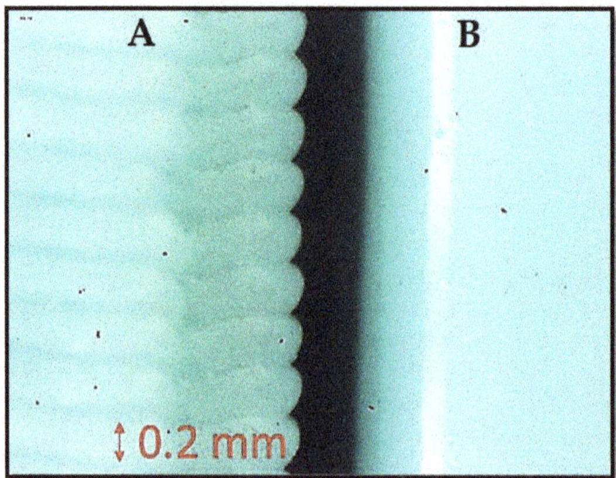

Figure 1. Effect of vapor smoothing, (**A**) the surface as printed, (**B**) same surface after chemical smoothing.

2.3. Case Study: Manufacturing of a Carbon Fiber Fuel Tap Protection for a Racing Motorbike

The need of light-weight carbon fiber protection is necessary to protect exposed components. This prevents debris or contact with other riders from causing the part to break or malfunction. The fuel cap is particularly exposed in the Husqvarna TC 85 motorbike, and it is therefore necessary to protect it, to avoid dangerous fuel leakage. The need is to produce protection to be installed on that motorbike for the European and World championships. This case study was followed by further components designed to be produced in a very limited series and for the exclusive usage of the team.

2.3.1. Mold Geometry

The starting point was the CAD drawing of the fuel tap guard. The software used is PTC Creo (PTC Inc., Boston, MA, USA), and the overall dimensions were acquired directly on the fuel tank by means of a caliper. Once the protection geometry was created, a Boolean approach was chosen for the construction of the mold. The CAD file of the protection was modified with the addition of material and draft angles to obtain the correct geometry for the slot on the mold. Finally, the addition of fittings made it possible to avoid ripples in the fabric that could rise to defects in the final component. The overall process is summarized in Figure 2.

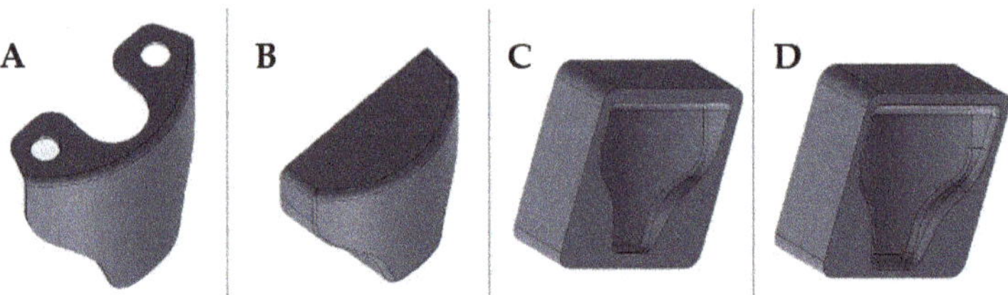

Figure 2. Steps followed during the design of the mold cad model, starting from: (**A**) the actual part, (**B**) making the solid block, (**C**) Cavity extrusion on Mold part, (**D**) Surface-optimized Mold.

2.3.2. Printing Strategy and Settings

The printing strategy adopted for this component could also be generalized to other parts with similar characteristics. Table 1 shows the printing parameters used to create the component. The first key point is the orientation of the part with respect to the build platform. Although it is a relatively simple component, there are four possible part orientations with respect to the print bed, as shown in Figure 3. The software used for slicing was Cura v4.9.1 (Ultmaker Inc., Zaltbommel, The Netherlands).

Table 1. Applied Slicing Printing Parameters in Cura.

Parameter	Value	Unit
Layer Height	0.22	mm
Line width	0.4	mm
Wall Line Count	3	-
Z seam position	Back Left	-
Top Layers	3	-
Bottom Layers	3	-
Infill Density	15	%
Infill Patten	Gyroid	-
Printing Temperature	205	°C
Build Plate Temperature	65	°C
Flow	100	%
Print Speed	80	mm/s
Travel Speed	250	mm/s
Retraction Distance	4	mm
Fan Speed	70	%
Regular Fan Speed at Height	0.2	mm
Support Structure	Tree	-
Support Overhang Angle	60	°
Adaptative Layers Maximum Variation	0.02	mm

Moreover, in the first case (Figure 3A), the mold is placed flat on the printing surface. The idea is to minimize the height of the printed component, thus reducing the total number of layers to be created. This printing mode allowed us to reduce the printing time, but it did not reproduce very well the curvature in the build direction of the component due to the so-called staircase-effect. The action of chemical smoothing can improve the surface roughness of the component and reduce the staircase-effect. In general, it is necessary to obtain the best possible surface prior to treatment in order to reduce the exposure time to solvent vapors. High exposure to the solvent could irreparably damage the surface.

In the second case (Figure 3B), the mold is placed vertically onto the build platform, leading to a time increase of 5% compared to the previous condition, but the staircase-effect problem was significantly improved. As a drawback, many supports were generated, meaning a waste of material and a poor surface finish on supported surfaces. Overall better mold finish quality could be achieved with the use of soluble supports at the expense of a significant price increase for the creation of the mold. Moreover, the result in case 3 (Figure 3C) was similar to the previous case, but the generation of supports was reduced, and the staircase-effect was still present at some points. Finally, the printing position used in the fourth case (Figure 3D) minimized the generation of supports and allowed the maximum resolution of the curvature of the mold.

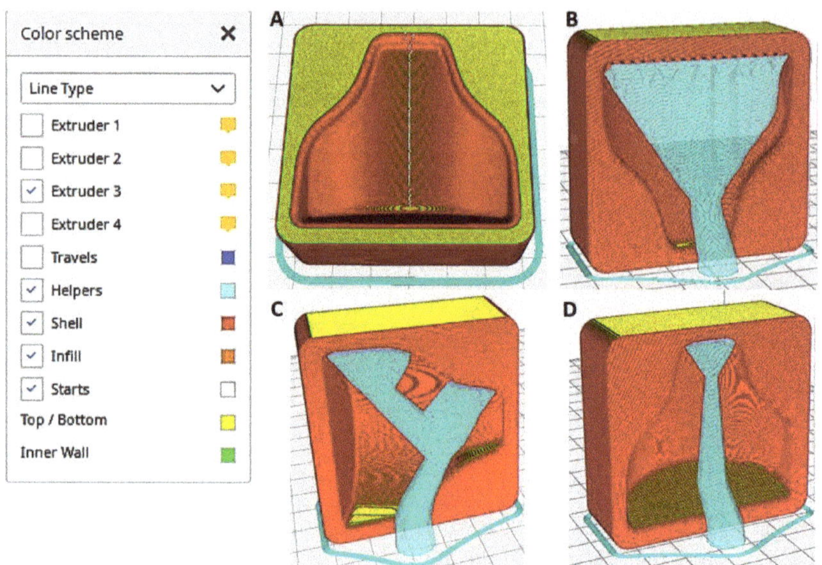

Figure 3. Part orientation on the buildplate, part helpers in blue.

However, the settings on Table 1 were adopted to further improve the quality of the mold before the chemical smoothing process to reduce possible imperfections in the cavity. The seam of the outermost wall was preferentially positioned in the rear corner of the mold to avoid seams in the cavity, as seen in Figure 4. Thereafter, it was decided to use a variable layer height, as seen in Figure 5, to further improve the fidelity of the curvature, whilst speed up the printing process at the same time. The molds were not 100% filled; a 20% Gyroid type infill approach with three contour lines were used. This allowed us to obtain molds that could withstand the vacuum lamination process and minimize the material used.

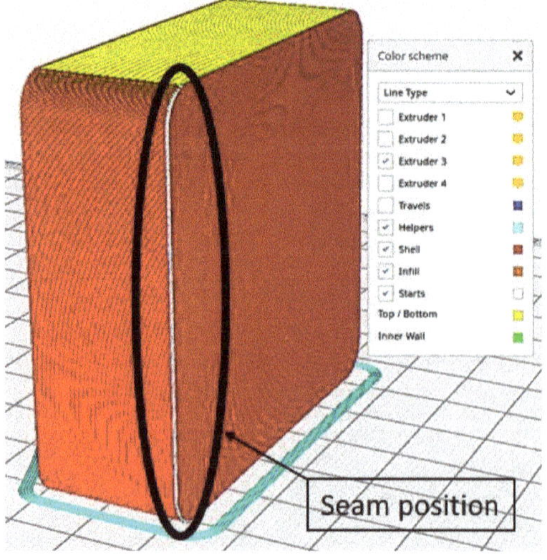

Figure 4. Seam positioning.

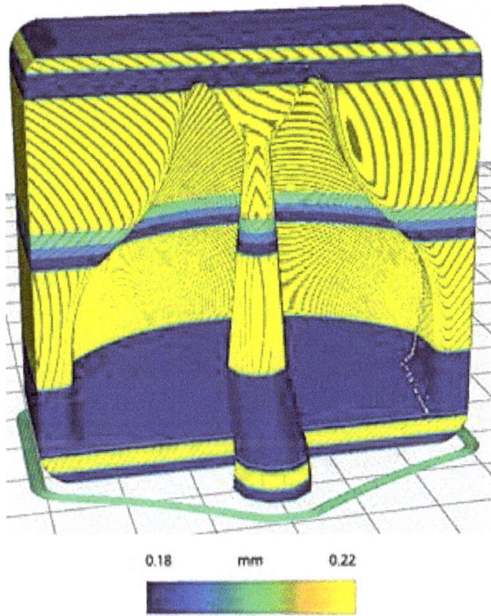

Figure 5. Preview of variable layer height.

2.3.3. Chemical Smoothing of the Mould Surface

The mold was initially placed inside the Polyshear device in the same position in which it was printed, and then a 10-min smoothing cycle was performed. Subsequently, it was turned upside down, and a second 10-min smoothing cycle was carried out. Subsequently, two molds were printed with the same printing parameters (thus the same gcode), and both sustained the smoothing process to verify the reproducibility of the process. Both were left to dry for 24 h at ambient temperature. Figure 6 shows the differences before and after the smoothing process.

Figure 6. (**A**) Mold immediately after printing; (**B**) mold after smoothing treatment. On the top of the (**B**) mold are small circular spots; these are due to the support surface during smoothing and the rotation of 180 degrees. In the lower part of the mold (**A**) it is possible to see the layers, due to the staircase effect; on the right they have disappeared thanks to the smoothing process.

3. Results

3.1. Dimensional Verification of a 3D-Printed Mold with an Optical 3D Scanner

A Faro 3D scanner was used to check the fidelity of the printed model compared to the designed CAD file. In Figure 7 it is possible to see the cloud of points obtained from the scanning of the first mold. In order to evaluate the reproducibility of this process, two molds with the same gcode were printed. Therefore, a comparison with the theoretical, CAD file could be appreciated in Figure 8, in which the matching was good altogether, as an absolute range of 0.05 mm was obtained in most of the mold. There were areas on the inner boundary highlighted in blue where the matching was not accurate. Nevertheless, the results shown how accurate the printing process was. On the other hand, the overall printing precision could be improved. A so-called loop optimization could be carried out to obtain greater fidelity with the cad file. It could be noticed that although the staircase effect was reduced to a minimum, its effect persisted at some points, where the differences between the cad file and the printed model were higher.

Figure 7. Cloud of point of the first mold.

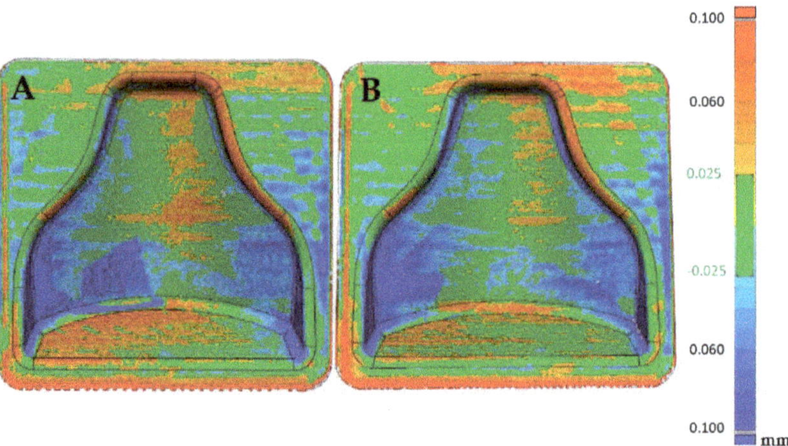

Figure 8. Comparison between the CAD model and first mold (**A**) and CAD model and second mold (**B**).

The visualization tool seen in Figure 9 allowed us to observe small oscillations on the surface of the two molds. These were probably due to the printing parameters and could be improved by reducing acceleration and jerk. These oscillations were less visible in Figure 8 because most result values were in the range of $-0.025 +0.025$ mm with only the peak values obtained outside this range.

Figure 9. Zebra stripes effect on the two molds after printing, (**left**) mold 1 and (**right**) mold 2.

Furthermore, Figure 10 shows the comparison between the two molds point-clouds with each other. It could be seen that the reproducibility of printing with this specific material was very high. In fact, the two molds could be superimposed, as can be seen from the almost complete green color of the image.

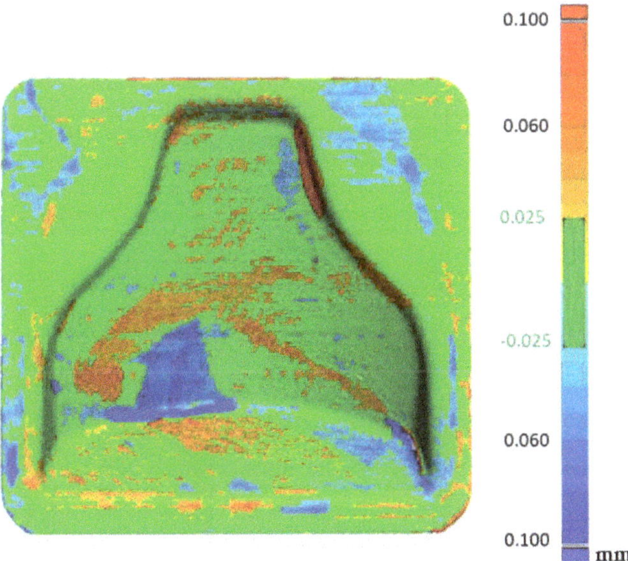

Figure 10. Comparison between the two molds after printing.

3.2. Dimensional Verification of the 3D-Printed, Chemical-Smoothed Mold with an Optical 3D Scanner

After the first dimensional verification, both molds were subjected to a chemical smoothing process, as described in Section 2.3.3. Afterwards, dimensional verification with respect to the CAD file gave the results visible in Figure 11. As expected, the chemical smoothing process reduced the peaks of innacuracies and filled the valleys, reducing the overall external dimensions of the component. Once again, the verification was carried out on both molds printed to evaluate the repeatability of the process, and a comparison was made with respect to the CAD file values and with respect to the scanning performed before the smoothing process.

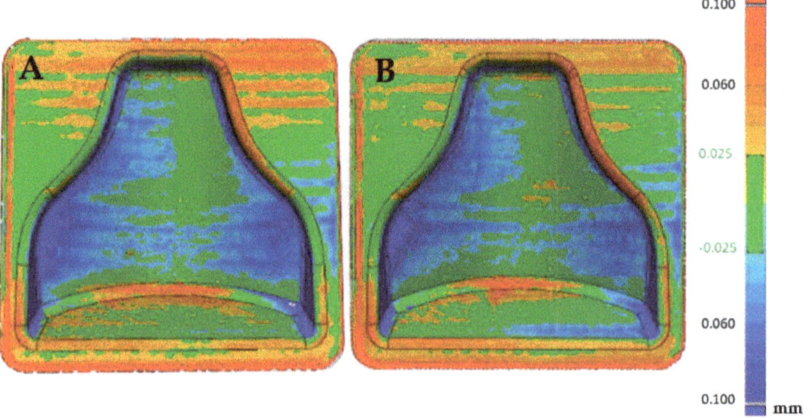

Figure 11. Comparison between the CAD model and first mold after chemical smoothing (**A**) and cad model and second mold after chemical smoothing (**B**).

Afterwards, thanks to the Geomagic visualization tool (3d systems Inc., Valencia, CA, USA), a comparison with Figure 9 was performed in Figure 12, and it was possible to see a greater homogenization of the surfaces, which can be seen from the zebra stripes. This was due to the smoothing process that turned the surfaces smoother and glossier. Additionally, it was possible to appreciate the effect of smoothing on the dimensional variation of the component in Figures 13 and 14. The molds before and after treatment were compared.

Figure 12. Zebra stripes effect on the two molds after printing. (**left**) mold 1 and (**right**) mold 2.

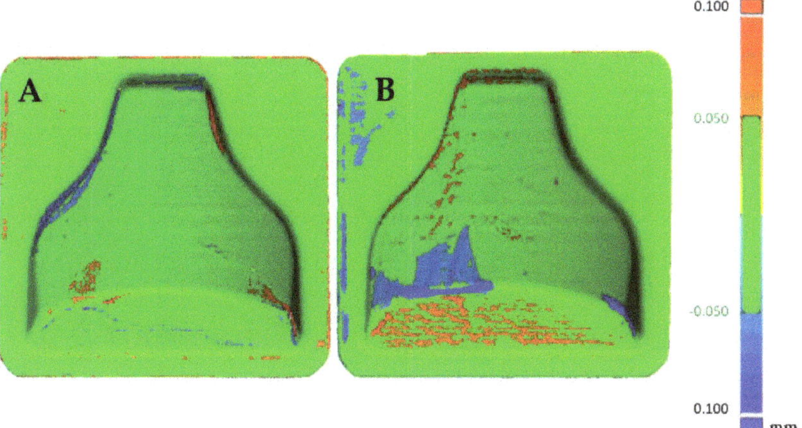

Figure 13. Comparison between the scan of mold 1 pre and post smoothing (**A**) and comparison of mold 2 pre and post smoothing (**B**).

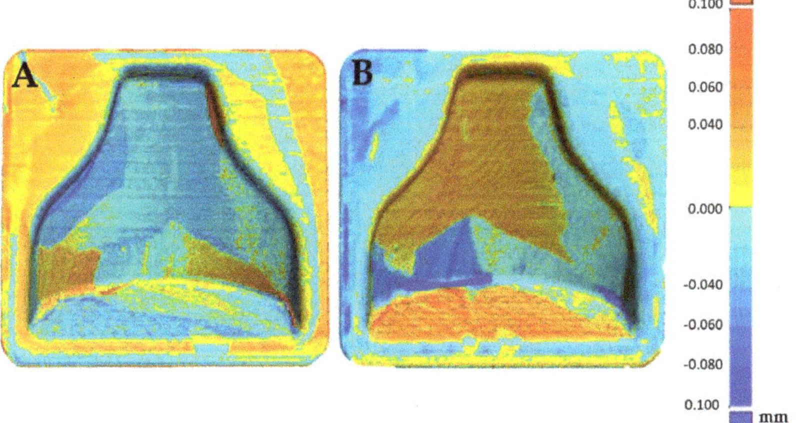

Figure 14. Comparison between the scan of mold 1 pre and post smoothing (**A**) and comparison of mold 2 pre and post smoothing (**B**).

From Figure 13 it could be deduced that overall, the dimensional tolerance of the component after smoothing reached an absolute value of 0.1 mm, with both molds completely colored green. Figure 14 shows that the treatment was uniform; positive or negative values also differed in the way in which the software superimposed the two scans and did not indicate removal or addition of material. Therefore, they must be understood in an absolute manner. In fact, the software tried to minimize the distance between the points of one mesh and those of the other, obtaining the result shown.

In Figure 15A the reproducibility of the process can be evaluated. In fact, it can be seen that the two post-treatment molds differed in an absolute value by less than 0.1 mm, with the likely analysis almost completely colored green. Figure 15B shows the areas that were slightly positive and those that were negative, but overall a good matching was obtained, making it possible to guarantee narrow tolerances.

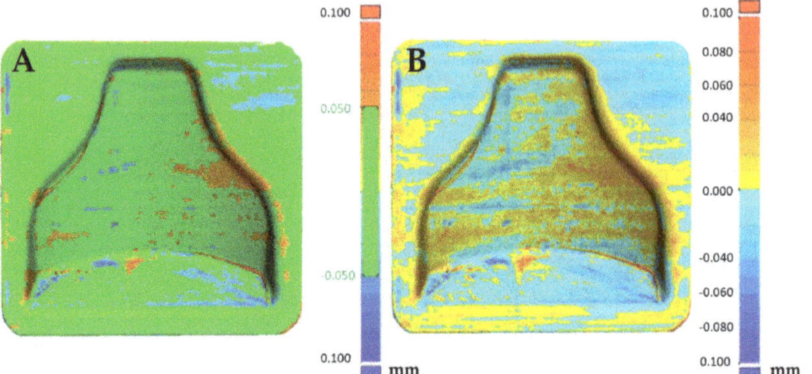

Figure 15. Comparison between the two molds (mold 1 (**A**), mold 2 (**B**)) after smoothing.

3.3. Carbon Fiber Vacuum Lamination Process

The mold in this study was used to create a part with the vacuum lamination process to make the component of Figure 2A. It is worth noticing that the FDM-produced mold did not require the use of a release agent, unlike conventional fiberglass and gelcoat or metal molds. Therefore, a resin-impregnated carbon fiber cloth was placed directly in the cavity of the mold, after which a layer of peel ply fabric and a layer of absorbent tissue were placed and the vacuum was made. The resin used was AERO68® (Rius Composites SRL, Italy)), suitable for wet-layup laminations at room temperature with carbon fiber, glass fiber, and Kevlar. The component was then extracted from the mold with the aid of a plastic wedge and then finished and mounted on the fuel tank. Figure 16 shows a series of images that summarize the production process of the component.

Figure 16. Carbon fiber component production steps: (**A**) component lamination and vacuum, (**B**) mold opened after resin curing, (**C**) component just extracted from the mold, (**D**) trimmed and finished component, (**E**) component mounted on the tank.

4. Conclusions

The procedure highlighted proved to be valid for creating carbon fiber components produced in very small numbers. The choice of using molds printed in FDM allowed us to reduce time and costs considerably compared to conventional methods. The need to obtain a smooth surface of the final component is also fundamental to guarantee optical and mechanical properties. The chemical smoothing allowed us to obtain a smooth surface, without using gelcoat or other products/techniques for reducing surface roughness. Overall, the quality of the final component was high, but there are margins for improvement, especially in terms of dimensional tolerances. By knowing the effect of the smoothing process at a dimensional level approximately, it is possible to modify the starting cad file to obtain more accurate final dimensions of the mold.

The printing parameters used for the creation of the mold could also be improved. The correct setting of printing parameters aimed at obtaining a better surface quality, therefore guaranteeing shorter smoothing times, and reaching higher dimensional tolerances. These ensure less alcohol absorption by the surface and therefore less time to get the component finished and ready to be used.

5. Future Developments

Future developments include further printing methodologies for this material, i.e., the adoption of a lower layer height, that would lead to much longer printing times but an even higher surface quality.

Additional tests of this element are needed, e.g., if it would be possible to make the entire mold in cheap PLA and only the outer layer in PVB, after verifying a good adhesion between the two materials. A possible alternative to PLA could be PETG, in order to further reduce the costs of a single mold and take full advantage of the existing technologies of 3D FDM printing and the smoothing technique.

Further research is needed for the vapor-process parameters and how to influence the surface roughness and the actual fabrications of parts with other materials. Mechanical tests could be carried out to evaluate the mechanical properties of the carbon fiber components obtained with this technology.

Author Contributions: Conceptualization, P.F. and G.M.S.; methodology, G.M.S., P.F.; validation, M.F., P.F.; formal analysis, P.F., C.L.-C.; investigation, P.F., C.L.-C.; resources, G.D., L.F.; data curation, P.F.; writing—original draft preparation, P.F., C.L.-C.; writing—review and editing, C.L.-C.; visualization, P.F., M.F.; supervision, L.F.; project administration, A.L. All authors have read and agreed to the published version of the manuscript.

Funding: This research received no external funding.

Institutional Review Board Statement: Not applicable.

Informed Consent Statement: Not applicable.

Conflicts of Interest: The authors declare no conflict of interest.

References

1. Shahrubudin, N.; Lee, T.; Ramlan, R. An Overview on 3D Printing Technology: Technological, Materials, and Applications. *Procedia Manuf.* **2019**, *35*, 1286–1296. [CrossRef]
2. Rajpurohit, S.R.; Dave, H.K. Effect of process parameters on tensile strength of FDM printed PLA part. *Rapid Prototyp. J.* **2018**, *24*, 1317–1324. [CrossRef]
3. Guessasma, S.; Belhabib, S.; Nouri, H. Effect of printing temperature on microstructure, thermal behavior and tensile properties of 3D printed nylon using fused deposition modeling. *J. Appl. Polym. Sci.* **2021**, *138*, 50162. [CrossRef]
4. Savu, I.D.; Savu, S.V.; Simion, D.; Sirbu, N.-A.; Ciornei, M.; Ratiu, S.A. PP in 3D Printing—Technical and Economic Aspects. *Mater. Plast.* **2019**, *56*, 931–936. [CrossRef]
5. Turner, B.N.; Strong, R.; Gold, S.A. A review of melt extrusion additive manufacturing processes: I. Process design and modeling. *Rapid Prototyp. J.* **2014**, *20*, 192–204. [CrossRef]
6. Bikas, H.; Stavropoulos, P.; Chryssolouris, G. Additive manufacturing methods and modelling approaches: A critical review. *Int. J. Adv. Manuf. Technol.* **2016**, *83*, 389–405. [CrossRef]

7. Tofail, S.A.; Koumoulos, E.P.; Bandyopadhyay, A.; Bose, S.; O'Donoghue, L.; Charitidis, C. Additive manufacturing: Scientific and technological challenges, market uptake and opportunities. *Mater. Today* **2018**, *21*, 22–37. [CrossRef]
8. Singh, R. Some investigations for small-sized product fabrication with FDM for plastic components. *Rapid Prototyp. J.* **2013**, *19*, 58–63. [CrossRef]
9. Kim, M.K.; Lee, I.H.; Kim, H.-C. Effect of fabrication parameters on surface roughness of FDM parts. *Int. J. Precis. Eng. Manuf.* **2018**, *19*, 137–142. [CrossRef]
10. Frizziero, L.; Donnici, G.; Dhaimini, K.; Liverani, A.; Caligiana, G. Advanced Design Applied to an Original Multi-Purpose Ventilator Achievable by Additive Manufacturing. *Appl. Sci.* **2018**, *8*, 2635. [CrossRef]
11. Bacciaglia, A.; Ceruti, A.; Liverani, A. Surface smoothing for topological optimized 3D models. *Struct. Multidiscip. Optim.* **2021**, 1–20. [CrossRef]
12. Casavola, C.; Cazzato, A.; Moramarco, V.; Pappalettera, G. Residual stress measurement in Fused Deposition Modelling parts. *Polym. Test.* **2017**, *58*, 249–255. [CrossRef]
13. Komineas, G.; Foteinopoulos, P.; Papacharalampopoulos, A.; Stavropoulos, P. Build time estimation models in thermal extru-sion additive manufacturing processes. *Procedia Manuf.* **2018**, *21*, 647–654. [CrossRef]
14. Tiwary, V.; Arunkumar, P.; Deshpande, A.S.; Rangaswamy, N. Surface enhancement of FDM patterns to be used in rapid investment casting for making medical implants. *Rapid Prototyp. J.* **2019**, *25*, 904–914. [CrossRef]
15. Harun, W.S.W.; Sharif, S.; Idris, M.H.; Kadirgama, K. Characteristic studies of collapsibility of ABS patterns produced from FDM for investment casting. *Mater. Res. Innov.* **2009**, *13*, 340–343. [CrossRef]
16. Mukhtarkhanov, M.; Perveen, A.; Talamona, D. Application of Stereolithography Based 3D Printing Technology in Investment Casting. *Micromachines* **2020**, *11*, 946. [CrossRef]
17. Stumpel, L.J. Deformation of Stereolithographically Produced Surgical Guides: An Observational Case Series Report. *Clin. Implant. Dent. Relat. Res.* **2010**, *14*, 442–453. [CrossRef]
18. Guo, H.; Zhou, Y.T. Research on the Impacting Factors of Dimensional Accuracy with Silicone Mold Casting Products Based on FDM. *Adv. Mater. Res.* **2014**, *912–914*, 309–313. [CrossRef]
19. Heidari-Rarani, M.; Rafiee-Afarani, M.; Zahedi, A. Mechanical characterization of FDM 3D printing of continuous carbon fiber reinforced PLA composites. *Compos. Part B Eng.* **2019**, *175*, 107147. [CrossRef]
20. Wang, P.H.; Kim, G.; Sterkenburg, R. Investigating the Effectiveness of a 3D Printed Composite Mold. *Int. J. Aerosp. Mech. Eng.* **2019**, *13*, 11. [CrossRef]
21. Sudbury, T.Z.; Springfield, R.; Kunc, V.; Duty, C. An assessment of additive manufactured molds for hand-laid fiber reinforced composites. *Int. J. Adv. Manuf. Technol.* **2016**, *90*, 1659–1664. [CrossRef]
22. Garg, A.; Bhattacharya, A.; Batish, A. Chemical vapor treatment of ABS parts built by FDM: Analysis of surface finish and mechanical strength. *Int. J. Adv. Manuf. Technol.* **2017**, *89*, 2175–2191. [CrossRef]
23. Singh, J.; Singh, R.; Singh, H. Investigations for improving the surface finish of FDM based ABS replicas by chemical vapor smoothing process: A case study. *Assem. Autom.* **2017**, *37*, 13–21. [CrossRef]
24. Kuo, C.C.; Wang, C.W.; Lee, Y.F.; Liu, Y.L.; Qiu, Q.Y. A surface quality improvement apparatus for ABS parts fabricated by ad-ditive manufacturing. *Int. J. Adv. Manuf. Technol.* **2017**, *89*, 635–642. [CrossRef]
25. Iwasaki, R.; Sato, C.; Latailladeand, J.L.; Viot, P. Experimental study on the interface fracture toughness of PVB (polyvinyl butyral)/glass at high strain rates. *Int. J. Crashworthiness* **2007**, *12*, 293–298. [CrossRef]
26. Liu, B.; Sun, Y.; Li, Y.; Wang, Y.; Ge, D.; Xu, J. Systematic experimental study on mechanical behavior of PVB (polyvinyl butyral) material under various loading conditions. *Polym. Eng. Sci.* **2012**, *52*, 1137–1147. [CrossRef]
27. Chen, Y.; Yao, X.; Zhou, X.; Pan, Z.; Gu, Q. Poly(lactic acid)/graphene nanocomposites prepared via solution blending using chloroform as a mutual solvent. *J. Nanosci. Nanotechnol.* **2011**, *11*, 7813–7819. [CrossRef]
28. Colpani, A.; Fiorentino, A.; Ceretti, E. Characterization of chemical surface finishing with cold acetone vapours on ABS parts fabricated by FDM. *Prod. Eng.* **2019**, *13*, 437–447. [CrossRef]
29. Bere, P.; Neamtu, C.; Udroiu, R. Novel Method for the Manufacture of Complex CFRP Parts Using FDM-based Molds. *Polymers* **2020**, *12*, 2220. [CrossRef] [PubMed]

 *polymers*

Article

3D Printing of Thermal Insulating Polyimide/Cellulose Nanocrystal Composite Aerogels with Low Dimensional Shrinkage

Chiao Feng [1] and Sheng-Sheng Yu [1,2,3,*]

[1] Department of Chemical Engineering, National Cheng Kung University, Tainan 70101, Taiwan; N36094136@gs.ncku.edu.tw
[2] Core Facility Center, National Cheng Kung University, Tainan 70101, Taiwan
[3] Program on Smart and Sustainable Manufacturing, Academy of Innovative Semiconductor and Sustainable Manufacturing, National Cheng Kung University, Tainan 70101, Taiwan
* Correspondence: ssyu@mail.ncku.edu.tw; Tel.: +886-6-2757575 (ext. 62628)

Abstract: Polyimide (PI)-based aerogels have been widely applied to aviation, automobiles, and thermal insulation because of their high porosity, low density, and excellent thermal insulating ability. However, the fabrication of PI aerogels is still restricted to the traditional molding process, and it is often challenging to prepare high-performance PI aerogels with complex 3D structures. Interestingly, renewable nanomaterials such as cellulose nanocrystals (CNCs) may provide a unique approach for 3D printing, mechanical reinforcement, and shape fidelity of the PI aerogels. Herein, we proposed a facile water-based 3D printable ink with sustainable nanofillers, cellulose nanocrystals (CNCs). Polyamic acid was first mixed with triethylamine to form an aqueous solution of polyamic acid ammonium salts (PAAS). CNCs were then dispersed in the aqueous PAAS solution to form a reversible physical network for direct ink writing (DIW). Further freeze-drying and thermal imidization produced porous PI/CNC composite aerogels with increased mechanical strength. The concentration of CNCs needed for DIW was reduced in the presence of PAAS, potentially because of the depletion effect of the polymer solution. Further analysis suggested that the physical network of CNCs lowered the shrinkage of aerogels during preparation and improved the shape-fidelity of the PI/CNC composite aerogels. In addition, the composite aerogels retained low thermal conductivity and may be used as heat management materials. Overall, our approach successfully utilized CNCs as rheological modifiers and reinforcement to 3D print strong PI/CNC composite aerogels for advanced thermal regulation.

Keywords: polyimide; aerogels; 3D printing

Citation: Feng, C.; Yu, S.-S. 3D Printing of Thermal Insulating Polyimide/Cellulose Nanocrystal Composite Aerogels with Low Dimensional Shrinkage. *Polymers* **2021**, *13*, 3614. https://doi.org/10.3390/polym13213614

Academic Editor: Houwen Matthew Pan

Received: 29 September 2021
Accepted: 19 October 2021
Published: 20 October 2021

Publisher's Note: MDPI stays neutral with regard to jurisdictional claims in published maps and institutional affiliations.

1. Introduction

Polyimide (PI) aerogels have received wide attention because of their low density, stable mechanical properties, and low thermal conductivity [1]. Aerogels are highly porous materials for various applications, but they are often limited by their poor mechanical strength and thermal stability [2]. The capability of aerogels may be expanded by using engineering plastics such as PI with high mechanical strength and thermal stability [3]. For example, poly(4,4'-oxydiphenylene pyromellitimide), also known by its trade name Kapton, is a fully aromatic PI that retains its performance over a wide range of temperatures for application in aerospace engineering, defense, and electronics [4]. Recently, there has been a growing interest in using porous PI as high-performance aerogels [5]. Typical PI aerogels are made by the chemical imidization of crosslinked polyamic acids, followed by supercritical CO_2 drying to remove organic solvents within the gels. The lightweight PI aerogels are especially attractive for thermal insulation [6,7], electromagnetic interference shielding [8], and aerospace engineering [9]. Nanocomposite PI aerogels can also be used as sensors [10] and nanogenerators [11]. However, the fabrication of PI aerogels is still limited to the traditional molding process. In addition, the large volumetric shrinkage of PI aerogels during preparation makes it difficult to control the shape of the final product precisely. It is desirable to achieve complex and customizable designs of PI aerogels to accurately assemble each component for advanced applications in thermal regulation and electronics.

Additive manufacturing (AM), also known as 3D printing, is an emerging tool for polymer processing. Through the layer-by-layer deposition of materials, geometrically complex objects from computer-aided design (CAD) can be achieved [12]. Light-based printing such as stereolithography (SLA) and digital light processing (DLP) utilized the photopolymerization of resins. On the other hand, fused deposition modeling (FDM) and direct ink writing (DIW) are extrusion-based printing. In particular, DIW is a low-cost and versatile technique compatible with a wide range of materials [13]. The DIW process prints the desired objects by extruding materials through fine nozzles. A successful layer-by-layer deposition is achieved by the proper viscoelastic response of the inks. The ink should display shear-thinning behavior for facile material extrusion under a high printing rate and high yield stress for good self-supporting ability [14,15].

Several strategies have been developed to 3D print PI-based materials. For example, photosensitive polyamic acids can be prepared by the reaction with common vinyl monomers [16–18]. Such inks can be used for SLA, DLP, and UV-assisted DIW. It is also possible to use the ionic interaction between polyamic acids and amine-containing vinyl monomers to prepare 3D printable inks for UV-assisted DIW and DLP [19,20]. The above approaches require the chemical modification of polyamic acids and a large quantity of organic solvents. The thermal imidization also induces a significant shrinkage of the printed object. Wang et al. demonstrated that DIW with a rapid drying apparatus enabled 3D printing of PI from poly(amic acid) ammonium salt (PAAS) hydrogels [21]. Water-soluble PAAS inks can be prepared by mixing polyamic acids and triethylamine (TEA). A careful selection of polyamic acids and the amines enables the formation of PAAS-based hydrogels through the potential hydrogen bonding and π–π stacking as physical networks [22]. In contrast to PI, the printing of PI aerogels has only been sporadically studied. It may be achieved by the freeze-drying of PAAS hydrogels [23] or using polymer microspheres as pore-forming agents and rheological modifiers for DIW [24].

Aside from the above methods, 3D printable ink may also be achieved by using nanoparticles as rheological modifiers. Recently, cellulose nanocrystals (CNCs) have been intensively studied because of their high elastic modulus and low density [25]. CNCs are rod-like particles (10 to 30 nm in diameter and few nanometers in length) with high crystallinity [26]. Indeed, CNCs have been included in chemically imidized PI aerogels [27] or PI thin films [28] as reinforcement. In addition, the dispersion of CNCs in water forms hydrogels with percolated networks. The CNC hydrogels exhibit strong shear-thinning behavior ideal for DIW [29,30]. Interestingly, it has also been demonstrated that the strong physical network of CNCs could significantly reduce the volumetric shrinkage of aerogels during the drying process [31].

Although several approaches have 3D printed PI-based materials successfully, the 3D printing of porous PI aerogels has only been sporadically studied. Furthermore, it is intriguing to utilize renewable nanomaterials such as CNCs to enhance the performance of PI aerogels further. In this work, we designed 3D printable inks to fabricate PI/CNC composite aerogels (Figure 1). The ink was mainly composed of water-soluble PAAS by the complexation of polyamic acids and TEA. CNCs were included in the aqueous solutions of PAAS to form physically crosslinked hydrogels for DIW. After printing, freeze-drying was employed as a simple and inexpensive method to make porous structures. Further thermal imidization of the freeze-dried PAAS/CNC aerogels leads to PI/CNC composite aerogels with improved mechanical strength and potential for thermal insulation. The physical network of CNCs effectively reduced the shrinkage of aerogels during preparation and enhanced the mechanical properties of the aerogels.

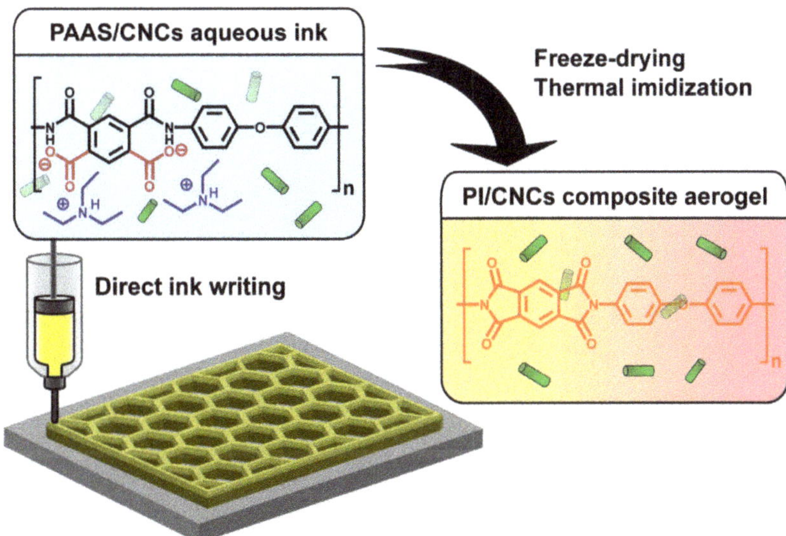

Figure 1. Synthesis and 3D printing of PI/CNC composite aerogels.

2. Materials and Methods

2.1. Materials

4,4-Oxydianiline (ODA, 98%) and triethylamine (TEA, 99%) were purchased from Acros Organics, Geel, Belgium. Pyromellitic dianhydride (PMDA, >98%) was obtained from Tokyo Chemical Industry, Tokyo, Japan. Cellulose nanocrystals (CNCs) were obtained from CelluForce Inc., Montreal, QC, Canada. The diameter and length of CNCs were 2.3 to 4.5 nm and 44 to 108 nm, respectively. N, N-Dimethylformamide (DMF, ACS grade) was offered by Avantor, Radnor, PA, USA, and was treated with molecular sieves for more than two days before synthesis. All other reagents were used as received.

2.2. Synthetic Procedures of PMDA-ODA Polyamic Acid

First, 4.806 g (24 mmol) ODA was added to a three-neck flask and mixed with 100 mL DMF under room temperature and a nitrogen atmosphere. A total of 5.235 g (24 mmol) PMDA was gradually added to form a polyamic acid precursor solution with a light yellow color. The solution was then stirred for another four hours to ensure the completeness of the reaction. Next, the viscous polyamic acid solution was diluted with another 100 mL DMF, and slowly poured into 600 mL deionized water under vigorous stirring for precipitation. The yellow polyamic acid was dried in a vacuum oven at 50 °C for more than two days before use. The detailed experimental procedure is summarized in Figure S1.

2.3. Preparation of PAAS/CNC Inks

First, 0.750 g polyamic acid, 0.363 g TEA, and 4.444 g deionized water were mixed in a glass vial. The solution was sonicated until all materials had dissolved to form the PAAS solution. Next, 700 mg of CNCs was dispersed in 4.444 g deionized water by a planetary mixer (MV-300S, CGT, Taiwan) at 1500 rpm for 10 min to form a uniform CNC hydrogel. Seven Teflon balls were added in the container to deagglomerate the CNC powder better. Finally, the aqueous PAAS solution was mixed with the CNC hydrogel by the planetary mixer at 1500 rpm for 30 min. The final concentration of polyamic acid and CNCs was 7.0 wt% and 6.5 wt%, respectively. For comparison, the amount of CNC was adjusted to make inks with 5.6 wt%, 6.5 wt%, and 7.4 wt% CNCs. Pure PI aerogels were prepared using a 7.5 wt% polyamic acid solution. Pure CNC aerogels were prepared from the inks containing only 6.5 wt% CNCs.

2.4. Rheological Test

The rheological behavior of inks was investigated by a rheometer (HR-2, TA instruments, New Castle, DE, USA) with parallel plates (diameter: 25 mm). The gap between the two parallel plates was set at 1000 μm. A steady-state shear flow test was used to measure the viscosity of the inks under different shear rates. The shear storage modulus (G′) and shear loss modulus (G″) were measured by an oscillatory stress sweep method at a frequency of 1 Hz. All rheological tests were performed at room temperature.

2.5. 3D Printing of PAAS/CNC Composite Ink

The as-prepared PAAS/CNC ink was transferred into a 10 mL disposable syringe. The bubbles within the ink were removed by the planetary mixer. A FDM printer (INFINITY X1Speed, INFINITY3DP, Kaohsiung, Taiwan) was modified for the DIW process by replacing the extruder with a 3D printed adapter. The syringe was then mounted on a syringe pump (NE-1010, New Era Instrument, Farmingdale, NY, USA) and connected to a 0.4 mm conical type plastic nozzle (TPND-22G-U, Musashi Engineering, Tokyo, Japan). The printer was controlled by Repetier software, and the G-codes were generated from the Slic3r toolpath generator. Typically, the printing speed was set at 20 mm/s, and the extrusion rate was 0.155 mL/min.

2.6. Preparation of PI, CNC, and PI/CNC Composite Aerogels

The printed objects were frozen in a −20 °C refrigerator overnight. The frozen samples were rapidly transferred to a freeze dryer and dried for more than two days. The dried samples further underwent a sequential temperature increase from 80, 120, 180, 240, to 300 °C under vacuum to form the PI/CNC composite aerogels. Each temperature was held for one hour, and the ramping rate was 1 °C/min in all cases.

2.7. Characterization

The compression strength was measured using a universal electromechanical testing machine (AGS-X 100kN, Shimadzu, Kyoto, Japan) equipped with a 10 kN load cell. All tests were performed following the ASTM-D695 standard.

The shrinkage ratio was calculated based on the following equation:

$$Shrinkage\ (\%) = 100\% \times \frac{V_{\mathrm{CAD}} - V_{\mathrm{s}}}{V_{\mathrm{CAD}}}$$

where V_{CAD} represents the volume of the cylindrical CAD models and V_{s} is the apparent volume of the aerogels. Bulk density was calculated by the weight and volume of each sample after imidization.

Fourier transform infrared spectroscopy (FTIR) was performed using a Nicolet 6700 Fourier transform infrared spectrometer (Thermo Scientific, Waltham, MA, USA). The scanning region was set from 650 to 4000 cm^{-1}. Resolution and scanning times were set at 2 cm^{-1} and 32 scans. Solid-state ^{13}C NMR spectra were recorded using a Bruker Avance III HD 400 MHz NMR spectrometer. Thermogravimetric analysis (TGA) was conducted on a TGA 4000 (PerkinElmer, Waltham, MA, USA) from 30 °C to 700 °C under a nitrogen atmosphere at a heating rate of 10 °C/min. The microstructures of the aerogels were observed using a high-resolution field emission scanning electron microscope (SU-8010, Hitachi, Tokyo, Japan). The accelerating voltage was 10 kV.

Thermal conductivity (λ) was measured by Hotdisk TPS 2500S (Hot Disk AB, Gothenburg, Sweden) with a 5465 f1 type Kapton® sensor and standard isotropic method under room temperature. The probing depth, heating power, and measuring time were set at 5 mm, 10 mW, and 10 s, respectively. Thermal images were taken by a FLIR C2 thermal imaging camera (FLIR, Wilsonville, OR, USA). The working distance was around 30 cm. Samples were placed on a 160 °C hotplate. A series of thermal images were taken before and after 10 min of heating treatment.

3. Results and Discussions

3.1. Rheological Analysis of the PAAS/CNC Composite Inks

We first explored the effect of CNCs on the rheological behavior of the PAAS/CNC inks. The concentration of CNCs varied from 5.6 wt% to 7.4 wt%, and polyamic acid was 7.0 wt%. Water-soluble PAAS was prepared by mixing PAA with TEA. TEA also catalyzes the imidization reaction of PAA to PI [32]. As mentioned earlier, CNCs form a reversible physical network in water, and a series of rheological analyses can evaluate the strength of the network. All inks demonstrated significant shear thinning behavior (Figure 2a). For example, the viscosity for the ink using 6.5 wt% CNCs was 18,134 Pa·s at 0.01 1/s, but it decreased to 228 Pa·s at 1 1/s. The reduction in viscosity at a high shear rate allows for the ink to be extruded at a high flow rate. The performance of DIW also depends on the storage modulus and yield stress. The ink should have a high storage modulus and yield stress, so the printed inks could stack successfully and even span through unsupported parts [33]. The yield stress of the ink could be obtained from the intersection between the storage (G') and loss modulus (G''). As shown in Figure 2b, all inks exhibited gel-like behavior with their storage modulus greater than loss modulus at small shear stress. The gel-like behavior of the inks at low shear stress was found in several works by studying the suspension of CNCs [29,30,34]. CNCs form a percolating network in ink with finite yield stress. The ink does not flow if the applied stress is lower than the yield stress. At high shear stress, the inks transited into a liquid-like state ($G'' > G'$) suitable for DIW because of the alignment of CNCs [34]. In addition, both storage modulus and yield stress increased with the concentrations of CNCs. Therefore, the strength of physical networks in the inks gradually increased with CNCs.

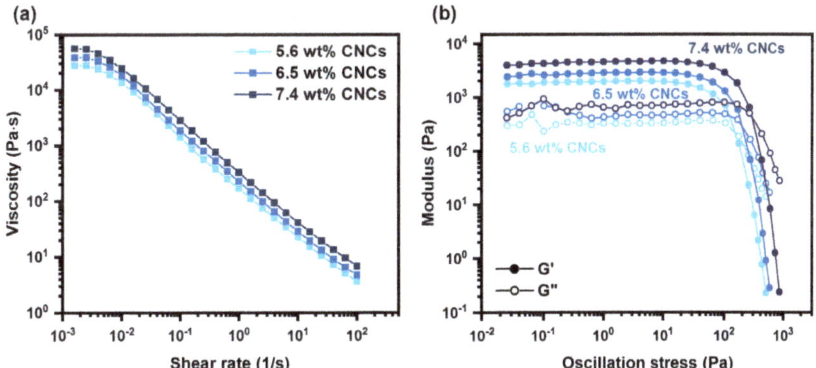

Figure 2. Rheological analysis of the PAAS/CNC inks under different concentrations of CNCs. (**a**) Steady-state shear viscosity of the inks. (**b**) Oscillatory rheological measurement.

From the above result, the inks using 6.5 or 7.4 wt% CNCs were sufficient for DIW because of their high yield stress (>200 Pa) [15]. Our previous work has shown that a higher concentration of CNCs (12 to 15 wt%) in water was needed to achieve similar rheological performance for DIW [35]. Therefore, PAAS may promote the formation of the physical network by CNCs and reduce the concentration of CNCs needed for printing. Similar behavior has been found for the dispersion of CNCs with anionic polymers such as carboxymethyl cellulose [36]. The rheological behavior of CNCs is strongly influenced by polymers, especially non-adsorbing polymers that have little affinity with the surface of CNC particles [37,38]. The presence of non-adsorbing polymers significantly reduced the gelation concentration of CNCs. This phenomenon can be attributed to the depletion force induced by the non-adsorbing polymers. The polymers lose their conformation entropy around the CNC particles and formed polymer-depleted regions near CNCs. Therefore, the difference in osmotic pressure between the depleted region and the bulk polymer solution provides entropically driven attraction to promote the gelation of CNCs. We hypothesized that the depletion-induced flocculation of CNCs also occurred in our PAAS/CNC inks. PAAS is an anionic polymer [22,32] that should have little adsorption on negatively charged CNCs. Therefore, PAAS may have a similar behavior to the non-adsorbing polymers [36]. The depletion effect of PAAS then reduced the concentration of CNCs required for DIW.

The effect of polyamic acid on the rheological behavior of the inks has also been investigated. A CNC of 6.5 wt% and the concentrations of polyamic acid ranged from 4.7 wt% to 9.7 wt%. All three samples had similar shear-thinning behavior for DIW (Figure S2a). Figure S2b further shows these inks had a similar response in storage modulus and yield stress. Although the presence of non-adsorbing polymers promotes the gelation of CNCs, the effect was less prominent at a high concentration of polymers [36,39]. Based on the above analysis, we concluded that adding more polyamic acids into the system did not significantly improve the printability of the ink.

The above results suggest that the viscosity and modulus were mainly determined by the concentration of CNCs rather than polyamic acids. A 6.5 wt% of CNCs was sufficient to achieve significant shear thinning behavior and high yield stress for DIW. Based on the above analysis, we then mainly focused on the effect of CNCs on the properties of the PI/CNC composite aerogels. We chose the ink with 6.5 wt% CNCs to demonstrate the printing performance (Figure 3). During the printing process, the ink could be successfully extruded and stacked layer-by-layer (Figure 3a,b). The printed object then underwent the freeze-drying process to remove water and form macropores. Further thermal treatment completed the PI/CNC composite aerogels. As shown in Figure 3d, the thermal imidization induced only small dimensional shrinkage. The printed object showed good shape-fidelity compared to our initial CAD design. Some cracks appeared in the imidized samples, possibly because of the drastic weight loss of the composite aerogels or thermal stress [40] during the thermal imidization. The presence of cracks may be mitigated by further adjusting the imidization process. From scanning electron microscopy (SEM), we found that the printed PI/CNC aerogels had high spatial resolution as expected (Figure 3e), and small pores could be observed on the surface of the aerogels (Figure 3f). Furthermore, a cylindrical model with a high aspect ratio was printed using the PAAS/CNC ink (Figure S3). The ink could successfully be stacked up to 70 layers (28.22 mm in height), indicating the high yield stress and the printability of the PAAS/CNC ink.

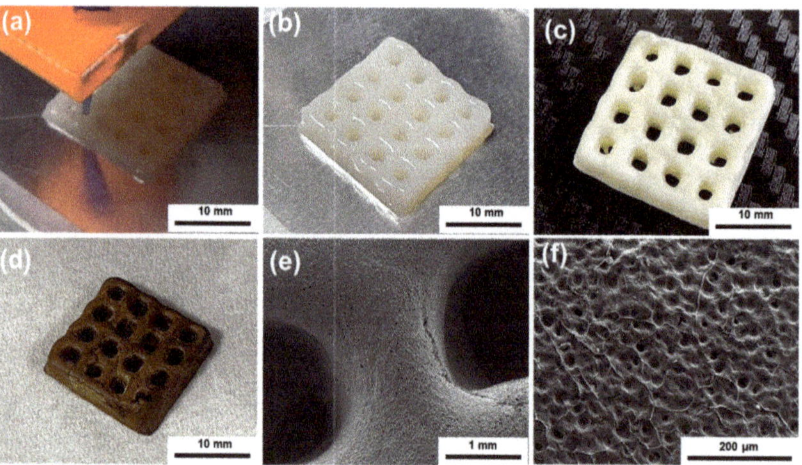

Figure 3. Performance of the PAAS/CNC ink for DIW. (**a**) The printing process. (**b**) The printed mesh structure. (**c**) The printed mesh after freeze-drying (**d**). The printed mesh after thermal imidization. (**e,f**) SEM images of the mesh after thermal imidization under different magnification.

3.2. Chemical Analysis of the PI/CNC Nanocomposite Aerogels

The PI/CNC aerogels were analyzed by FTIR to probe the degree of imidization (Figure 4a). The PI aerogel without CNCs and the pure CNCs aerogel were used for a comparison. Pure PI aerogels had two peaks at 1776 cm^{-1} and 1721 cm^{-1}, corresponding to the asymmetrical and symmetrical C=O stretching of the imide ring, respectively [8]. The peaks at 1500 cm^{-1} and 1375 cm^{-1} came from the C=C stretching of the aromatic ring and the C–N stretching of the imide ring. Besides these characteristic peaks of PI, the absence of signals at 1540 and 1660 cm^{-1} (amide II and C=O stretching of amide) also implied a successful imidization of the pure PI aerogels. As shown in Figure S4, the pure CNC aerogel had a broad peak around 3400 cm^{-1}, corresponding to the hydroxyl groups on CNCs. The FTIR spectrum of the PI/CNC aerogels was similar to that of the pure PI aerogel, suggesting that the imidization was not affected by CNCs. Mixing CNCs with PAAS did not affect the final cyclization process of PAAS to PI. In addition, the PI/CNC aerogels also showed the characteristic peak of the hydroxyl group at 3400 cm^{-1}. Therefore, a significant amount of CNCs remained in the composite aerogels, even after the thermal imidization process.

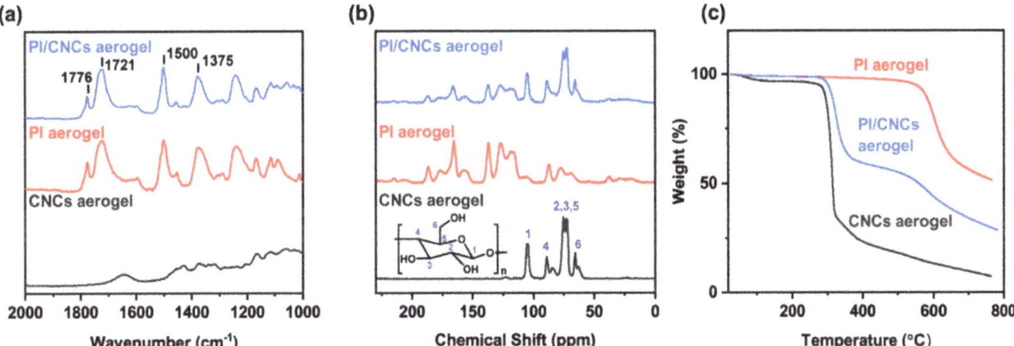

Figure 4. (**a**) FTIR spectra of the aerogels. (**b**) Solid-state ^{13}C NMR spectra of the aerogels. (**c**) TGA curves of aerogels. The PI/CNC aerogel was prepared by using 6.5 wt% CNCs.

Solid-state ^{13}C NMR was also used to probe the interaction or chemical bonding between PI and CNCs in the composite aerogels (Figure 4b). The characteristic signals of CNCs were found at 65.6 ppm (C6 crystalline), 70–75 ppm (C2, C3, C5), 84–89 ppm (C4), and 104.9 ppm (C1) [41]. In both pure PI and PI/CNC aerogels, we found the peaks of C=O imide moiety (165.6 ppm), phenoxy carbon (156.2 ppm), and the peaks of aromatic rings (115–134 ppm). These signals were similar to early reported results [42,43]. The peaks around 170–190 and 70–90 ppm were attributed to the spinning sidebands by the magic angle spinning [44]. The main signals of CNCs could also be found in the composite aerogels, suggesting that most CNCs were not degraded by thermal imidization. In addition, esterification might also occur between the carboxylic acid groups of PAA and the surface hydroxyl groups of CNCs during the thermal treatment. As we did not find any new signals in the PI/CNC composite aerogels, the reaction between PAA and CNCs should be minimal and the CNCs mainly acted as rheological modifiers and reinforcing agents. The above results also suggest that complete imidization was achieved in the composite aerogels and a significant amount of CNCs still remained.

Besides FTIR and solid-state NMR, we also used TGA to confirm the completeness of the thermal imidization process (Figure 4c). Aromatic polyimides typically possess great stability in a high-temperature environment. For the pure PI aerogel, no significant weight loss was found until 550 °C, and the final char yield was 51%. The onset temperature of degradation was approximately 576 °C, indicating a high-temperature tolerance of the pure PI aerogel. On the other hand, the pure CNCs aerogel decomposed around 290 °C and showed a char yield of 7 wt% at 760 °C, similar to the previous report [27]. The PI/CNC composite aerogels had two degradation temperatures around 300 °C and 540 °C. The first onset of decomposition at 300 °C was the degradation of CNCs, and the second one at 540 °C should be the thermal decomposition of PI. We also found that the thermal stability of CNCs increased slightly in the PI/CNC aerogels, but the decomposition temperature of PI decreased slightly. The improved thermal stability of CNCs in the composite aerogels may come from the hydrogen bonding between the surface hydroxyl group of CNCs and PI [45]. Although CNCs may degrade around the thermal imidization temperature (300 °C), a large quantity of CNCs still remained in our PI/CNC aerogels based on the results of TGA. We then recorded the weight of each freeze-dried sample before and after the thermal imidization process to understand the degradation behavior of CNCs (Figure S5). For the pure PI aerogel without CNCs, a significant loss of weight to 42.3% was observed. This result could be attributed to the loss of water by imidization and the evaporation of TEA. If a similar degree of weight loss occurs in the composite aerogels, the 6.5 wt% CNCs sample should only lose 25.8% of its weight. However, the observed weight loss was 32.4%, indicating that some CNCs degraded during the thermal imidization process. The degradation of CNCs was more significant in the aerogels with 5.6 wt% CNCs than the other composite aerogels. Fortunately, a significant amount of CNCs remained in the aerogels prepared by 6.5 wt% and 7.4 wt% CNCs. Based on the above results, we expect that the weight loss of CNCs should be minimal, even at 300 °C if there is a high concentration of CNCs. This result may come from the enhanced thermal stability of CNCs in the composites, as indicated by our TGA analysis. Although lowering the imidization temperature could preserve CNCs in the composite aerogels, the mechanical properties of PI may be compromised [46].

3.3. Internal Morphology of the PI/CNC Nanocomposite Aerogels

We further analyzed the internal morphologies of the aerogels prepared using different concentrations of CNCs. Figure 5a,c shows that the PI/CNC aerogels prepared by 6.5 wt% CNCs had a lamellar structure. This morphology was consistent with the previous work studying CNC aerogels prepared under a slow freezing rate [47]. The paralleled pores came from the anisotropic growth of ice crystals along the freezing front. The pure PI aerogels generally showed a smaller pore distribution, but the lamellar morphology was less pronounced. In addition, the morphology of the aerogels using different concentrations of CNCs was similar (Figure S6), indicating that the presence of CNCs was sufficient to form the lamellar structure. We also examined the morphology of the freeze-dried pure PAAS and pure CNC aerogels (Figure S7). The PAAS aerogels also contained some parallel sheet-like structures, indicating that anisotropic growth of ice crystals still occurred. However, such structures generally disappeared after thermal imidization, potentially because of the shrinkage of the aerogels. On the other hand, the freeze-dried CNC aerogels formed pores much larger than the composite aerogels. This result was expected as a low concentration of CNCs was not enough to form strongly packed networks. The pure CNCs aerogel was also too fragile for further studies. The results of the microstructures further support our rheological analysis. The physical network of CNCs was facilitated by the presence of PAAS in the inks. Although the pure PI aerogels contained smaller and denser pores than the composite aerogels, the lamellar structures induced by CNCs may further enhance the thermal insulation performance of the aerogels. Some studies have attempted to control the growth of ice crystals to modulate the thermal insulation of aerogels in different directions [6,9]. We are currently investigating the possibility of integrating 3D printing and directional freezing.

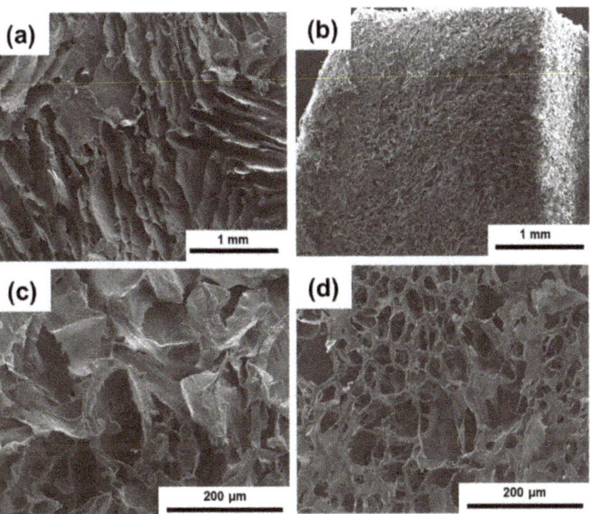

Figure 5. Cross-sectional SEM images of the aerogels after the thermal imidization process. (**a,c**) The PI/CNC composite aerogel using 6.5 wt% CNCs. (**b,d**) Pure PI aerogel.

As implied in the internal morphologies, CNCs should significantly reduce the shrinkage of the aerogels during preparation. The degree of shrinkage is summarized in Figure 6a. As the concentration of CNCs increased, the volumetric shrinkage generally decreased, suggesting that the aerogels had higher shape-fidelity after imidization. The PI/CNC composite aerogels only had about a 10% decrease in volume after freeze-drying. Further imidization reduced the volume of the composite aerogels, but they still retained more than 70% of the initial volume.

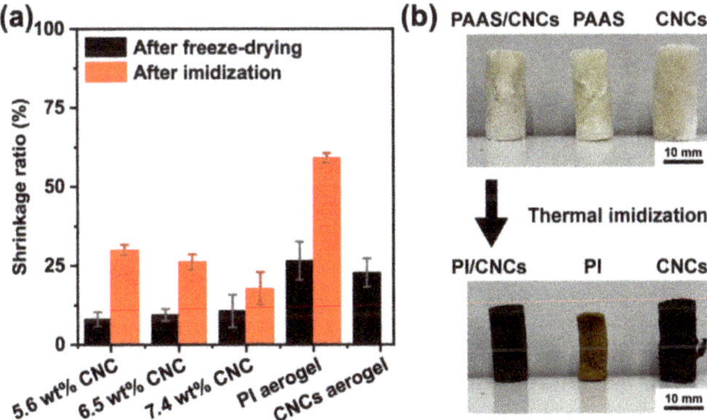

Figure 6. (**a**) Volumetric shrinkage ratio of aerogels after freeze-drying and after thermal imidization. (**b**) The appearance of aerogels before and after thermal imidization.

In contrast to the composite aerogels, the pure PI aerogels had a much higher shrinkage ratio, up to 60%. These results demonstrate that CNCs could strengthen the polymer matrices during freeze-drying and thermal imidization. As shown in Figure 6b, the pure PI aerogel shrank and wrapped significantly after thermal imidization. In contrast, the composite aerogel and the pure CNC aerogel only shrank slightly. However, the color of the pure CNC aerogels became darker, indicating that some CNCs were thermally degraded. Figure S8 shows the result of the bulk density. The bulk density of the composite aerogels increased with the concentration in CNCs because of the increased solid content. The severe shrinkage of the pure PI aerogel also led to a high bulk density.

3.4. Performance of the PI/CNC Composite Aerogels

The mechanical properties of the PI/CNC composite aerogels were studied, as shown in Figure 7. All samples followed the typical compression stress–strain curves of aerogels including an elastic zone, a platform zone, and a dense zone [48]. The pure PI aerogel had the lowest compression modulus. On the other hand, the compression modulus and maximum stress increased with the concentrations of CNCs. In general, the modulus of aerogels increases with their density [49]. As mentioned earlier, the pure PI aerogels generally had smaller and denser pores than the composite aerogels because of the shrinkage during thermal imidization. As a result, the density of the PI aerogel was similar to the composite aerogels. Under a similar bulk density, CNCs greatly reinforced the aerogels during compression. Although the pure PI aerogels contained dense and small pores that are generally desirable for strong aerogels, the effect of CNCs was more significant. As shown in Figure 7b, our PI/CNCs demonstrated a high compression modulus ranging from 6.40 to 13.64 MPa. The mechanical stiffness was higher than the previous works using a lower concentration of nanofillers [6,8,48]. Therefore, a sufficient concentration of CNCs provides both 3D printability and high mechanical strength.

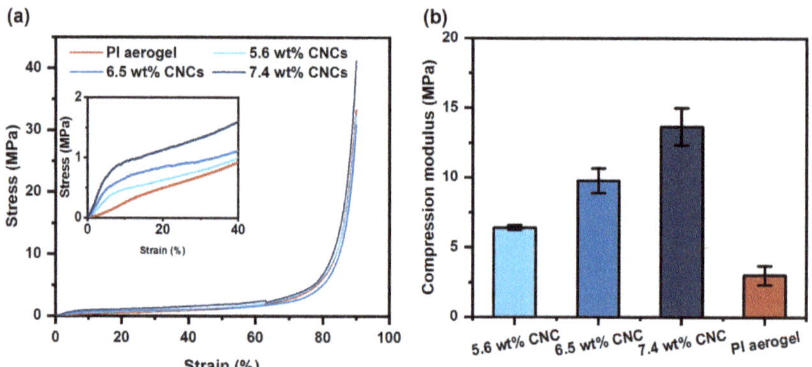

Figure 7. Mechanical properties of the aerogels using different concentrations of CNCs. (**a**) Compressive stress–strain curves. (**b**) Comparison of compressive modulus.

Finally, we compared the thermal insulation ability of the aerogels. Figure 8a shows a series of thermal images for pure PI, pure CNCs, and three composite aerogels with a cuboid shape. These aerogels were placed on a hot plate at 160 °C. After 10 min, we found that all samples retained a temperature below 80 °C, suggesting they all had low thermal conductivity. The concentration of CNCs appeared to have minimal effects on the thermal conductivity of the aerogels based on the thermal images. Indeed, the thermal conductivities of the composite aerogels with 5.6 wt% and 6.5 wt% CNCs were 89.39 ± 0.087 and 92.27 ± 1.03 mW/m·K, respectively. This result was comparable to the thermal conductivities of the PI aerogels prepared by the supercritical CO_2 drying [50] and bidirectional freezing [6]. We also attempted to measure the thermal conductivity of the pure PI, pure CNCs, and the composite aerogels with 7.4 wt% CNCs. However, the accurate measurement could not be achieved because these samples typically had low mechanical integrity or high surface roughness.

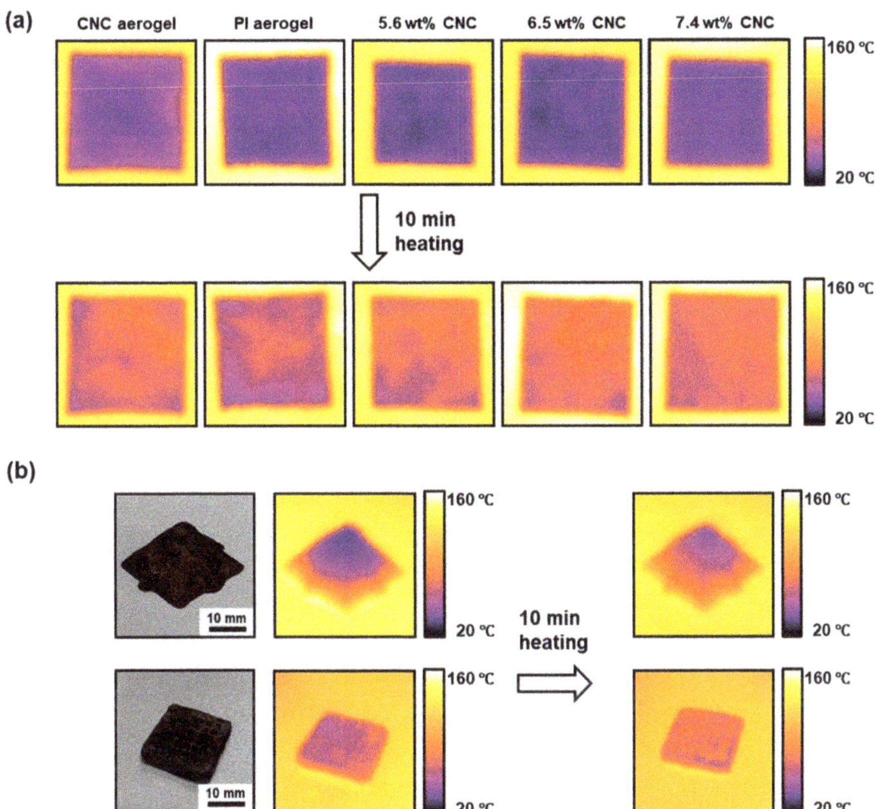

Figure 8. The thermal insulation performance of aerogels. (**a**) Infrared images of pure CNC aerogel, pure PI aerogel, and PI/CNC aerogels with different concentrations of CNCs. The sample size was $2.8 \times 2.8 \times 0.4$ cm^3. (**b**) 3D printed pyramid and grid of PI/CNC aerogels.

In addition to common cuboid structures, we also placed a 3D printed pyramid on the hotplate (Figure 8b). The pyramid was 1 cm in height with 30 printed layers, indicating the robust printability of inks. We found that there was a gradient distribution of temperature from the hot plate to the top of the printed pyramid. Therefore, the 3D printed sample still had good thermal insulation ability, as expected. A grid with a finer surface structure than the mesh in Figure 3 was also printed for the thermal image analysis. As shown in Figure 8b, the temperature distribution of the grid depended on the macrostructure enabled by 3D printing. Therefore, 3D printing could provide an opportunity to locally control the temperature distribution and modulate the overall thermal conductivity of the aerogels. Although some effective designs could also be fabricated by molding, 3D printing is a rapid test of different macrostructures. The combination of thermal insulation capability and 3D printability of our PI/CNC composite aerogels may enable the easy production of a customizable design for advanced applications.

Our results indicate that CNCs can be used as renewable fillers to 3D print PI/CNC composite aerogels with enhanced mechanical strength. Although our PAAS/CNC ink was printed under room temperature, several previous works preheated the inks to reduce the viscosity of the ink, facilitate material extrusion, and extend the printable time [21,23,51]. The printing speed and resolution may be further improved by selecting a suitable heating parameter. Further work is being conducted to understand the effects of temperature and heating time on the rheological behavior of the PAAS/CNCs inks. In addition, DIW can be used to align CNCs by controlling the shear stress during printing [29,34]. The aligned CNCs in the composites exhibited anisotropic mechanical response and optical properties. The alignment of nanomaterials can also be achieved by high-resolution aerosol jet printing [52]. Therefore, the mechanical stiffness and thermal insulating properties of the PI/CNC aerogels may also be controlled by the printing process.

The low degradation temperature of CNCs limits the overall thermal stability of the aerogels compared to PI. The thermal resistance of the PI/CNC aerogels may be improved by the surface modification of CNCs [53,54]. In addition, the thermal insulation capability of the PI/CNC aerogels was affected by the distribution and the direction of pores [6]. Although our work only showed the 3D printability of the PI/CNC composite aerogels, we envision that a further combination of 3D printing with other pore-forming strategies may produce robust PI-based aerogels with enhanced thermal regulation performance.

4. Conclusions

In this study, we utilized CNCs as renewable nanofillers to prepare water-based inks for the 3D printing of PI composite aerogels. Freeze-drying and thermal imidization of the printed objects produced porous PI/CNC composite aerogels. The potential depletion effect by PAAS promoted the gelation of CNCs and reduced the required concentration of CNCs for DIW. In addition, the strong physical networks of the CNCs increased the thermal stability of the porous structures and reduced the shrinkages of the aerogels after thermal imidization. Furthermore, CNCs reinforced the PI/CNC composite aerogels to provide high mechanical strength. Our composite aerogels exhibited low thermal conductivities for thermal regulation. Overall, our work demonstrated a simple way to print PI aerogels with customizable shapes, low density, and high mechanical strength. The 3D printing of high-performance composite aerogels may further enable advanced and rapid design of thermal insulating materials to precisely control the local temperature distribution of the target device.

- High-resolution printing by DIW can be achieved by using the ink containing only 6.5 wt% CNCs.
- Chemical analysis showed that a significant amount of CNCs remained in the composite after thermal imidization, and CNCs displayed improved thermal stability.
- CNCs significantly reduce the dimensional shrinkage of the aerogels during preparation and enable the 3D printing of aerogels with high shape fidelity.
- The PI/CNC composite aerogels demonstrated increased mechanical stiffness and sufficient thermal insulating capability.

Supplementary Materials: The following are available online at https://www.mdpi.com/article/10.3390/polym13213614/s1, Figure S1: The experimental procedure of the 3D printed PI/CNC composite aerogels; Figure S2: Rheological analysis of the inks under different concentrations of polyamic acid; Figure S3: A 3D printed cylinder that shows the thickness achievable by the PAAS/CNC ink; Figure S4: Full FTIR spectra of the aerogels after thermal imidization; Figure S5: Weight loss of the freeze-dried samples after thermal imidization; Figure S6: Cross-sectional SEM images of the composite aerogels after the thermal imidization process; Figure S7: Cross-sectional SEM images of pure PAAS aerogels and pure CNC aerogels after freeze-drying; Figure S8: Bulk densities of PI/CNC composite aerogels, CNC aerogels, and PI aerogels.

Author Contributions: Data curation, C.F.; Formal analysis, C.F. and S.-S.Y.; Funding acquisition, S.-S.Y.; Investigation, S.-S.Y.; Methodology, C.F. and S.-S.Y.; Project administration, S.-S.Y.; Visualization, S.-S.Y.; Writing—original draft, S.-S.Y.; Writing—review & editing, S.-S.Y. All authors have read and agreed to the published version of the manuscript.

Funding: This research was funded by the Ministry of Science & Technology in Taiwan, grant No. 108-2636-E-006-005, 109-2636-E-006-005 and 110-2222-E-006-009-MY3.

Institutional Review Board Statement: Not applicable.

Informed Consent Statement: Not applicable.

Data Availability Statement: The data presented in this study are available on request from the corresponding author.

Acknowledgments: We are grateful for the valuable discussion and inputs from Mitsuru Ueda. We thank the Ministry of Science & Technology in Taiwan for financial support (Young Scholar Fellowship Program, MOST 108-2636-E-006-005 and 109-2636-E-006-005). We also thank the support by Higher Education Sprout Project, Ministry of Education to the Headquarters of University Advancement at National Cheng Kung University (NCKU). The authors gratefully acknowledge the use of Bruker Avance III HD 400 NMR spectrometer (NMR000800) of MOST 108-2731-M-006-001 belonging to the Core Facility Center of National Cheng Kung University.

Conflicts of Interest: The authors declare no conflict of interest.

References

1. Shi, B.; Ma, B.; Wang, C.; He, H.; Qu, L.; Xu, B.; Chen, Y. Fabrication and applications of polyimide nano-aerogels. *Compos. Part A Appl. Sci. Manuf.* **2021**, *143*, 106283. [CrossRef]
2. Feng, J.; Su, B.-L.; Xia, H.; Zhao, S.; Gao, C.; Wang, L.; Ogbeide, O.; Feng, J.; Hasan, T. Printed aerogels: Chemistry, processing, and applications. *Chem. Soc. Rev.* **2021**, *50*, 3842–3888. [CrossRef] [PubMed]
3. Liaw, D.-J.; Wang, K.-L.; Huang, Y.-C.; Lee, K.-R.; Lai, J.-Y.; Ha, C.-S. Advanced polyimide materials: Syntheses, physical properties and applications. *Prog. Polym. Sci.* **2012**, *37*, 907–974. [CrossRef]
4. Ni, H.-J.; Liu, J.-G.; Wang, Z.-H.; Yang, S.-Y. A review on colorless and optically transparent polyimide films: Chemistry, process and engineering applications. *J. Ind. Eng. Chem.* **2015**, *28*, 16–27. [CrossRef]
5. Gu, W.; Wang, G.; Zhou, M.; Zhang, T.; Ji, G. Polyimide-based foams: Fabrication and multifunctional applications. *ACS Appl. Mater. Interfaces* **2020**, *12*, 48246–48258. [CrossRef] [PubMed]
6. Zhang, X.; Zhao, X.; Xue, T.; Yang, F.; Fan, W.; Liu, T. Bidirectional anisotropic polyimide/bacterial cellulose aerogels by freeze-drying for super-thermal insulation. *Chem. Eng. J.* **2020**, *385*, 123963. [CrossRef]
7. Wang, Y.; Cui, Y.; Shao, Z.; Gao, W.; Fan, W.; Liu, T.; Bai, H. Multifunctional polyimide aerogel textile inspired by polar bear hair for thermoregulation in extreme environments. *Chem. Eng. J.* **2020**, *390*, 124623. [CrossRef]
8. Wang, Y.-Y.; Zhou, Z.-H.; Zhou, C.-G.; Sun, W.-J.; Gao, J.-F.; Dai, K.; Yan, D.-X.; Li, Z.-M. Lightweight and robust carbon nanotube/polyimide foam for efficient and heat-resistant electromagnetic interference shielding and microwave absorption. *ACS Appl. Mater. Interfaces* **2020**, *12*, 8704–8712. [CrossRef]
9. Cheng, Y.; Zhang, X.; Qin, Y.; Dong, P.; Yao, W.; Matz, J.; Ajayan, P.M.; Shen, J.; Ye, M. Super-elasticity at 4 K of covalently crosslinked polyimide aerogels with negative poisson's ratio. *Nat. Commun.* **2021**, *12*, 4092. [CrossRef]
10. Zhao, X.; Wang, W.; Wang, Z.; Wang, J.; Huang, T.; Dong, J.; Zhang, Q. Flexible PEDOT:PSS/polyimide aerogels with linearly responsive and stable properties for piezoresistive sensor applications. *Chem. Eng. J.* **2020**, *395*, 125115. [CrossRef]
11. Zheng, Q.; Fang, L.; Guo, H.; Yang, K.; Cai, Z.; Meador, M.A.B.; Gong, S. Highly porous polymer aerogel film-based triboelectric nanogenerators. *Adv. Funct. Mater.* **2018**, *28*, 1706365. [CrossRef]
12. Ligon, S.C.; Liska, R.; Stampfl, J.; Gurr, M.; Mülhaupt, R. Polymers for 3D printing and customized additive manufacturing. *Chem. Rev.* **2017**, *117*, 10212–10290. [CrossRef] [PubMed]
13. Truby, R.L.; Lewis, J.A. Printing soft matter in three dimensions. *Nature* **2016**, *540*, 371–378. [CrossRef] [PubMed]
14. Lewis, J.A. Direct ink writing of 3D functional materials. *Adv. Funct. Mater.* **2006**, *16*, 2193–2204. [CrossRef]
15. Peng, E.; Zhang, D.; Ding, J. Ceramic robocasting: Recent achievements, potential, and future developments. *Adv. Mater.* **2018**, *30*, 1802404. [CrossRef] [PubMed]
16. Hegde, M.; Meenakshisundaram, V.; Chartrain, N.; Sekhar, S.; Tafti, D.; Williams, C.B.; Long, T.E. 3D printing all-aromatic polyimides using mask-projection stereolithography: Processing the nonprocessable. *Adv. Mater.* **2017**, *29*, 1701240. [CrossRef] [PubMed]
17. Guo, Y.; Xu, J.; Yan, C.; Chen, Y.; Zhang, X.; Jia, X.; Liu, Y.; Wang, X.; Zhou, F. Direct ink writing of high performance architectured polyimides with low dimensional shrinkage. *Adv. Eng. Mater.* **2019**, *21*, 1801314. [CrossRef]
18. Guo, Y.; Ji, Z.; Zhang, Y.; Wang, X.; Zhou, F. Solvent-free and photocurable polyimide inks for 3D printing. *J. Mater. Chem. A* **2017**, *5*, 16307–16314. [CrossRef]
19. Herzberger, J.; Meenakshisundaram, V.; Williams, C.B.; Long, T.E. 3D printing all-aromatic polyimides using stereolithographic 3D printing of polyamic acid salts. *ACS Macro Lett.* **2018**, *7*, 493–497. [CrossRef]
20. Rau, D.A.; Herzberger, J.; Long, T.E.; Williams, C.B. Ultraviolet-assisted direct ink write to additively manufacture all-aromatic polyimides. *ACS Appl. Mater. Interfaces* **2018**, *10*, 34828–34833. [CrossRef]
21. Wang, C.; Ma, S.; Li, D.; Zhao, J.; Zhou, H.; Wang, D.; Zhou, D.; Gan, T.; Wang, D.; Liu, C.; et al. 3D printing of lightweight polyimide honeycombs with the high specific strength and temperature resistance. *ACS Appl. Mater. Interfaces* **2021**, *13*, 15690–15700. [CrossRef]
22. Zhang, L.; Wu, J.; Sun, N.; Zhang, X.; Jiang, L. A novel self-healing poly(amic acid) ammonium salt hydrogel with temperature-responsivity and robust mechanical properties. *J. Mater. Chem. A* **2014**, *2*, 7666–7668. [CrossRef]

23. Qin, S.; Jiang, Y.; Ji, Z.; Yang, C.; Guo, Y.; Zhang, X.; Qin, H.; Jia, X.; Wang, X. Three-dimensional printing of high-performance polyimide by direct ink writing of hydrogel precursor. *J. Appl. Polym. Sci.* **2021**, *138*, 50636. [CrossRef]
24. Yang, C.; Jiang, P.; Qin, H.; Wang, X.; Wang, Q. 3D printing of porous polyimide for high-performance oil impregnated self-lubricating. *Tribol. Int.* **2021**, *160*, 107009. [CrossRef]
25. Mariano, M.; El Kissi, N.; Dufresne, A. Cellulose nanocrystals and related nanocomposites: Review of some properties and challenges. *J. Polym. Sci. Part B Polym. Phys.* **2014**, *52*, 791–806. [CrossRef]
26. Frka-Petesic, B.; Guidetti, G.; Kamita, G.; Vignolini, S. Controlling the photonic properties of cholesteric cellulose nanocrystal films with magnets. *Adv. Mater.* **2017**, *29*, 1701469. [CrossRef]
27. Nguyen, B.N.; Cudjoe, E.; Douglas, A.; Scheiman, D.; McCorkle, L.; Meador, M.A.B.; Rowan, S.J. Polyimide cellulose nanocrystal composite aerogels. *Macromolecules* **2016**, *49*, 1692–1703. [CrossRef]
28. Lee, H.-G.; Kim, G.-H.; Ha, C.-S. Polyimide/amine-functionalized cellulose nanocrystal nanocomposite films. *Mater. Today Commun.* **2017**, *13*, 275–281. [CrossRef]
29. Siqueira, G.; Kokkinis, D.; Libanori, R.; Hausmann, M.K.; Gladman, A.S.; Neels, A.; Tingaut, P.; Zimmermann, T.; Lewis, J.A.; Studart, A.R. Cellulose nanocrystal inks for 3D printing of textured cellular architectures. *Adv. Funct. Mater.* **2017**, *27*, 1604619. [CrossRef]
30. Lai, C.-W.; Yu, S.-S. 3D printable strain sensors from deep eutectic solvents and cellulose nanocrystals. *ACS Appl. Mater. Interfaces* **2020**, *12*, 34235–34244. [CrossRef] [PubMed]
31. Heath, L.; Thielemans, W. Cellulose nanowhisker aerogels. *Green Chem.* **2010**, *12*, 1448–1453. [CrossRef]
32. Yang, J.; Lee, M.-H. A water-soluble polyimide precursor: Synthesis and characterization of poly(amic acid) salt. *Macromol. Res.* **2004**, *12*, 263–268. [CrossRef]
33. Smay, J.E.; Cesarano, J.; Lewis, J.A. Colloidal inks for directed assembly of 3-D periodic structures. *Langmuir* **2002**, *18*, 5429–5437. [CrossRef]
34. Hausmann, M.K.; Rühs, P.A.; Siqueira, G.; Läuger, J.; Libanori, R.; Zimmermann, T.; Studart, A.R. Dynamics of cellulose nanocrystal alignment during 3D printing. *ACS Nano* **2018**, *12*, 6926–6937. [CrossRef]
35. Lai, P.-C.; Yu, S.-S. Cationic cellulose nanocrystals-based nanocomposite hydrogels: Achieving 3D printable capacitive sensors with high transparency and mechanical strength. *Polymers* **2021**, *13*, 688. [CrossRef]
36. Oguzlu, H.; Dobyrden, I.; Liu, X.; Bhaduri, S.; Claesson, P.M.; Boluk, Y. Polymer induced gelation of aqueous suspensions of cellulose nanocrystals. *Langmuir* **2021**, *37*, 3015–3024. [CrossRef]
37. Oguzlu, H.; Boluk, Y. Interactions between cellulose nanocrystals and anionic and neutral polymers in aqueous solutions. *Cellulose* **2017**, *24*, 131–146. [CrossRef]
38. Oguzlu, H.; Danumah, C.; Boluk, Y. Colloidal behavior of aqueous cellulose nanocrystal suspensions. *Curr. Opin. Colloid Interface Sci.* **2017**, *29*, 46–56. [CrossRef]
39. Laurati, M.; Egelhaaf, S.U.; Petekidis, G. Nonlinear rheology of colloidal gels with intermediate volume fraction. *J. Rheol.* **2011**, *55*, 673–706. [CrossRef]
40. Liu, F.; Liu, Z.; Gao, S.; You, Q.; Zou, L.; Chen, J.; Liu, J.; Liu, X. Polyimide film with low thermal expansion and high transparency by self-enhancement of polyimide/sic nanofibers net. *RSC Adv.* **2018**, *8*, 19034–19040. [CrossRef]
41. Li, M.-C.; Mei, C.; Xu, X.; Lee, S.; Wu, Q. Cationic surface modification of cellulose nanocrystals: Toward tailoring dispersion and interface in carboxymethyl cellulose films. *Polymer* **2016**, *107*, 200–210. [CrossRef]
42. Rahnamoun, A.; Engelhart, D.P.; Humagain, S.; Koerner, H.; Plis, E.; Kennedy, W.J.; Cooper, R.; Greenbaum, S.G.; Hoffmann, R.; van Duin, A.C.T. Chemical dynamics characteristics of kapton polyimide damaged by electron beam irradiation. *Polymer* **2019**, *176*, 135–145. [CrossRef]
43. Waters, J.F.; Likavec, W.R.; Ritchey, W.M. 13C CP–MAS NMR study of absorbed water in polyimide film. *J. Appl. Polym. Sci.* **1994**, *53*, 59–70. [CrossRef]
44. Qiu, W.; Leisen, J.E.; Liu, Z.; Quan, W.; Koros, W.J. Key features of polyimide-derived carbon molecular sieves. *Angew. Chem. Int. Ed.* **2021**, *60*, 22322–22331. [CrossRef]
45. Cui, Y.; Jin, R.; Zhang, Y.; Yu, M.; Zhou, Y.; Wang, L.-Q. Cellulose nanocrystal-enhanced thermal-sensitive hydrogels of block copolymers for 3D bioprinting. *Int. J. Bioprinting* **2021**, *7*, 397. [CrossRef]
46. Chen, W.; Chen, W.; Zhang, B.; Yang, S.; Liu, C.-Y. Thermal imidization process of polyimide film: Interplay between solvent evaporation and imidization. *Polymer* **2017**, *109*, 205–215. [CrossRef]
47. Li, V.C.-F.; Dunn, C.K.; Zhang, Z.; Deng, Y.; Qi, H.J. Direct ink write (DIW) 3D printed cellulose nanocrystal aerogel structures. *Sci. Rep.* **2017**, *7*, 8018. [CrossRef] [PubMed]
48. Yu, Z.; Dai, T.; Yuan, S.; Zou, H.; Liu, P. Electromagnetic interference shielding performance of anisotropic polyimide/graphene composite aerogels. *ACS Appl. Mater. Interfaces* **2020**, *12*, 30990–31001. [CrossRef] [PubMed]
49. Guo, H.; Meador, M.A.B.; McCorkle, L.; Quade, D.J.; Guo, J.; Hamilton, B.; Cakmak, M. Tailoring properties of cross-linked polyimide aerogels for better moisture resistance, flexibility, and strength. *ACS Appl. Mater. Interfaces* **2012**, *4*, 5422–5429. [CrossRef]
50. Feng, J.; Wang, X.; Jiang, Y.; Du, D.; Feng, J. Study on thermal conductivities of aromatic polyimide aerogels. *ACS Appl. Mater. Interfaces* **2016**, *8*, 12992–12996. [CrossRef] [PubMed]

51. Tan, J.J.Y.; Lee, C.P.; Hashimoto, M. Preheating of gelatin improves its printability with transglutaminase in direct ink writing 3D printing. *Int. J. Bioprint.* **2020**, *6*, 296. [CrossRef] [PubMed]
52. Goh, G.L.; Agarwala, S.; Yeong, W.Y. Aerosol-jet-printed preferentially aligned carbon nanotube twin-lines for printed electronics. *ACS Appl. Mater. Interfaces* **2019**, *11*, 43719–43730. [CrossRef] [PubMed]
53. Ben Azouz, K.; Ramires, E.C.; Van den Fonteyne, W.; El Kissi, N.; Dufresne, A. Simple method for the melt extrusion of a cellulose nanocrystal reinforced hydrophobic polymer. *ACS Macro Lett.* **2012**, *1*, 236–240. [CrossRef]
54. Nagalakshmaiah, M.; El Kissi, N.; Dufresne, A. Ionic compatibilization of cellulose nanocrystals with quaternary ammonium salt and their melt extrusion with polypropylene. *ACS Appl. Mater. Interfaces* **2016**, *8*, 8755–8764. [CrossRef]

 *polymers*

Article

Representative Volume Element (RVE) Analysis for Mechanical Characterization of Fused Deposition Modeled Components

Patrich Ferretti, Gian Maria Santi, Christian Leon-Cardenas *, Elena Fusari, Giampiero Donnici and Leonardo Frizziero *

Department of Industrial Engineering, Alma Mater Studiorum, University of Bologna, I-40136 Bologna, Italy; patrich.ferretti2@unibo.it (P.F.); gianmaria.santi2@unibo.it (G.M.S.); elena.fusari2@studio.unibo.it (E.F.); giampiero.donnici@unibo.it (G.D.)
* Correspondence: christian.leon2@unibo.it (C.L.-C.); leonardo.frizziero@unibo.it (L.F.)

Abstract: Additive manufacturing processes have evolved considerably in the past years, growing into a wide range of products through the use of different materials depending on its application sectors. Nevertheless, the fused deposition modelling (FDM) technique has proven to be an economically feasible process turning additive manufacture technologies from consumer production into a mainstream manufacturing technique. Current advances in the finite element method (FEM) and the computer-aided engineering (CAE) technology are unable to study three-dimensional (3D) printed models, since the final result is highly dependent on processing and environment parameters. Because of that, an in-depth understanding of the printed geometrical mesostructure is needed to extend FEM applications. This study aims to generate a homogeneous structural element that accurately represents the behavior of FDM-processed materials, by means of a representative volume element (RVE). The homogenization summarizes the main mechanical characteristics of the actual 3D printed structure, opening new analysis and optimization procedures. Moreover, the linear RVE results can be used to further analyze the in-deep behavior of the FDM unit cell. Therefore, industries could perform a feasible engineering analysis of the final printed elements, allowing the FDM technology to become a mainstream, low-cost manufacturing process in the near future.

Keywords: FEM; FDM; additive manufacturing; microstructure behavior; linear analysis; RVE

Citation: Ferretti, P.; Santi, G.M.; Leon-Cardenas, C.; Fusari, E.; Donnici, G.; Frizziero, L. Representative Volume Element (RVE) Analysis for Mechanical Characterization of Fused Deposition Modeled Components. *Polymers* **2021**, *13*, 3555. https://doi.org/10.3390/polym13203555

Academic Editors: Houwen Matthew Pan and Eric Luis

Received: 16 July 2021
Accepted: 13 October 2021
Published: 15 October 2021

Publisher's Note: MDPI stays neutral with regard to jurisdictional claims in published maps and institutional affiliations.

1. Introduction

The process of three-dimensional (3D) printing, known as additive manufacturing (AM) has achieved an unexpected evolution. The ability of producing most kinds of complex, irregularly shaped geometries is an asset for this technology. Moreover, the rapid increase in design software makes 3D printing ideal for manufacturing custom components impossible to be produced at the industrial level by using standard processes. Nowadays, many AM technologies for polymers offer high levels of material and aesthetics quality such as stereolithography (SLA), selective laser sintering (SLS), digital light processing (DLP), and ink-deposition technologies including the Polyjet (patented by Stratasys [1–5]. In contrast, these methodologies may result expensive due to the uniqueness of each process, materials availability, and the need of a more expensive, specialized equipment [6].

Fortunately, the fused deposition modelling (FDM) technology have evolved substantially over the last years, with the arrival of a wide range of filaments and materials, leading to higher manufacturability and aesthetic quality. In addition, subsequent advances on printing machines have led to multi-material deposition and finer surface quality. Software also offers a wide range of possibilities from open sources to commercial solutions underlying the interest in G-code optimization. To catch the opportunity of making FDM available for industrial applications, it is necessary to deeply understand and predict the behavior of printed components. The derived mechanical tensile response and the elongation percentage of the part would help designers to close an important gap to product industrialization. The studies currently available in the literature present limited results or focus on a single component and are difficult to be extended, such as the study of Somireddy et al. that considered the behavior of a foil section [7]. Although the proposed model is built on an image of a printed section, it does not allow reducing the size of voids and therefore a variable model was not obtained. Furthermore, the tests conducted used beam elements, but the number of elements was enormously high, and it was a difficult approach to apply on large components. The model presented by Bhandari et al. is very interesting, because it divided the model into two well-defined areas, i.e., the infill and the contour lines [8]. The infill is schematized as a series of beams, and the contour act as shell elements. Again, it is difficult to extend the model to complex components, but it is very interesting from a mathematical point of view. The study of Garg and Bhattacharya proposes a methodology that faithfully reproduces the layer and line deposition typical of the FDM-printed model [9]. However, it is extremely computationally demanding to mesh each individual line with enough elements. Following this approach, even the simplest geometry requires extremely high computing power and time. This proposal builds a model based on the image analysis of the mesostructure of the molded component. In this way, the representative volume element (RVE) model is not fixed in its dimensions, but every time the material or printing parameters are changed based on the image analysis. Thanks to the RVE model, it is then possible to use macroelements that summarize the mesoscopic properties of the component and allow complex components to be studied with relative ease.

1.1. FDM Part Micromechanics

It is known that AM is a technology in which its understanding relates to the high variability of the process and environment characteristics. This variability can be translated into correct parameter definition issues, which otherwise would lead to an internal anisotropy, making this type of resulting material very complex to analyze. Extruded material irregularities due to FDM processes are discussed previously by Lee et al. [10] and Kotlinski et al. [11], which agree that anisotropy leads to limitations in obtaining feasible prototype properties. Currently, only a few studies currently have started to analyze the internal behavior of FDM-produced parts. An FEM modelling of the mesostructure of FDM-printed parts was discussed by Somireddy et al. [12]. The FEM analysis was used to find the elastic modulus of a single printed layer of a unidirectional (UD), polymer-extruded material. The authors have laid down an understanding in order to gather an FEM analysis that took into account the anisotropic behavior of the 3D printed part. Additionally, the study of Bhandari and Lopez-Anido underlies a distinctive, rather interesting approach to material anisotropy [8], as the FEM analysis was performed in a lattice-like internal structure, with an about 20% difference in Poisson's ratios with respect to the test values.

Nevertheless, the accurate analysis of the mesostructure of the printed material is therefore needed to gather an FEM methodology that could allow the analysis of whole structures manufactured by FDM. By understanding the variability characteristics of the printed structure in a single, homogeneous element helps to further study the part's macroscopical behavior. The analysis of composites structures as an unique element in FEM could assess accurately changes by any of their components, being able to differentiate different resins analogous to those in the study by Croccolo et al. [13]. A second study by Somireddy et al. established a numerical homogenization procedure for a more efficient material modeling of the printed parts [14]. This model claimed to gather the influence of printing variables such as build orientation, printing direction, and layer thickness. A layer deposition influence is noted as well, and it has been stated that the material responses to different parts of the modelled structure are dependent on the build orientations and thicknesses of the parts.

1.2. RVE FEM Analysis

Computational analysis tools have given engineers the capacity of create new methodologies able to enhance the understanding of physical phenomena in modern, multimaterial, anisotropic problems. The state-of-the-art developments of complex structures have evolved thanks to the introduction of composite materials. Nowadays, some analysis of yield and failure theories in polymer–matrix composites (PMCs) are possible [15]. This is due to a unified theory that can be used to predict their nonlinearity and strength by also considering the anisotropy and tension-compression asymmetry simultaneously. Another one-parameter yield function was proposed earlier by Sun and Chen to establish a nonlinear plastic model for UD PMCs [16,17]. Additionally, Mellinger et al. described polymer–air composites, and they showed that they are elastically softer due to the air content and in relation to the size and shape of the polymer walls [18]. This behavior defines foam structures as described by Ashby in [19]. Because of these considerations, the influence of air is relevant with respect to its volume, and it should be considered for FDM-made component simulations. Using FEM software is indeed possible to calculate mechanical behaviors of anisotropic materials, i.e., composites, using a micromechanical RVE [20]. RVE is the smallest volume in which a measurement can be made in order to homogenize the entire domain. In the studied case, a periodic unit cell is a simpler choice. ANSYS software elaborates the area of analysis by means of the material cross-section geometry. The RVE study of Bhaskara et al. started from periodic boundary conditions, applied to the RVE to calculate elastic modulus of composite elements, which are characterized as an anisotropic behavior [21]. Most micromechanical models are applied to a composition of fibers immersed in a matrix, so that an RVE or unit cell can be insulated. This methodology is used to study composite materials (e.g., UD carbon fibers in an epoxy matrix) and anisotropic materials in general. Because of the similarity between long-fiber RVE and FDM deposition, the authors choose this approach to analyze the resulting mesostructure of a 3D printed component. Printed lines are similar to UD carbon fibers composites. The voids between lines resemble the structure of porous materials such as foams. FDM-printed parts are in fact made of air and polymeric materials and can be considered "composites".

The RVE methodology has proven to show good results for the analysis of anisotropic materials.

2. Materials and Methods

2.1. Microstructure Definition

It is necessary to define the minimum volume to be analyzed by means of the RVE. Because of the previous literature findings and the study performed by Grimal et al. [22], it is possible to state that the structure of a 3D printed material can be defined as periodic. The aim of this research was to create a methodology to study the behavior of 3D printed components. This model would start from a linear, discrete model and then be extended to a model that can consider diverse material and process variabilities that deliver such nonlinearities to polymeric materials due to the 3D printing process. Furthermore, to guarantee the continuity between both models, it was decided to use a unit cell with an actual size slightly greater than that of a single line. However, Section 3.1 highlights this aspect in the linear field. Considering overlapping lines revealed no loss or addition of information by varying the size of the RVE.

2.2. Model Considerations

- The material used to validate this model was polylactic acid (PLA). The average material values used for the Young's modulus, Poisson's ratio, and density came from specimens made through injection molding [23–25].
- The stress–strain behavior of the material was approximated as linear. This allowed simplifying the problem but limits the validity of the model to small deformations, while generally the behavior of polymers is nonlinear.

- The model does not take into consideration that there is a preferential orientation of the polymer chains along the direction of extrusion during the 3D printing process. [26]. The value of the mesoscopic geometry plays a key role, given the fact that such geometry defines the anisotropy of the structure. Different geometry schematizations led to different results from both a numerical and a physical point of view.
- The condition of "perfect bonding" was set between various layers, so the assumptions that the adhesion is perfect at the interface between one layer and the next and between one line and the adjacent one was taken. This hypothesis allowed keeping the model linear, avoiding problems of nonliner contact behavior. Due to this hypothesis, the results reported higher values for E_{22} and E_{33} (Young's moduli along directions perpendicular to the fiber direction) and the shear modules, according to the reference system, as seen in Figure 1. However, E_{11} (Young's modulus along the "fiber" direction) should remain unchanged because it is not affected by the layers' adhesion.

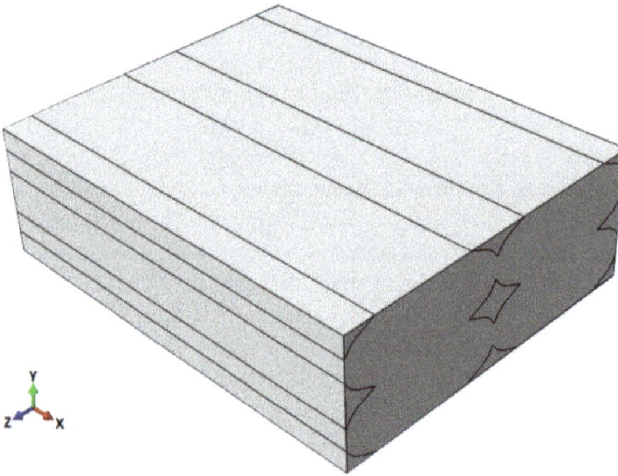

Figure 1. The representative volume element (RVE) and the reference system.

However, the possibility of an optimization of the microstructure in order to reduce voids, as previously presented in [27], was not taken into consideration. This is due to the hypotheses considered in this study beforehand that established a perfect bonding with no lines or layer boundaries. An optimization process will lead to no appreciable result.

The specimens were made with an E3D Tool-changer 3D printer, and the toolpath (G-Code) was obtained using Cura 4.9.1 (Ultrecht, The Netherlands) free slicing software.

The main printing parameters are reported in the Table 1; it should be noted that different printers, with the same type of material but from a different supplier (brand), would give different results in terms of microstructure. An image analysis on the specimen failure section is always necessary to evaluate the obtained geometry and therefore to remodel the RVE accordingly. To sum up, different volume-area ratios of voids produce different behaviors in various directions.

Table 1. Slicing parameters.

Slicing Parameter	Value
Layer height	0.15
Extrusion multiplier	100%
Line width	0.4 mm
Cooling	100
Print temperature	205 °C
Bed temperature	65 °C
Default printing speed	60 mm/s
Line direction	90° *

* Lines were printed along the X direction, considering the reference system reported in both Figures 1 and 2.

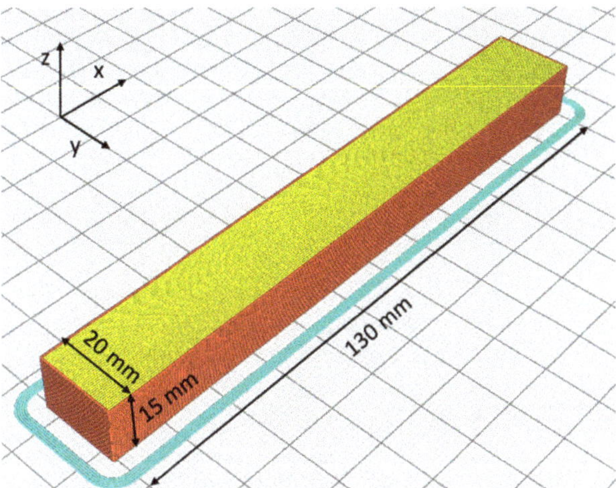

Figure 2. Dimensions of the specimen at the end of the slicing process in Cura.

2.3. Additional Image Analysis Considerations

The specimen microstructure was obtained using an optical microscope with a $20\times$ magnification. The specimen must be long enough to allow the printing speed and extrusion rate to stabilize. To simplify the geometry of the RVE model, the specimen must be printed with lines parallel to the largest dimension of the specimen, as shown in Figure 2, in order to be able to clearly and uniquely highlight the microstructure, once broken at $90°$ with respect to the largest dimension. The choice of creating a parallel line RVE model is dependent by the following reasons:

- Overlapping parallel lines are always found on the contour (perimeter) of the component. An example is thinner wall parts in general; in this case, the walls often form the entire structure of the printed component. This is not the conventional way of 3d printing, but it is still significant for certain type of parts.
- On tests carried out on FDM-printed components, where bending and torsional loads were predominant, compared to tensile loads, it was observed by the authors that the initialization and propagation of cracks started from the external surface of the component. This is a further incentive to create an RVE model with parallel lines to understand better the behavior in that location.
- The presented model is linear with a constraint of the perfect adhesion between layers, leading to a simplified model for tensile tests. In any case, it is possible to create a $45°$–$45°$ RVE model by using the same methodology but needs the study of a new RVE model.

When the printer started extruding a new line, its speed started from zero and grew up to the cruise value in accordance with the acceleration set in the Firmware or directly in the slicing software.

Furthermore, the extruded material flow could vary in areas where the speed was not constant, changing the volume/area ratio of the voids. There are effective solutions to adjust the material flow rate during the acceleration and deceleration phases, for example "coast at end" and "retraction extra prime amount" in the slicing software or even directly in the Firmware such as "Lin advance" in Marlin firmware. The possibility of a nonperfect matching between the general and the local microstructures in these areas may still exist.

The specimen must be broken in a brittle manner, and the failure must be carried out in the central area of the sample. In many cases, the need for a heated bed platform is essential to ensure adhesion to the build plate.

The height of the first layer was extremely dependent on the calibration of the distance between the nozzle and the build plate. The combination of these two aforementioned factors can produce a series of layers with variations in microstructure. The image necessary to create the RVE model must therefore be taken in the medium–high parts of the specimen, as shown in Figure 3, to avoid influences on the height of the first layer and the temperature of the printing surface.

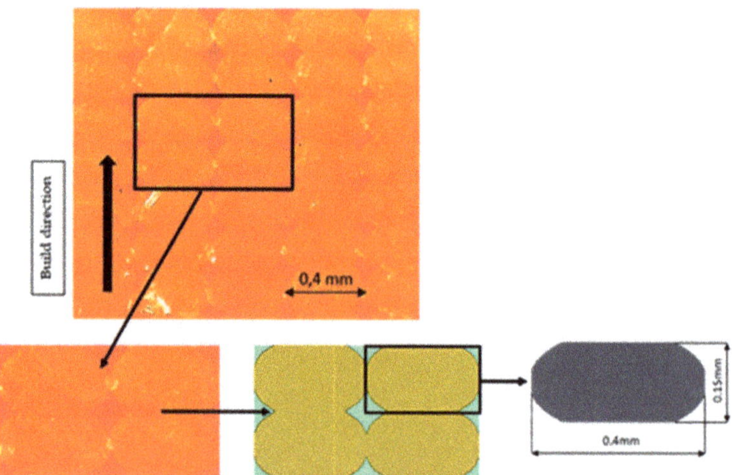

Figure 3. Workflow of geometry selection.

2.4. Input Data

The geometry used was similar to a UD-fiber composite RVE and was therefore composed by PLA and "Air". The second one is necessary for the definition of the elementary volume which needs to have parallel plane faces with the same number of nodes on the opposite surfaces. This type of mesh is called periodic mesh. The reference values for these materials are given in Table 2.

Table 2. Material properties.

Material	Young's Modulus (MPa)	Shear Modulus (MPa)	Poisson's Ratio
PLA	3000	740	0.35
Air	1×10^{-6}	–	0

Moreover, the values of the "Air" material were assumed from the Ansys database, and they had no structural meaning in the simulation. The influence of this material on the mechanical behavior of the structure was in fact irrelevant. It could have a nonzero influence in the case of the thermal analysis of cooling or heating of the component being air as an insulating material. However, this analysis is beyond the scope of this study, and therefore for the simplicity and clarity, the "air" material was considered "linear elastic" to maintain compatibility in the simulation. The air material results, however, were mandatory to perform the RVE analysis, since the calculation of the various modules was weighed on the lateral area of the RVE. Therefore, for the simplicity of calculation of the software, it is a good practice to keep a parallelepipedal shape.

Moreover, since it is currently a linear model, it is only valid for small deformations. However, if a large deformations analysis is performed with nonlinearities or thermal analyses, the contribution of air would be crucial for validating the idea of implementing air in the system.

2.5. Geometry

The geometry followed the observations under the microscope. The dimensions were obtained by analyzing the section image of the specimen. It was decided to consider a single unit of length for the "Z" direction. The schematization of the printed lines was made considering an ellipsoidal section, while the contact areas were defined as continuous between adjacent lines and adjacent layers. Figure 4 shows that four lines were chosen in the definition of the RVE.

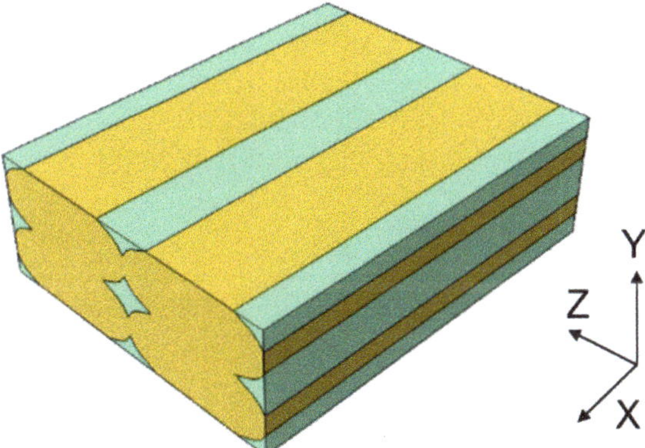

Figure 4. RVE model. The polylactic acid (PLA) mesostructure is indicated in yellow, and air is indicated in green.

2.6. Loads and Constraints

In order to evaluate the characteristic modules of the homogenized material, it is necessary to apply particular boundary conditions to the fundamental volume. Specifically, Neumann conditions were not applied to the structure, but only Dirichlet conditions. Since the elementary volume was a part of the total volume, its behavior will be symmetrical with respect to the opposite surfaces. This involves the application of periodic boundary conditions to the model studied [28]. To clarify this, equations of the two-dimensional (2D) version are reported as Equation (1). The obtained 3D version of Figure 5 is an extension of these conditions. The displacement of the N-th node in the X and Y directions, $U_{(x,y)}^{N*}$, were defined as follows:

$$\begin{aligned}
U_{(x,y)}^{N_B} - U_{(x,y)}^{N_A} - U_{(x,y)}^{N_2} + U_{(x,y)}^{N_1} = 0 \\
U_{(x,y)}^{N_C} - U_{(x,y)}^{N_D} - U_{(x,y)}^{N_4} + U_{(x,y)}^{N_1} = 0
\end{aligned} \tag{1}$$

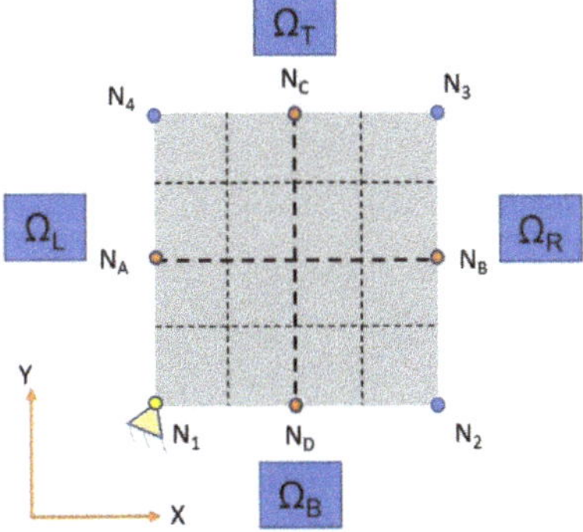

Figure 5. Two-dimensional (2D) RVE model for the application of boundary conditions.

Subsequently, in order to find out the mechanical characteristics, particular conditions of the imposed displacement were studied afterwards. Figure 6 represents a schematization of various load cases. In Figure 7, there is the detailed 2D schematization of how the various shear moduli were calculated. Taking the G12 case and element characterization as examples, Young's modulus E, Poisson's ratio v, and Shear modulus G were calculated according to [29]:

$$E = \frac{Stress}{Axial\ strain} \tag{2}$$

$$v = \frac{-Transvers\ strain}{Axial\ strain} \tag{3}$$

$$E_{11} = \frac{\frac{\sum Front\ surface\ nodal\ forces_{in\ 1-Direction}}{Front\ surface\ area\ (H \times W)}}{\frac{\Delta L}{L}} \tag{4}$$

$$G_{12} = \frac{\frac{\sum Top\ surface\ nodal\ forces_{in\ 1-Direction}}{Top\ surface\ area\ (L \times W)}}{\frac{\Delta 1}{H} + \frac{\Delta 2}{L}} \tag{5}$$

$$G = \frac{Shear\ stress}{Tensors\ of\ shear\ strain} \tag{6}$$

$$v_{12} = \frac{\frac{\Delta H}{H}}{\frac{\Delta L}{L}} \tag{7}$$

$$v_{13} = \frac{\frac{\Delta W}{W}}{\frac{\Delta L}{L}} \tag{8}$$

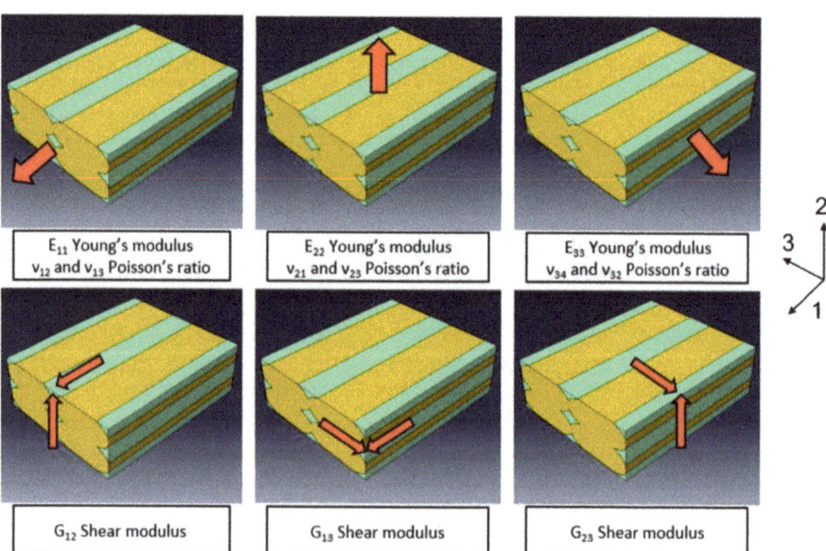

Figure 6. Load cases in the RVE model. The red arrows indicate the directions of the imposed displacement on the RVE. Software axes 1, 2, and 3 indicate the X, Y, and Z directions.

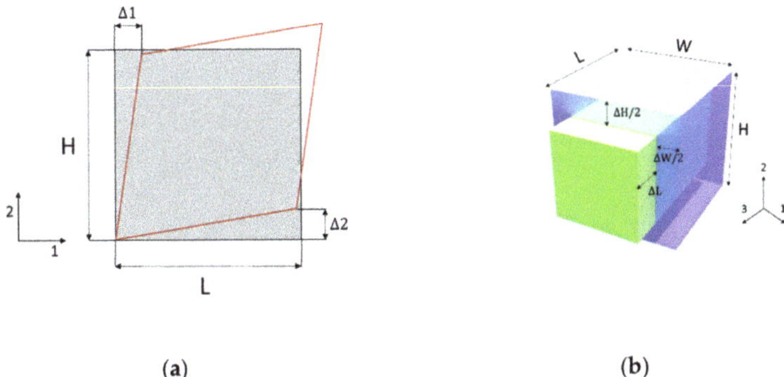

(a)   (b)

Figure 7. (**a**) Deformed part schematization of displacement to evaluate G_{12}; (**b**) 3D schematization of the load case to evaluate E_{11} and Poisson's ratios ν_{12} and ν_{13}.

3. Results

The obtained simulation results defined in the precedent chapter are displayed in Table 3.

Table 3. Homogenized elastic properties.

Property	Value	Unit
E_{11}	2729	MPa
ν_{12}	0.35	/
ν_{13}	0.35	/
E_{22}	2193	MPa
ν_{21}	0.28	/
ν_{23}	0.29	/
E_{33}	2259	MPa
ν_{31}	0.29	/
ν_{32}	0.3	/
G_{12}	870	MPa
G_{13}	897	MPa
G_{23}	790	MPa

As expected, there was a decrease in values in all directions due to the presence of voids in the structure. As for E_{11}, its value was closely related to the resistant area, regardless of the geometry drawn. Along the direction of the lines, Young's modulus was in fact the highest, similar to the behavior of UD composites. The E_{22} and E_{33} responses depended on the contact area between adjacent lines and layers. By varying the printing parameters, it is possible to modify the geometry and its modules in the Y and Z directions. In the presented case study, the RVE model was linear, and there was a constraint of perfect adhesion between one layer and the next, i.e., continuity between the material of one line and the adjacent and overlying one. There would be a rather linear increase in the various moduli (Young's and shear moduli) by varying the size of the voids and therefore increasing the contact area between the lines, until there were no more voids in the structure. Three kinds of RVE with different void dimensions were considered and tested, as shown in Figure 8. The same dimension of the RVE and the basic dimensions of the lines were maintained as reported in Figure 3. An increase of the contact area between the printed lines was obtained by keeping the same dimension of the RVE and reducing the voids dimension.

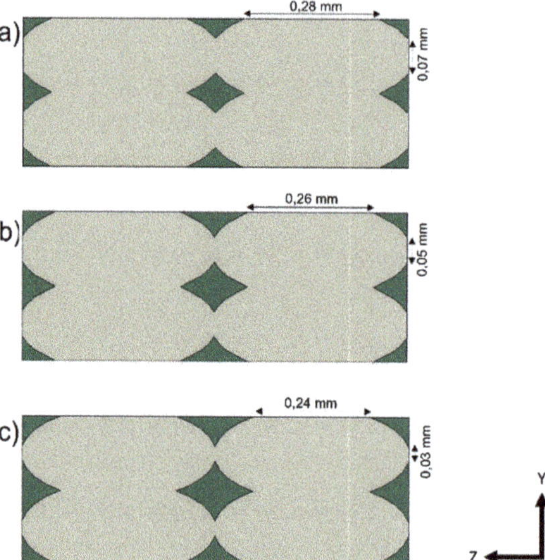

Figure 8. Increased RVE void length dimensions in Y and Z directions, respectively: (**a**) lengths of 0.07 and 0.28 mm; (**b**) lengths of 0.05 and 0.26 mm; (**c**) lengths of 0.03 and 0.24 mm.

Such reductions could be demonstrated by using the methodology proposed by Patrich et al. [27]. However, the variations of Young's modulus and the Shear modulus values with respect to the relative density could be seen from the graphs reported in Figure 9. The data are shown in a similar manner as the reported research on a foam-like material by Imwinkelried [30] and Goods et al. [31], as the values increased in a linear way.

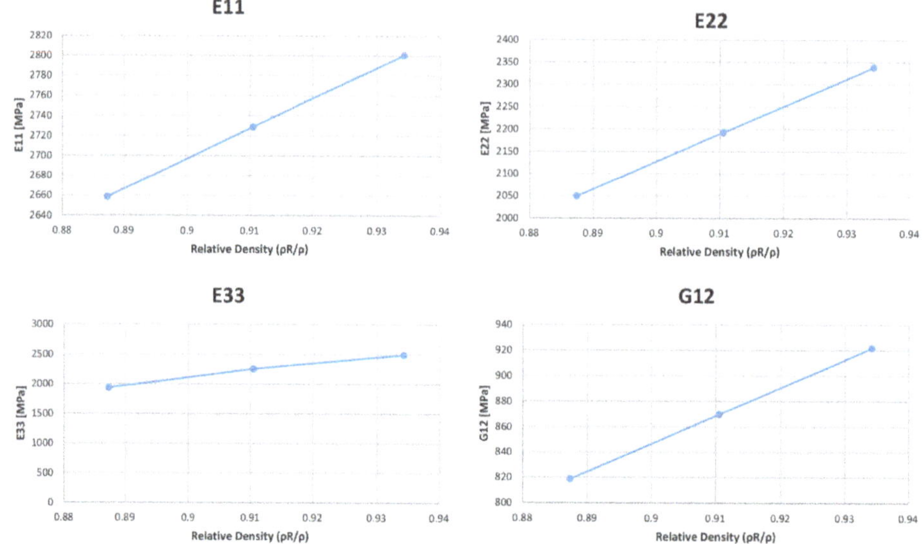

Figure 9. *Cont.*

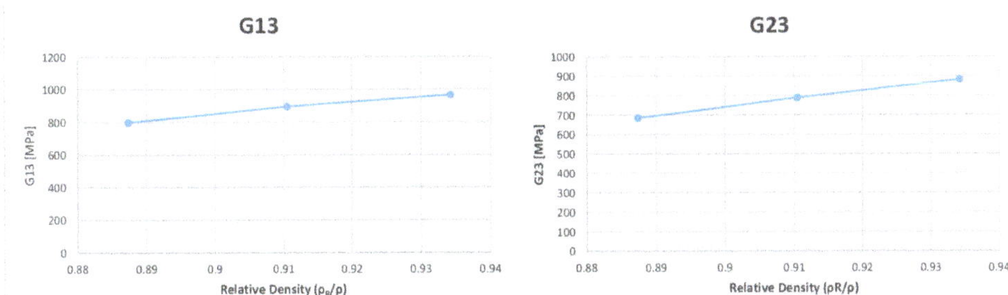

Figure 9. Variations of Young's modulus and the Shear modulus with respect to the relative density.

Consequently, shear moduli G_{12}, G_{13}, and G_{23} were reduced compared to those in the case of a homogeneous isotropic material and strictly depended on the contact behavior between layers and lines. The displacements and directional stresses of all the load cases to show the symmetric behavior of the RVE are reported in Figures 10–15.

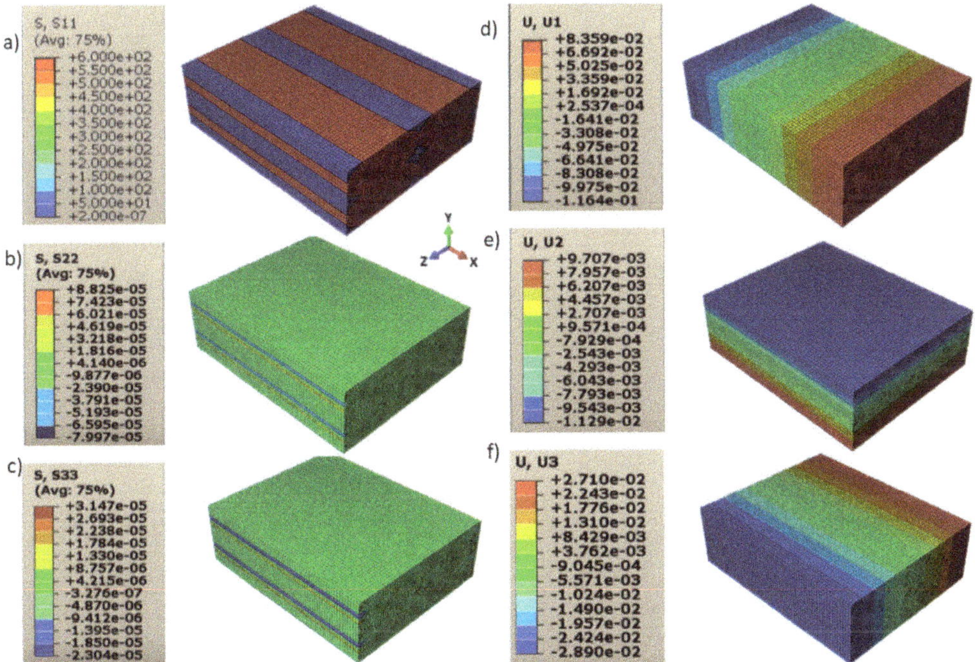

Figure 10. Load cases to evaluate E_{11}: (**a**) stress in direction 11; (**b**) stress in direction 22; (**c**) stress in direction 33; (**d**) displacement in direction 11; (**e**) displacement in direction 22; and (**f**) displacement in direction 33.

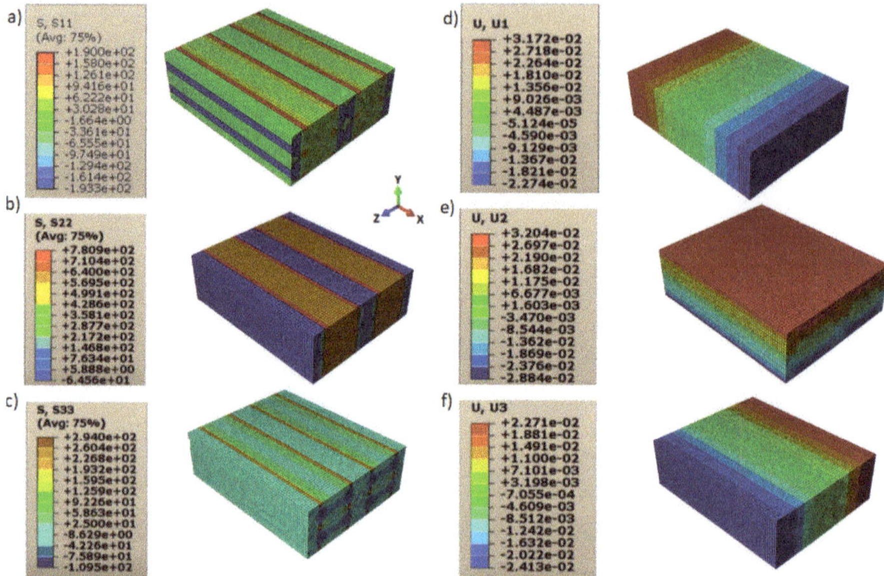

Figure 11. Load cases to evaluate E_{22}: (**a**) stress in direction 11; (**b**) stress in direction 22; (**c**) stress in direction 33; (**d**) displacement in direction 11; (**e**) displacement in direction 22; (**f**) displacement in direction 33.

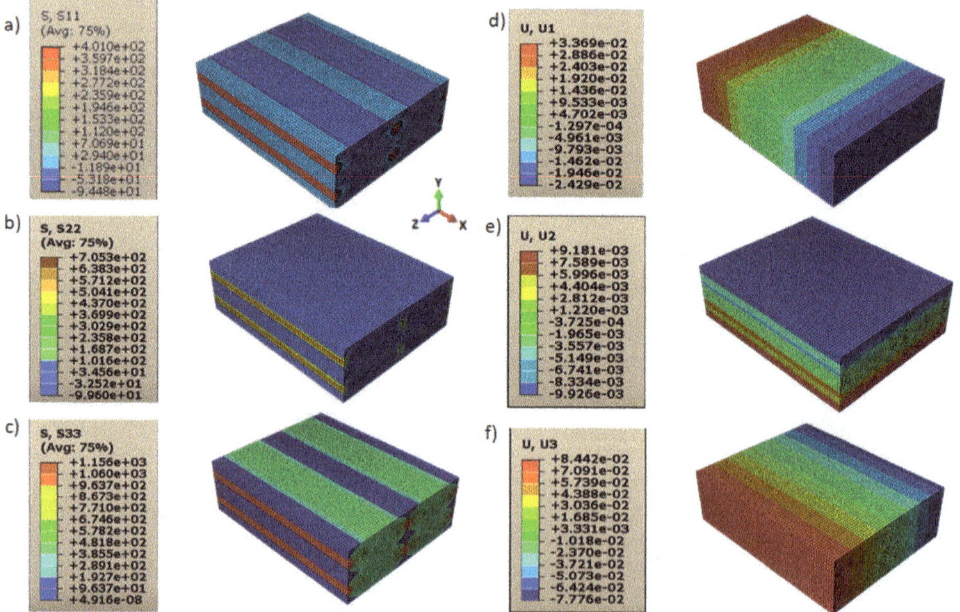

Figure 12. Load cases to evaluate E_{33}: (**a**) stress in direction 11; (**b**) stress in direction 22; (**c**) stress in direction 33; (**d**) displacement in direction 11; (**e**) displacement in direction 22; (**f**) displacement in direction 33.

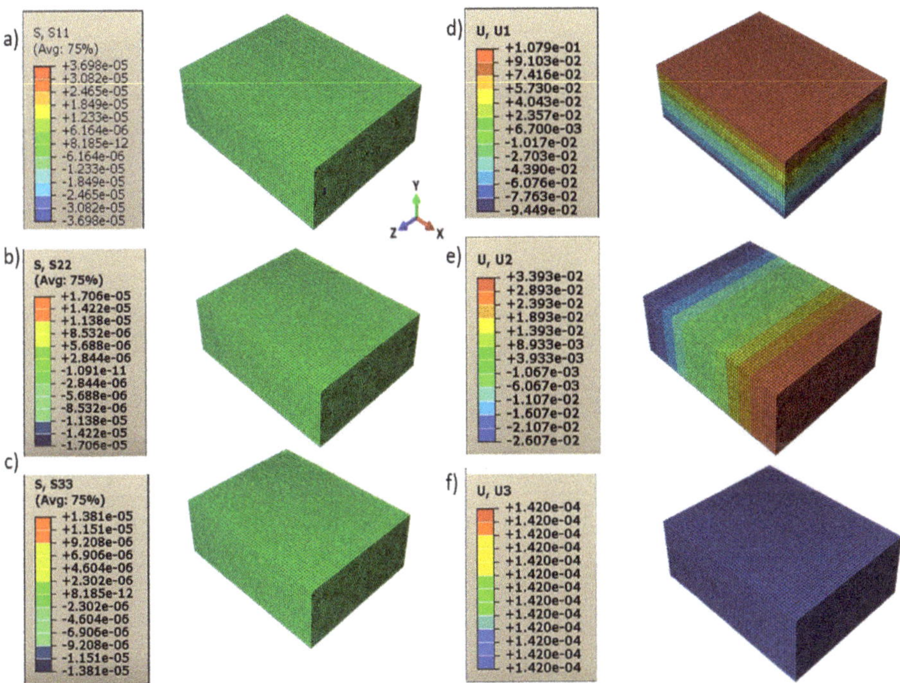

Figure 13. Load cases to evaluate G_{12}: (**a**) stress in direction 11; (**b**) stress in direction 22; (**c**) stress in direction 33; (**d**) displacement in direction 11; (**e**) displacement in direction 22; (**f**) displacement in direction 33.

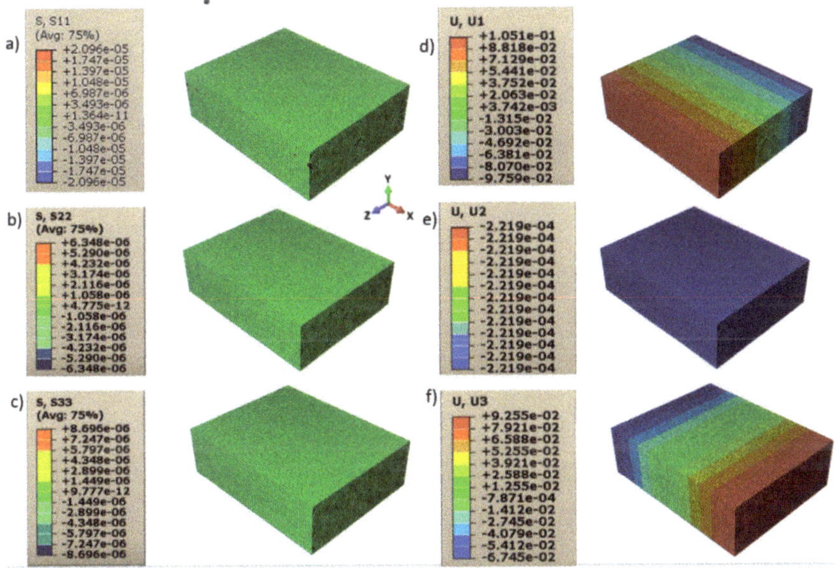

Figure 14. Load cases to evaluate G_{13}: (**a**) stress in direction 11; (**b**) stress in direction 22; (**c**) stress in direction 33; (**d**) displacement in direction 11; (**e**) displacement in direction 22; (**f**) displacement in direction 33.

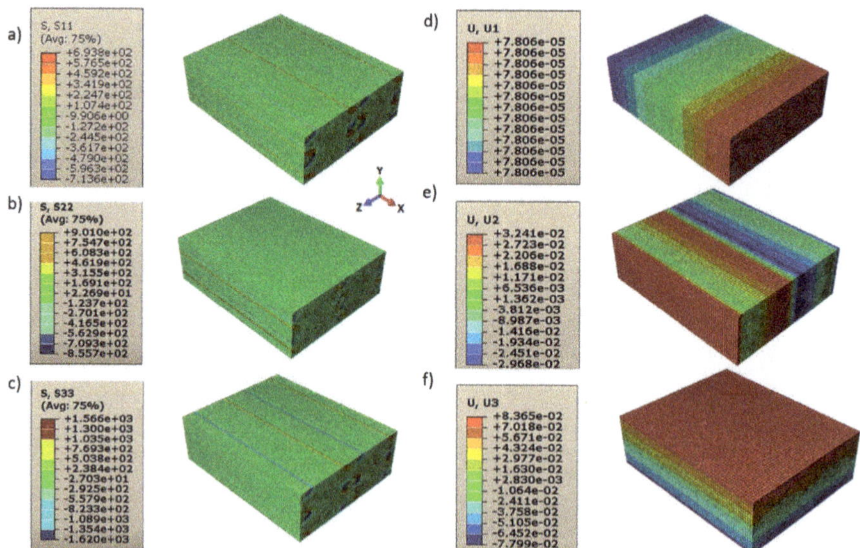

Figure 15. Load cases to evaluate G_{23}: (**a**) stress in direction 11; (**b**) stress in direction 22; (**c**) stress in direction 33; (**d**) displacement in direction 11; (**e**) displacement in direction 22; (**f**) displacement in direction 33.

It can also be noted in Figures 16 and 17 that the layer deposition was generated from the areas in which there was a concentration of the stresses; in particular, this always happened between one layer and the subsequent one. This is particularly evident especially in cases where the imposed deformation was not parallel to the direction of the fibers (load cases to evaluate G_{13} and G_{23}).

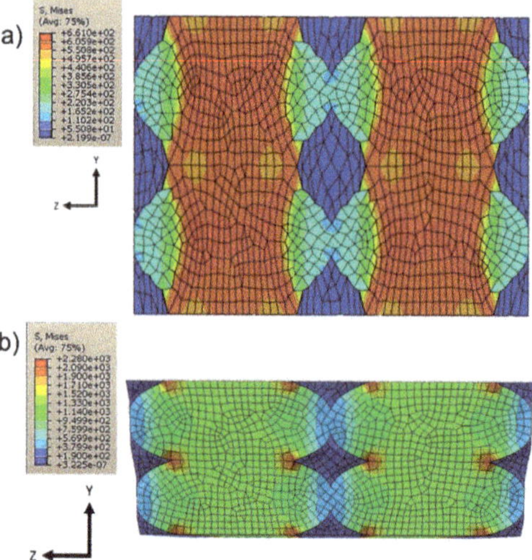

Figure 16. Von Mises stresses for E_{22} (**a**) and G_{12} (**b**).

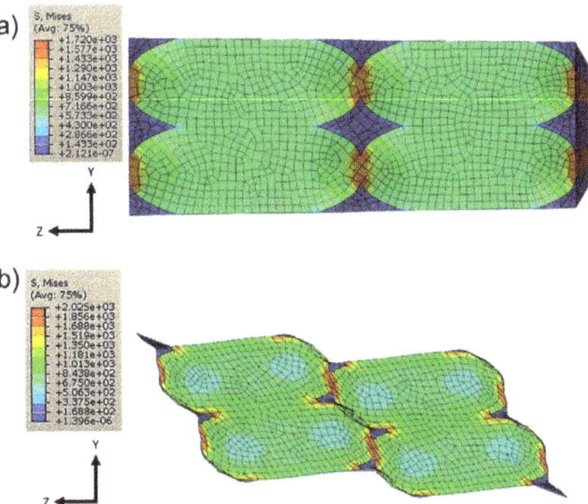

Figure 17. Von Mises stresses for G_{13} (**a**) and G_{23} (**b**).

3.1. Consideration about the RVE Dimension

Different RVE dimensions were considered, as shown in Figure 18, to prove that for a linear model with parallel lines the dimension of RVE can be as small as a single line. Moreover, for a widen understanding, the results of the Young's and Shear moduli of all the models considered were basically the same of the ones reported in Tables 3 and 4. The variation between the results was less than the 1% and could be related to the discretization of the model. This verified that additional information was not given for a linear model with a linear elastic material by enlarging the RVE dimension. Moreover, when the problem is not linear, the choice of the RVE dimension has a great influence on the modules value, as reported by Okereke and Akpoyomare [32,33]. For this reason, it was decided to use a larger unit cell, which could lead into further research that develops this model.

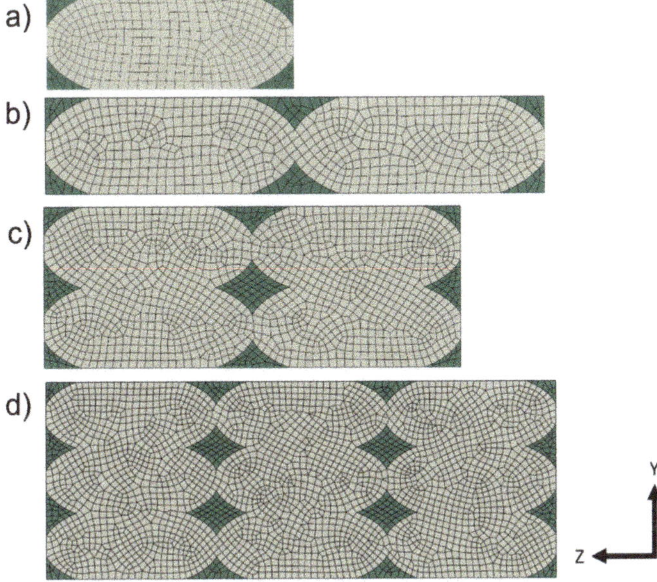

Figure 18. Different RVE dimensions given by number of printed lines used to evaluate the effect of the size influence. (**a**) one line;(**b**) two lines; (**c**) four lines;(**d**) nine lines.

Table 4. Mechanical properties for all the RVE sizes.

	a	b	c	d	Unit
E_{11}	2729	2729	2729	2729	MPa
ν_{12}	0.35	0.35	0.35	0.35	/
ν_{13}	0.35	0.35	0.35	0.35	/
E_{22}	2196	2194	2193	2195	MPa
ν_{21}	0.28	0.28	0.28	0.28	/
ν_{23}	0.29	0.29	0.29	0.29	/
E_{33}	2261	2261	2259	2257	MPa
ν_{31}	0.29	0.29	0.29	0.29	/
ν_{32}	0.30	0.30	0.3	0.29	/
G_{12}	871	871	870	871	MPa
G_{13}	897	896	897	896	MPa
G_{23}	791	791	790	791	MPa

4. Conclusions

The analysis carried out involved the construction of a linear RVE model in order to predict the macroscopic behavior of 3D printed geometries. Particular attention was paid to the geometry of the printed mesostructure observed under a microscope after printing. The given reference material for the model was PLA, which is low-cost and easy to print through FDM methodologies, but with discrete performing mechanical characteristics compared to other thermoplastic polymers. As the similarity between the structure of a 3D FDM-printed component and a UD composite has allowed evaluating the behavior of an FDM-printed sample through a linear elastic RVE.

Subsequently, the construction of a linear elastic model allows obtaining valid results only in the field of small deformations. The condition of the good adhesion between one layer and another is an ideal assumption and constitutes a reference point in the study of these microstructures. The model presented this condition by defining the intersection areas between adjacent lines as a continuous, perfect bonding behavior. This guaranteed a perfect match with the theoretical model and simplified the numerical model.

Furthermore, studies of a complex model in which an adhesion condition is implemented between various printed lines are desirable in order to accurately evaluate intralayer phenomena that are the main cause of failure. The analysis presented showed how the mesoscopic description of the geometry influenced the macroscopic properties of the material, effectively inducing a geometric anisotropy to be considered for the construction of complex components. Future developments are aimed at validating the macroscopic behaviors of the parameters calculated with reference to experimental tests.

5. Future Developments

Future developments will consider nonlinear elastic–plastic behaviors common to most polymers to develop a valid model over a wider range of applications. Considerations will also be made about the adhesion between one layer and the next and adjacent layers and its effect on the properties of an RVE. In addition, an experimental setting should be planned to demonstrate the effectiveness of this model.

Author Contributions: Conceptualization, P.F. and G.M.S.; methodology, L.F.; software, L.F.; validation, P.F., E.F. and C.L.-C.; formal analysis, G.M.S. and G.D.; investigation, E.F. and C.L.-C.; resources, P.F.; data curation, P.F.; writing—original draft preparation, C.L.-C.; writing—review and editing, C.L.-C. and E.F.; visualization, P.F.; supervision, G.M.S.; project administration, L.F. All authors have read and agreed to the published version of the manuscript.

Funding: This research received no external funding.

Institutional Review Board Statement: Not applicable.

Informed Consent Statement: Not applicable.

Data Availability Statement: Not applicable.

Conflicts of Interest: The authors declare no conflict of interest.

References

1. Hofstätter, T.; Pedersen, D.B.; Tosello, G.; Hansen, H.N. State-of-the-art of fiber-reinforced polymers in additive manufacturing technologies. *J. Reinf. Plast. Compos.* **2017**, *36*, 1061–1073. [CrossRef]
2. Frascio, M.; Marques, E.A.d.S.; Carbas, R.J.C.; da Silva, L.F.M.; Monti, M.; Avalle, M. Review of Tailoring Methods for Joints with Additively Manufactured Adherends and Adhesives. *Materials* **2020**, *13*, 3949. [CrossRef]
3. Zarringhalam, H.; Hopkinson, N.; Kamperman, N.F.; de Vlieger, J.J. Effects of processing on microstructure and properties of SLS Nylon 12. *Mater. Sci. Eng. A* **2006**, *435–436*, 172–180. [CrossRef]
4. Linares-Alvelais, J.A.R.; Figueroa-Cavazos, J.O.; Chuck-Hernandez, C.; Siller, H.R.; Rodríguez, C.A.; Martínez-López, J.I. Hydrostatic high-pressure post-processing of specimens fabricated by DLP, SLA, and FDM: An alternative for the sterilization of polymer-based biomedical devices. *Materials* **2018**, *11*, 2540. [CrossRef] [PubMed]
5. Ali, Z.; Türeyen, E.B.; Karpat, Y.; Çakmakcı, M. Fabrication of polymer micro needles for transdermal drug delivery system using DLP based projection stereo-lithography. *Procedia CIRP* **2016**, *42*, 87–90. [CrossRef]
6. Schmid, M.; Amado, A.; Wegener, K. Polymer powders for selective laser sintering (SLS). In Proceedings of the AIP Conference Proceedings, Cleveland, OH, USA, 6–12 June 2014; AIP Publishing LLC: College Park, MD, USA, 2015; Volume 1664, p. 160009.
7. Somireddy, M.; De Moraes, D.A.; Czekanski, A. Flexural behavior of fdm parts: Experimental, analytical and numerical study. In Proceedings of the 28th Annual International Solid Freeform Fabrication Symposium—An Additive Manufacturing Conference, Austin, TX, USA, 7–9 August 2017; pp. 7–9.
8. Bhandari, S.; Lopez-Anido, R. Finite element analysis of thermoplastic polymer extrusion 3D printed material for mechanical property prediction. *Addit. Manuf.* **2018**, *22*, 187–196. [CrossRef]
9. Garg, A.; Bhattacharya, A. An insight to the failure of FDM parts under tensile loading: Finite element analysis and experimental study. *Int. J. Mech. Sci.* **2017**, *120*, 225–236. [CrossRef]
10. Lee, C.S.; Kim, S.G.; Kim, H.J.; Ahn, S.H. Measurement of anisotropic compressive strength of rapid prototyping parts. *J. Mater. Process. Technol.* **2007**, *187–188*, 627–630. [CrossRef]
11. Kotlinski, J. Mechanical properties of commercial rapid prototyping materials. *Rapid Prototyp. J.* **2014**, *20*, 499–510. [CrossRef]
12. Somireddy, M.; Czekanski, A. Mechanical characterization of additively manufactured parts by FE modeling of mesostructure. *J. Manuf. Mater. Process.* **2017**, *1*, 18.
13. Croccolo, D.; De Agostinis, M.; Fini, S.; Liverani, A.; Marinelli, N.; Nisini, E.; Olmi, G. Mechanical characteristics of two environmentally friendly resins reinforced with flax fibers. *Stroj. Vestnik-J. Mech. Eng.* **2015**, *61*, 227–236. [CrossRef]
14. Somireddy, M.; Czekanski, A.; Singh, C.V. Development of constitutive material model of 3D printed structure via FDM. *Mater. Today Commun.* **2018**, *15*, 143–152. [CrossRef]
15. Chen, Y.; Zhao, Y.; He, C.; Ai, S.; Lei, H.; Tang, L.; Fang, D. Yield and failure theory for unidirectional polymer-matrix composites. *Compos. Part B Eng.* **2019**, *164*, 612–619. [CrossRef]
16. Chen, J.L.; Sun, C.T. A plastic potential function suitable for anisotropic fiber composites. *J. Compos. Mater.* **1993**, *27*, 1379–1390. [CrossRef]
17. Sun, C.T.; Chen, J.L. A simple flow rule for characterizing nonlinear behavior of fiber composites. *J. Compos. Mater.* **1989**, *23*, 1009–1020. [CrossRef]
18. Mellinger, A.; Wegener, M.; Wirges, W.; Mallepally, R.R.; Gerhard-Multhaupt, R. Thermal and temporal stability of ferroelectret films made from cellular polypropylene/air composites. *Ferroelectrics* **2006**, *331*, 189–199. [CrossRef]
19. Ashby, M.F.; Evans, A.; Fleck, N.A.; Gibson, L.J.; Hutchinson, J.W.; Wadley, H.N.G. *Metal Foams: A Design Guide*; Butterworth-Heinemann: Oxford, UK, 2000; 251p, ISBN 0-7506-7219-6. [CrossRef]
20. Li, S.; Sitnikova, E. Representative Volume Elements and Unit Cells: Concepts, Theory, Applications and Implementation. In *Representative Volume Elements and Unit Cells: Concepts, Theory, Applications and Implementation*; Elsevier: Amsterdam, The Netherlands, 2019; pp. 1–468. [CrossRef]
21. Devireddy, S.B.R.; Biswas, S. Effect of Fiber Geometry and Representative Volume Element on Elastic and Thermal Properties of Unidirectional Fiber-Reinforced Composites. *J. Compos.* **2014**, *2014*, 629175. [CrossRef]
22. Grimal, Q.; Raum, K.; Gerisch, A.; Laugier, P. A determination of the minimum size of representative volume elements for the prediction of cortical bone elastic properties. *Biomech. Model. Mechanobiol.* **2011**, *10*, 925–937. [CrossRef]
23. *Fillamentum Technical Datasheet-PLA*; Fillamentum Manufacturing Czech s.r.o.: Brzezie, Poland, 2020.
24. *Formfutura Technical Datasheet-PLA*; Formfutura BV: Nijmegen, The Netherlands, 2020.
25. *Fiberology Technical Datasheet-PLA*; Fiberlab S.A.: Brzezie, Poland, 2020.
26. Aranda, S. *3D Printing Failures: How to Diagnose & Repair all Desktop 3D Printing Issues*; Feeney, D., Ed.; CreateSpace: Scotts Valley, CA, USA, 2019; p. 289.
27. Ferretti, P.; Leon-Cardenas, C.; Santi, G.M.; Sali, M.; Ciotti, E.; Frizziero, L.; Donnici, G.; Liverani, A. Relationship between FDM 3D Printing Parameters Study: Parameter Optimization for Lower Defects. *Polymers* **2021**, *13*, 2190. [CrossRef]
28. Santi, G.M.; Francia, D.; Cesari, F. Effect of coriolis force on vibration of annulus pipe. *Appl. Sci.* **2021**, *11*, 58. [CrossRef]

29. Omairey, S.L.; Dunning, P.D.; Sriramula, S. Development of an ABAQUS plugin tool for periodic RVE homogenisation. *Eng. Comput.* **2019**, *35*, 567–577. [CrossRef]
30. Imwinkelried, T. Mechanical properties of open-pore titanium foam. *J. Biomed. Mater. Res. Part A* **2007**, *81*, 964–970. [CrossRef] [PubMed]
31. Goods, S.H.; Neuschwanger, C.L.; Whinnery, L.L.; Nix, W.D. Mechanical properties of a particle-strengthened polyurethane foam. *J. Appl. Polym. Sci.* **1999**, *74*, 2724–2736. [CrossRef]
32. Okereke, M.I.; Akpoyomare, A.I. A virtual framework for prediction of full-field elastic response of unidirectional composites. *Comput. Mater. Sci.* **2013**, *70*, 82–99. [CrossRef]
33. Okereke, M.; Keates, S. *Finite Element Applications: A Practical Guide to the FEM Process*; Springer: Cham, Switzerland, 2018; ISBN 3319671251.

 *polymers*

Review

3D/4D Printing of Polymers: Fused Deposition Modelling (FDM), Selective Laser Sintering (SLS), and Stereolithography (SLA)

Abishek Kafle [1], **Eric Luis** [2], **Raman Silwal** [1], **Houwen Matthew Pan** [3], **Pratisthit Lal Shrestha** [1,*] **and Anil Kumar Bastola** [4,*]

[1]   Design Lab, Department of Mechanical Engineering, Kathmandu University, Dhulikhel 45200, Nepal; abishek.kafle@ku.edu.np (A.K.); ramansilwal24@gmail.com (R.S.)

[2]   Faculty of Medicine, Macau University of Science and Technology, Avenida Wai Long, Macau SAR, China; laguntureric@must.edu.mo

[3]   Division of Chemistry and Biological Chemistry, School of Physical and Mathematical Sciences, Nanyang Technological University, 21 Nanyang Link, Singapore 637371, Singapore; matthew.pan@u.nus.edu

[4]   Centre for Additive Manufacturing (CfAM), School of Engineering, University of Nottingham, Nottingham NG8 1BB, UK

*   Correspondence: pratisthit@ku.edu.np (P.L.S.); anilkuma001@e.ntu.edu.sg (A.K.B.)

Abstract: Additive manufacturing (AM) or 3D printing is a digital manufacturing process and offers virtually limitless opportunities to develop structures/objects by tailoring material composition, processing conditions, and geometry technically at every point in an object. In this review, we present three different early adopted, however, widely used, polymer-based 3D printing processes; fused deposition modelling (FDM), selective laser sintering (SLS), and stereolithography (SLA) to create polymeric parts. The main aim of this review is to offer a comparative overview by correlating polymer material-process-properties for three different 3D printing techniques. Moreover, the advanced material-process requirements towards 4D printing via these print methods taking an example of magneto-active polymers is covered. Overall, this review highlights different aspects of these printing methods and serves as a guide to select a suitable print material and 3D print technique for the targeted polymeric material-based applications and also discusses the implementation practices towards 4D printing of polymer-based systems with a current state-of-the-art approach.

Keywords: 3D printing; 4D printing; fused deposition modelling; selective laser sintering; stereolithography; polymers

Citation: Kafle, A.; Luis, E.; Silwal, R.; Pan, H.M.; Shrestha, P.L.; Bastola, A.K. 3D/4D Printing of Polymers: Fused Deposition Modelling (FDM), Selective Laser Sintering (SLS), and Stereolithography (SLA). *Polymers* **2021**, *13*, 3101. https://doi.org/10.3390/polym13183101

Academic Editor: Geoffrey R. Mitchell

Received: 22 August 2021
Accepted: 9 September 2021
Published: 15 September 2021

Publisher's Note: MDPI stays neutral with regard to jurisdictional claims in published maps and institutional affiliations.

1. Introduction

3D printing or additive manufacturing (AM) is a digital manufacturing process, in which the materials are added layer by layer to create 3D objects directly from the computer-aided design (CAD) models [1–8]. 3D printing has gained significant popularity in the last two decades due to a number of appealing advantages such as the limitless design freedom and capability to produce low cost and multifunctional objects with highly delicate/complex structures in a short period of time [9]. For example, 3D printing of concrete materials possesses the potential to reduce construction waste by 30–60%, labour cost by 50–80%, and construction time by 50–70% [10,11]. Therefore, 3D printing has become a suitable manufacturing technique in both rapid prototyping as well as in various engineering fields such as mechanical engineering, civil engineering, aerospace, electronics, biomedical, etc. [5,6,9,12–19].

A variety of AM methods are available to 3D print a wide range of materials including metals [20–23], polymers [24–29], polymer composites [30–33], ceramics [34–39], and cement [40–43]. The ASTM (ISO/ASTM 52900:2015) has classified the range of AM processes into seven general categories. This classification is made on the basis of the fundamental principle of operation, and it includes material jetting, binder jetting, vat photopolymerization, powder bed fusion, material extrusion, direct energy deposition, and sheet lamination [5]. Furthermore, according to the type of base material used, AM can be grouped into three different categories i.e., solid-based, powder-based, and liquid-based (Figure 1). The solid-based AM is further classified into laminated object manufacturing (LOM), fused deposition modelling (FDM), wire and arc additive manufacturing (WAAM), and electron beam free form fabrication (EBF3). Powder-based additive manufacturing can be classified into selective laser sintering (SLS), electron beam melting (EBM), selective laser melting (SLM), and laser metal deposition (LMD). The liquid-based methods mostly include material jetting (MJ) and vat-based printing such as stereolithography (SLA) and digital light processing (DLP). We refer to these excellent review articles to get a comprehensive insight into the above-mentioned AM techniques, LOM [44–46], FDM [47–52], WAAM [53,54], EBF3 [55], SLS [56,57], EBM [58,59], SLM [39,59,60], LMD [61,62], SLA [63–65], DLP [66–68], and MJ [69–71].

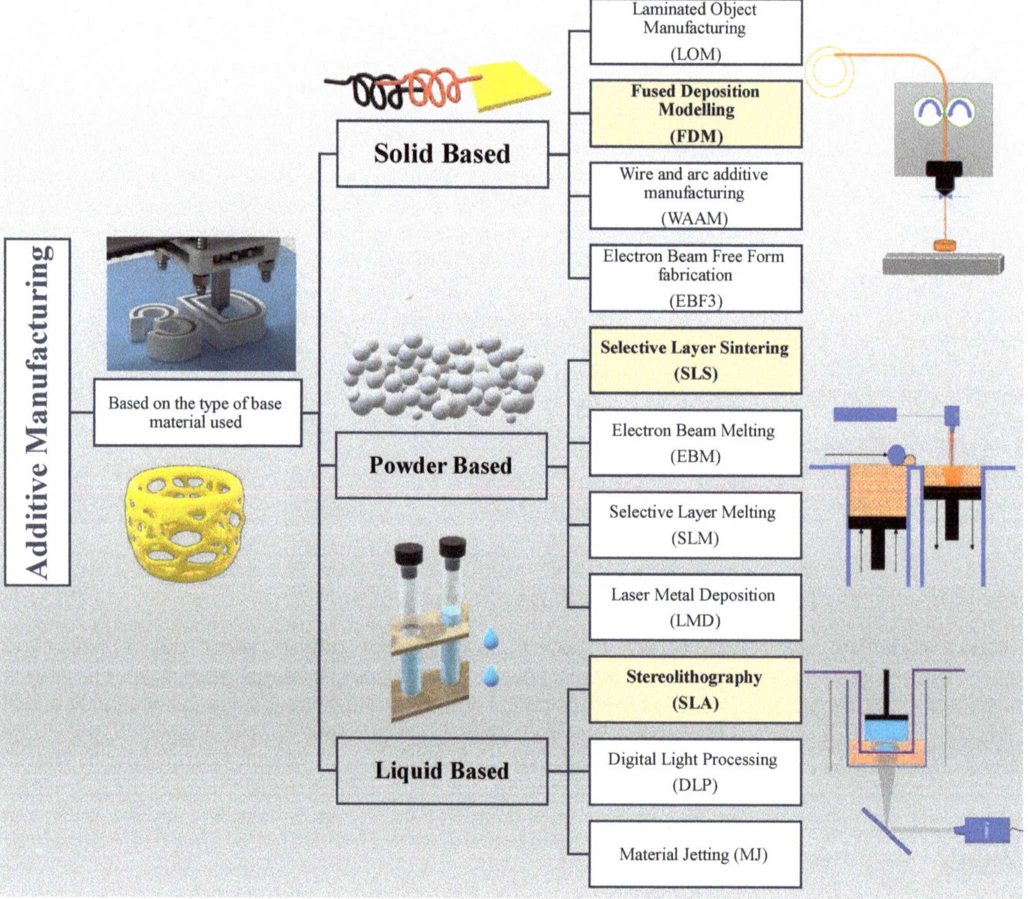

Figure 1. Classification of AM techniques based on the type of base materials used and the scope of the current review as highlighted (FDM, SLS, and SLA). The 3D printing image is taken from [72].

A polymer is a substance or material consisting of very large molecules, or macromolecules, composed of many repeating subunits [73]. Polymers are one of the prominent materials in a number of different applications due to their wide range of mechanical, thermal, electrical, fire-resistant, and biocompatible properties. According to the Web of Science (accessed on 2 August 2021), more than 60% of AM studies are focused on polymer printing. The polymers can be 3D printed with all three (i.e., solid-based, powder-based, and liquid-based) AM techniques [74]. FDM is the most conventional and widely used solid-based 3D printing technique to create polymer parts. On the other hand, SLS is a prominent AM technique to produce polymer parts using polymeric powders as a base material, while a vat-based technique, SLA, is another widely used early adopted AM technique to create polymer parts by processing the polymeric liquid as a base material. The details of these print methods are discussed in Section 2.

Although there are a number of review articles available in the literature focusing on various aspects of polymer printing based on FDM [47–52,75–77], SLS [1,78–81], and SLA [82–85], to the best of the authors' knowledge, a comprehensive study focusing on correlating the material-process-properties for these techniques is not available for both conventional 3D printing and emerging 4D printing techniques. In this article, we aim to provide the correlation of material-process-properties for these three most conventional yet widely adopted polymer-based 3D printing techniques; fused deposition modelling (FDM), selective laser sintering (SLS), and stereolithography (SLA). Furthermore, we also briefly cover how these methods are adopted towards the 4D printing (3D printing of smart materials) of polymer-based materials giving an example of 4D printing of magnetic field responsive polymers.

2. Printing Process

The fundamental process of 3D printing is the formation of parts by printing successive layers of materials that are formed on top of each other. The workflow of the 3D printing process is illustrated in Figure 2. Firstly, the CAD model of the object to be developed is created, then the standard tessellation language (.stl) file of the CAD is generated. The STL file creation process mainly converts the continuous geometry in the CAD file into small triangles [86]. The .stl file is then exported in a model slicing software which creates a tool path for the 3D printer. Here, the 3D model is translated to 2D slices that contain the information of cross-sections [87]. The 3D printer then starts the material processing and layering process. The final product is then taken out of the printer. There are a number of different factors in the printing process that determine the overall quality of the printed parts. For example, the .stl file can influence the overall quality of the printed objects. The finer the size of the triangles in .stl file, the better the printed object/part shape fidelity. The part orientation during the printing process is responsible for the mechanical properties while the environmental factors such as temperature and humidity also play the role on the overall quality of the final product [88].

Fused deposition modelling (FDM) also known as fused filament fabrication (FFF) is a process of depositing thermoplastic filaments layer by layer on a build platform [30,49,89–95]. The polymer filament is heated to a semi-solid state and deposited on the print bed or heated platform. The nozzle follows the path as of the final object in the given layer. For the next consequent layer, the platform moves one step lower, or the nozzle moves one step upward, and the material is extruded and again the nozzle follows the path of the object in the given layer. To generate the path that the nozzle follows, a slicer slices the model layer by layer and produces a G-code (computer numerical control programming language), which is followed by the nozzle in each layer. The height that the nozzle travels after each layer is the layer thickness of the model. The nozzle temperature, bed temperature, and layer height are the responsible parameters for the fractional behaviour of the 3D printed parts [96].

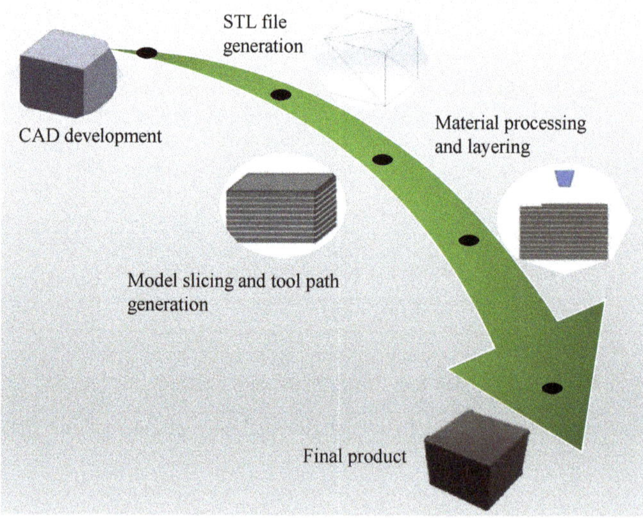

Figure 2. The workflow of the 3D printing process.

Print parameters can be grouped as machine parameters and process parameters for each printing technique (Figure 3).

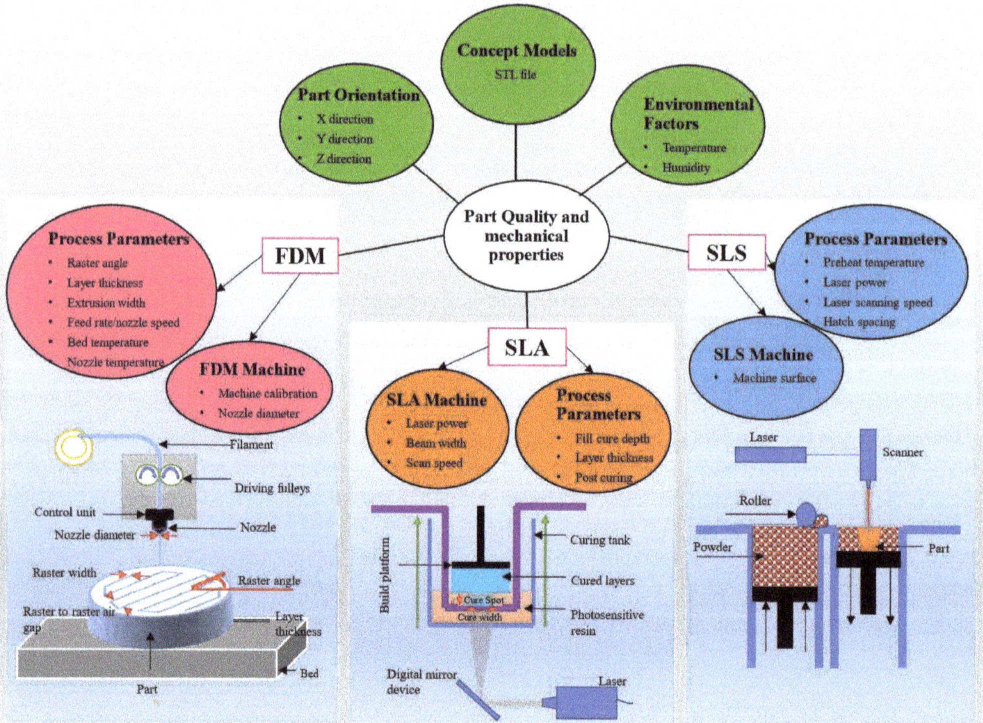

Figure 3. Overview of the process parameters of three different print methods: FDM (**left**), SLS (**right**), and SLA (**middle**).

The machine parameters for FDM printing are bed calibration and nozzle diameter while the process parameters are nozzle temperature, bed temperature, extrusion width, and raster angle [52,97], see Figure 3 (left) for the schematic illustration. The bed calibration is one of the most important considerations. FDM is a contact print method as the nozzle is used to directly deposit the material layer by layer, therefore, the distance between the nozzle and bed should be at a standard distance and constant throughout the bed. An improper bed calibration leads to the uneven distance between the nozzle and the bed at two (or more) different points on the print platform/bed, which causes warpage and also leads to the printer hitting the bed and the prints. The diameter of the 3D printing nozzle can be changed/replaced which however impacts the part quality and production time. The use of a nozzle with a large hole diameter accelerates the part production time [98]. It has also been reported that increasing the nozzle diameter increases the part quality and mechanical properties in FDM 3D printing [99]. The other important parameter is the ambient temperature which causes part warpage. For example, PLA part warpage of about 50%, 30%, and 10% at 10 °C, 15 °C, and 20 °C (ambient print environment temperature) respectively, is reported [100]. The process parameters affecting part properties in FDM are raster angle, extrusion width, extrusion rate, bed temperature, nozzle temperature, and nozzle speed. The raster angle is the angle between the direction of the nozzle and the x-axis (or y-axis, depending on notation) of the printing platform [101]. The extrusion rate is the rate at which the filament is extruded from the nozzle onto the build platform. Bed temperature refers to the temperature of the build platform. The bed temperature is required to maintain the adhesion between the build platform and the print part and avoid warpage [102]. Nozzle temperature is the temperature at which the material is melted and extruded from the nozzle. It has a high influence on the mechanical properties and microstructure of the 3D printed parts [99]. For instance, the increase in relative density (from 89.9% to 92.8% for PEEK) with the increase in nozzle temperature from 370 °C to 390 °C is reported [103]. Nozzle speed is the speed at which the nozzle moves while depositing/printing the filament melt from the nozzle onto the build platform. It greatly influences the dimensional precision of the printed parts although print time is reduced. For example, the increase in wall thickness of ring-shaped design from 2.00 mm to 2.17 mm with the increase in nozzle speed from 30 mm/s to 90 mm/s is reported [104].

Selective laser sintering (SLS), a variant of powder bed fusion and widely used AM technique, is a process used to produce objects from powdered materials using one or more lasers to selectively fuse the particles at the surface, layer upon layer, in an enclosed chamber [57,105–108]. The powders can be fused together with different particle binding mechanisms namely solid-state sintering, chemically induced binding, liquid phase sintering (partial melting), and full melting [109]. The working schematic of SLS is described in Figure 3 (right), also see Figure 4 for SLS process parameters. The printing system consists of a laser supply source, scanning system, roller, powder supply platform, and a sintering platform. Usually, the powders are fused by molecular diffusion under the influence of a high-power laser. After the first layer of powders is fused the sintering platform moves a step downwards and the next layer of powders are fused [110]. The process continues until the top layer of the final product is fused. The movement of the laser is determined again by the G-code generated from the slicer like in FDM. After the sintering process is completed, the un-sintered powder is removed, and the part is extracted from the platform.

The energy density (Equation (1)), in SLS, is the most vital parameter that is responsible for the overall process and part property. It is the amount of energy stored in a given system or region of space per unit volume.

$$ED = \frac{P}{v.h} \times \frac{d}{h} \tag{1}$$

where ED is energy density, P is the laser power, d is laser beam diameter, v is scan velocity, and h is the hatch spacing. The hatch spacing, laser scanning speed, laser power, and preheat temperature are therefore determining process parameters responsible for the part properties of SLS printed objects [105,111,112], see Figure 4. Laser power is the input power set as the ratio of the total permissible power as per the requirement of a given material and layer thickness [113]. Laser scanning speed is the rate at which the laser beam is moved along the hatching or contour lines. It influences the maximum energy at a point of material and the total time required to complete a product [114]. Hatch spacing is also known as scan spacing is the distance between two consecutive laser beams. Preheating temperature is another important parameter that affects the part property in SLS. A powder that is not preheated requires a higher-power laser beam source to melt. Furthermore, higher preheating also reduces the temperature gradient between the sintered and un-sintered parts—contributing to the elimination of thermal stress and avoiding distortion [115].

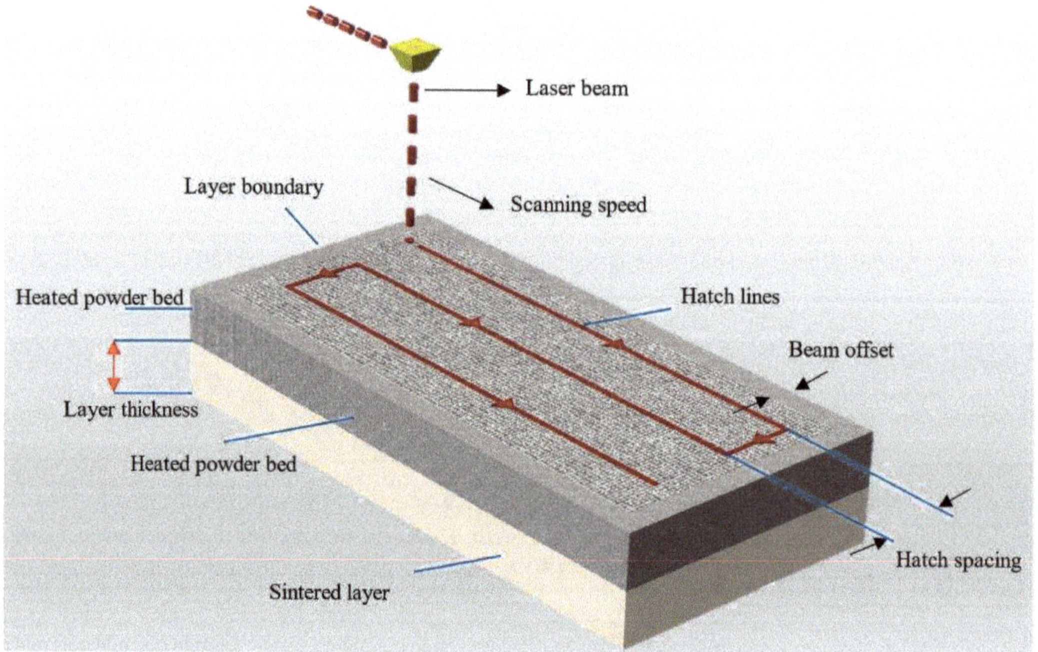

Figure 4. Process parameters for SLS printing, redrawn from [105].

Stereolithography apparatus (SLA), a vat-based and the early adopted AM technique, works on the process of 3D printing by using photopolymerization in which the photocurable resin is solidified through photopolymerization initiated by absorbing light [82,84,116–119]. Photopolymerization refers to a technique that uses rays of light to propagate a chain polymerization process which results in the photo-crosslinking of the pre-existing macromolecules [116]. The crosslinker is another component/material that links one polymer chain to another by the covalent or an ionic bond. The photopolymerization results in the solidification of a pattern inside the resin layer in order to hold the subsequent layers. A photoinitiator or photoinitiator system is required to convert photolytic energy into the reactive species (radical or cation) which can drive the chain growth via radical or cationic mechanism [116]. The measurement of attenuation of light by a chemical species at a given wavelength is given by the molar attenuation coefficient. The molar attenuation coefficient is a measurement of how strongly a chemical species attenuates light at a given wavelength. Typically, photoinitiators with molar attenuation coefficients at a short wavelength (UV < 400 nm) are used to initiate the photochemical reaction [120]. Using a computer-controlled laser beam, a pattern is illuminated on the surface of a resin. The area in the resin where the light beam strikes solidify. This principle is used repeatedly layer by layer to solidify the resin and form each layer of a product in SLA 3D printing. The thickness of the layer is controlled by the energy of the light source and exposure time [64].

The major process parameters that influence the quality of SLA printed parts are fill-cure depth, layer thickness, and post-curing, see Figure 3 (middle) for the illustration of the process parameters. The cure depth depends on the energy of the light being exposed to the resin. The energy is controlled by the laser power and the time the resin is being exposed to the light. The curing depth (C_d) should be high enough to avoid excessive fabrication time. However, the curing depth must be low enough to avoid over polymerization resulting in the over-cured part with poor resolution. Curing depth is given by an equation based on the Beer–Lambert equation (Equation (2)):

$$Cd = Dp \, log \frac{E}{E_c} \tag{2}$$

where D_p is the penetration depth (m), E is the light exposure (J m^{-2}), and E_c is the critical light exposure (J m^{-2}) [121]. The wavelength of the laser light being used is another important consideration. The wavelength of the UV light reported in the literature is in the range of 300 nm to 400 nm [63]. Usually, in SLA, objects/parts need to be post-cured after printing. Post-curing is performed to enhance the mechanical properties of the printed objects/parts. For example, a post-curing time up to 60–90 min for SLA 3D printed dental parts such as crown and bridge materials is reported [122].

The printing parameters required for FDM, SLS, and SLA 3D printing are collectively summarized in Figure 3. The initial printing parameters such as quality of the .stl file, part orientation, and environmental factors are common printing process parameters in all three techniques. However, due to the variation in the structure formation technique, the printing process and a number of process parameters differ in each of these methods. The printing techniques can be chosen according to the requirement of the simplicity of printing, mechanical properties, printing time and layer resolution. For instance, SLA has the capability of printing high-resolution parts of up to 10 μm [33], while the minimum layer resolution of the SLS printed part is 20 μm [123] and the FDM only has the capability of printing high-resolution parts of up to 40 μm [124]. The value for the print resolution should be considered as a comparative guide only because the exponential growth of the 3D printing industry is continuously offering optimized versions of the 3D printers. The print resolution of some common commercially available 3D printers is listed in Table 1. Based on the data provided by the manufacturer, the print resolution up to 25 μm [125], 50 μm [126], 1 μm [127] for FDM, SLS, and SLA, respectively, is also claimed. On the other hand, in terms of the process simplicity in printing, the FDM is the most suitable because the process is as simple as heating the filament polymer to a semi-solid state and depositing it directly on the print bed. SLS is a comparatively complicated process among others, it requires the movement of two systems: roller and laser light. A list of commercially available FDM, SLS, and SLA 3D printers along with their material and print specifications is summarized in Table 1. This table provides an overview for selecting the desired 3D printer on the basis of the required print volume, material, and print process. From Table 1, it is also evident that the FDM provides a wide range of materials for 3D printing while SLA provides a high print resolution. Further discussion based on the material and print part properties is presented in Sections 3 and 4.

Table 1. A directory of commercially available industrial-grade FDM, SLS, and SLA 3D printers and their specifications, information is collected from the supplier's website. The green highlight indicates the print material best suitable for the given printer according to the supplier. For SLA, the materials are categorized based on their type/specific properties as claimed by the supplier.

AM Technology	Name	Dimension of Printer (mm)	Print Volume (cm³)	Layer Thickness (mm)	Available Material/Type					Ref.
FDM	Stratasys F900	914 × 609 × 914	508,756.16	0.127–0.5	PLA	ABS	PEEK	Nylon	ULTEM	[128]
	Essentium HSE 280i HT	695 × 495 × 600	206,415	0.1–0.55	PLA	ABS	PEEK	Nylon	ULTEM	[129]
	CreatBot PEEK-300	300 × 300 × 400	36,000	0.04–0.4	PLA	ABS	PEEK	Nylon	ULTEM	[130]
	Anisoprint ProM IS 500	600 × 420 × 300	75,600	0.06	PLA	ABS	PEEK	Nylon	ULTEM	[131]
	3DGence F420	380 × 380 × 420	60,648	0.05	PLA	ABS	PEEK	Nylon	ULTEM	[132]
	Roboze Argo 500	500 × 500 × 500	125,000	0.025–0.2	PLA	ABS	PEEK	Nylon	ULTEM	[125]
	WASP 4070 Tech	400 × 400 × 700	112,000	0.1	PLA	ABS	PEEK	Nylon	ULTEM	[133]
	Cincinnati MAAM	1050 × 1015 × 1015	1,081,736.25	0.2	PLA	ABS	PEEK	Nylon	ULTEM	[134]
	Tractus 3D T850P	280 × 280 × 400	31,360	0.01–0.8	PLA	ABS	PEEK	Nylon	ULTEM	[135]
	AON-M2+	450 × 450 × 640	129,600	0.05–0.5	PLA	ABS	PEEK	Nylon	ULTEM	[136]
	Kumovis R1	180 × 180 × 150	4860	0.1–0.4	PLA	ABS	PEEK	Nylon	ULTEM	[137]
	Ultimaker S5	330 × 340 × 300	33,660	0.02–0.25	PLA	ABS	PEEK	Nylon	ULTEM	
SLS	Sintratec Kit	100 × 100 × 100	1000	0.05–0.15	PA 12	PA 11	TPU	TPE	PP	[138]
	Red Rock 3D	180 × 180 × 180	5832	0.1	PA 12	PA 11	TPU	TPE	PP	[139]
	Sinterit Lisa Pro	110 × 160 × 245	4312	0.05	PA 12	PA 11	TPU	TPE	PP	[140]
	Formlabs Fuse 1	165 × 165 × 300	8167.5	0.1	PA 12	PA 11	TPU	TPE	PP	[141]
	Sintratec S2	⌀160 × 400	8038.4	0.1	PA 12	PA 11	TPU	TPE	PP	[142]
	Sharebot SnowWhite 2	100 × 100 × 100	1000	0.05	PA 12	PA 11	TPU	TPE	PP	[143]
	Wematter Gravity	300 × 300 × 300	27,000	0.1	PA 12	PA 11	TPU	TPE	PP	[144]
	XYZ printing MfgPro230 xS	230 × 230 × 230	12,167	0.08–0.2	PA 12	PA 11	TPU	TPE	PP	[145]
	Nexa3D QLS350	350 × 350 × 400	49,000	0.05–0.2	PA 12	PA 11	TPU	TPE	PP	[146]
	Shining 3D EP-P3850	380 × 380 × 500	72,200	0.08–0.3	PA 12	PA 11	TPU	TPE	PP	[147]
	Prodways Promaker P1000	300 × 300 × 360	32,400	0.06–0.12	PA 12	PA 11	TPU	TPE	PP	[148]
	EOS Formiga P 110 Velocis	200 × 250 × 330	16,500	0.06–0.12	PA 12	PA 11	TPU	TPE	PP	[149]
	3D Systems ProX SLS 6100	381 × 330 × 460	57,835.8	0.08–0.15	PA 12	PA 11	TPU	TPE	PP	[150]
	Farsoon eForm	250 × 250 × 320	20,000	0.06–0.3	PA 12	PA 11	TPU	TPE	PP	[151]
SLA	Nyomo's Minny	44 × 28 × 70	86.24	0.01	Standard resin	Castable	Bio compatible	Flexible	Clear	[152]
	Asiga's Pico 2	51 × 32 × 76	124.032	0.001	Standard resin	Castable	Bio compatible	Flexible	Clear	[127]
	XYZprinting's Nobel 1.0 A	128 × 128 × 200	3276.8	0.025–0.1	Standard resin	Castable	Bio compatible	Flexible	Clear	[153]
	Formlabs Form 2	145 × 145 × 175	3679.375	0.025–0.2	Standard resin	Castable	Bio compatible	Flexible	Clear	[154]
	Photocentric's Liquid Crystal	121 × 68 × 160	1316.48	0.05	Standard resin	Castable	Bio compatible	Flexible	Clear	[155]
	Nexa3D's the NXV	220 × 120 × 380	10,032	0.03	Standard resin	Castable	Bio compatible	Flexible	Clear	[126]
	DWS's XPRO S	300 × 300 × 300	27,000	0.01	Standard resin	Castable	Bio compatible	Flexible	Clear	[156]
	UnionTech's RSPro 800	800 × 800 × 550	352,000	0.07–0.25	Standard resin	Castable	Bio compatible	Flexible	Clear	[157]
	3D Systems' ProX 950	1500 × 750 × 550	618,750	0.01	Standard resin	Castable	Bio compatible	Flexible	Clear	[158]

3. Print Materials

The common print materials available for FDM, SLS, and SLA 3D printing are given in Table 2. The material for FDM and SLS are thermoplastic polymers. Due to the process requirements, for FDM, the material is in filament form, while for SLS, it is in powder form. The thermoplastic polymers can be classified into amorphous and semi-crystalline thermoplastic. Amorphous thermoplastic polymers have a glass transition temperature (T_g) above which they soften and transform into a glassy state. They do not have a fixed melting temperature, while semi-crystalline polymers have a fixed glass transition temperature (T_g) and melting temperature (T_m). The melt viscosity of the semi-crystalline thermoplastics decreases with the increase in temperature above the melting temperature (T_m)—allowing flowability [159]. Apart from the pure polymers, the use of several modified FDM filaments has also been reported for various applications such as printed electronics by incorporating different materials such as carbon-black, graphene, and copper [160], carbon nanotube incorporated filaments for textile [161], carbon nanotubes incorporated capacitive and piezoresistive actuators [162], etc. A few approaches of the modification of materials for FDM/SLS printing by giving an example of magnetic materials is discussed in Section 5.

Table 2. A general classification of available materials (commercial and laboratory-grade) for FDM, SLS, and SLA printing.

AM Technique	Material		
FDM	Thermoplastic filament	Semi-crystalline	PEEK
			PVDF
			PP
			PLA
			TPU
			TPE
			PPS
			PCL
			PLGA
			PEVA
			PA6
			PA12
			POM
			PET
		Amorphous	PEI
			PAI
			PPSU
			PC
			PVA
			HIPS
			PEKK
			ASA
			ABS
			PMMA
			PS

Table 2. *Cont.*

AM Technique	Material		
SLS	Thermoplastic powder	Semi-crystalline	PA12
			PA11
			PA6
			PET
			PLA
			PCL
			TPU
			POM
			PEEK
			PEK
			PEKK
		Amorphous	PC
			PMMA
			PS
			PI
			PSU
			PES
			PVA
SLA	Resins	Polyester	PPF
			PLA
			PCL
			PCL/PEG/Chitosan
		Polycarbonate	PTMC
			PTMC/Gelatin
			Trimethylolpropane Carbonate
		Polyether	PEG
			PEG/Chitosan
			PEO/PEG
			Poly tetrahydrofuran ether

The materials used in SLA are photosensitive thermoset polymers. Thermoset is also known as a thermosetting polymer and is a polymer that is obtained by irreversibly polymerizing/curing a soft solid or viscous liquid prepolymer (resin). The curing, also sometimes known as solidification or vulcanization or polymerization, is achieved via photopolymerization in the presence of UV light. An SLA resin usually contains several components including monomer/oligomer, diluent, chain transfer agent, and photoinitiator [84,163]. Monomers/oligomers are reactive prepolymers that are primarily responsible for the part properties after undergoing a polymerization reaction. Diluents are low-molecular weight, low-viscosity compounds used to modify the viscosity of a resin or enhance the solubility of a resin. A chain transfer agent is essential to modify the crosslinking agent while photoinitiator is necessary to trigger the photopolymerization. The widely used resins are polyester or polycarbonate or polyether-based polymers in SLA or the vat-based printings [24,84,163].

3.1. Material Requirements

The material requirement majorly depends on the process of 3D printing. Figure 5 provides a pictorial summary of the material requirements for successful printability for all three print methods. Rheological properties are a common requirement of print materials in all three processes. It mostly includes the viscosity of the print material (polymer melt or resin). The thermal properties of the material include heat capacity, coefficient of thermal expansion, crystallinity, and conductivity of heat of the print materials. The thermal properties are an important consideration in FDM and SLS. On the other hand, in SLS and SLA, as both of these processes deals with absorption of energy through laser light, optical properties including reflection, absorption, transmission, and scattering are of utmost importance. In FDM, the mechanical properties of the filament including its elastic modulus and strain at yield is also considered. In SLS, extrinsic properties such as powder shape and powder surface are a considerable requirement. For SLA, chemical properties such as active centre stability, molecular weight, functional group, and degree of functionality of the resin play a key role. The details of material requirements and their significance are discussed in the following sections. Here, we first discuss individual process requirements of FDM, SLS, and SLA respectively and followed by the common requirements.

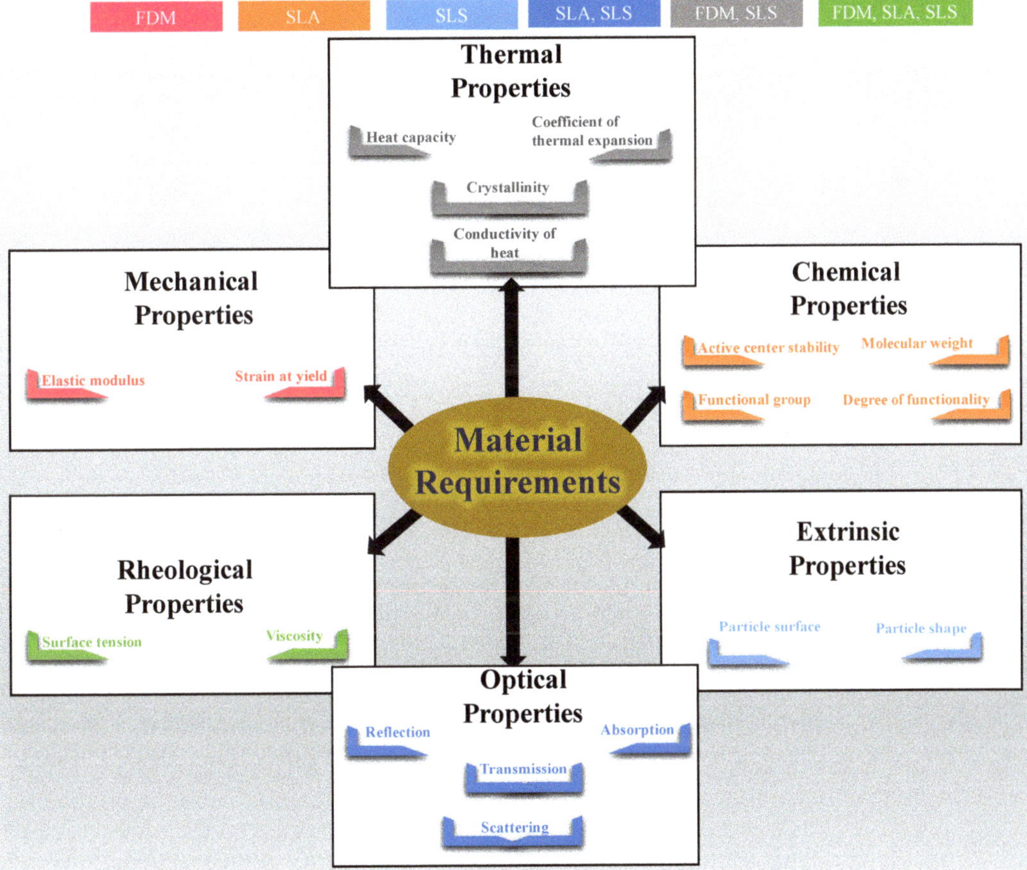

Figure 5. A pictographic summary of the various properties of print materials demanded for successful printability via FDM, SLS, and SLA 3D printing.

3.1.1. Mechanical Properties

In FDM, the mechanical properties of thermoplastic filament are one of the major material properties to be understood. The column strength of the solid filament is significant for thermoplastic polymers [164]. In the printing process, the solid filament serves as a piston under compression; therefore, the column strength should be sufficient enough to avoid buckling between the driving pulley and the melt chamber [52,165]. The critical load for buckling is given by the formula derived from Euler's buckling as given in Equation (3).

$$P_{cr} = \frac{\pi^2 E d^2}{16L^2}$$

(3)

where E is the elastic modulus, d is the diameter, and L is the length of the solid filament between the driving pulley and the melt chamber [165]. The filaments should however be flexible enough to allow their spooling and despoiling during printing and thus maximum strain at yield is recommended to be about 5% [166,167].

3.1.2. Extrinsic Properties

In SLS, particle shape, size, and distribution have a considerable influence on the overall flow behaviour and powder density. It is therefore an important consideration in SLS as it influences the thin, dense, and smooth layer of powders—influencing the quality of the produced part. The SLS powder particle distribution is reported to be between 20 μm and 80 μm [79], some particles have a higher size but mostly a d_{50} of around 60 μm [168]. The particle shape is required to be ideally spherical. Schmidt et al. [169] used the tensile strength to determine the powder flowability and concluded that the increase in powder flowability resulted in the decrease of tensile strength for the spherical particles.

3.1.3. Chemical Properties

In SLA, the chemical properties of the resin are substantial material properties to be understood as the process comprises the photopolymerization completely driven by chemical reaction to convert the liquid into a solid object in the presence of UV light. Herein, curing kinetics is the most important consideration [170]. The curing kinetics is influenced by the degree of functionality, steric effect, and the stability of radical or cationic active centres, for more detail please refer to [171]. A moderate curing rate is required to enable the faster part fabrication and at the same time provide sufficient time for interlayer adhesion. A significant difference in static and dynamic properties has been reported for the curing time of 5 min in comparison to that of 25 and 30 min [172]. The curing degree increases with an increase in light intensity. For instance, the increase in curing degree from 3.1% to 87.7% with the increase in light intensity from 5 mW/cm^2 to 40 mW/cm^2 is reported [173]. The same study [173] also presents the influence of exposure time on the curing degree. The curing degree significantly increases from 26.85% to 70.98% for the same increase in light intensity. On the other hand, the curing degree only increases from 70.98% to 81.74% with the increase in exposure time from 3 s to 12 s.

3.1.4. Thermal Properties

The thermal properties of the print material are highly influential material properties and need to be understood in-depth for successful printing via the SLS and FDM techniques.

In SLS, the requirement of the laser energy intensity also depends on the temperature of the powder. The polymer powders are heated to a temperature close to the melting temperature for semi-crystalline powders and up to glass transition temperature for amorphous powders to lower the required laser energy and reduce the temperature gradient which also decreases the non-uniform shrinkage in the printed parts [174,175]. The preheating temperature should be close to melting temperature but should not be greater than the onset melting temperature to avoid premature melting of the powders. For general polymer materials, a preheating temperature 5–10 °C lower than the glass transition temperature is suggested [115]. Thus, the processing temperature must be precisely controlled between onset melting and onset crystallization temperature, and this metastable thermodynamic region is called the sintering window (Figure 6a) [176]. The sintering window can be characterized using a differential scanning calorimeter at a fixed heating and cooling rate, for example, 10 °C/min [168]. The sintering window, however, depends on the polymer being used. A wider sintering window is usually preferred in SLS. Figure 6b shows the thermo-analytical results (DSC measurement) of a commercial injection moulding PA12 grade in comparison with a commercial PA12 for SLS processing [168,177]. The stretch in the sintering window of SLS powders (red curves) can clearly be marked [168].

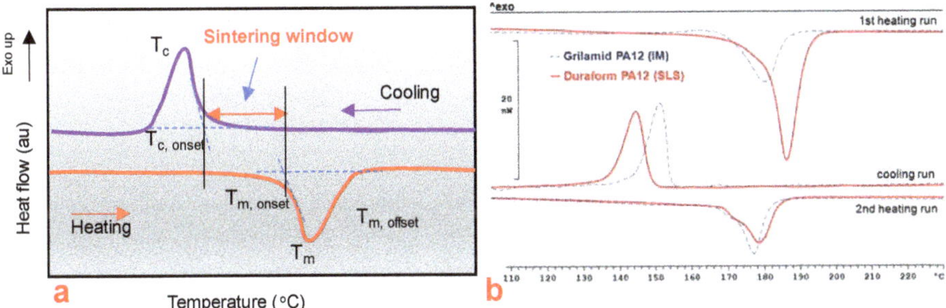

Figure 6. (**a**) Illustration of a dynamic DSC curve of a polymer. (**b**) Comparison of a commercial injection moulding PA12 grade and commercial PA12 for SLS processing, adapted from [168].

In FDM, the filaments are heated at a temperature a few degrees above the melting temperature at the nozzle. Reduced viscosity with increased temperature facilitates the polymer melt extrusion. The thermal properties of thermoplastic filament, moreover, influence the part shrinkage after the polymer melt deposition. The thermal properties include a coefficient of thermal expansion, heat capacity, heat conductivity and crystallinity of the polymer, please refer to these articles for more detail [178,179]. Similar to SLS, here, the thermal gradient leads to uneven shrinkage of printed parts [166]. For FDM, usually, the amorphous polymer filaments are favoured in comparison to semi-crystalline. The amorphous thermoplastics possess a low coefficient of thermal expansion—as a result, lower shrinkage, warpage, and distortion of the printed parts [180]. Another important consideration for FDM filaments is the printing temperature. This becomes an even more important consideration specifically for customized filaments with sensitive ingredients [181]. Usually, the printing temperature has to be above the melting point and should be always lower than the thermal degradation temperature of the print material. Thermogravimetric analysis (TGA) is used to characterize the filament material thermal stability by monitoring the weight change that occurs as the sample is heated at a constant rate.

See Tables 3 and 4 for a summary of the thermal properties of a few common thermoplastic filaments for FDM, and powders for SLS.

3.1.5. Optical Properties

Optical properties are a key requirement of print material for its successful use in SLS and SLA 3D printing as materials absorb light in both processes.

The absorption of the energy from the laser source by the material is dependent on its optical properties. In SLS, a process involving the melting of polymer powders in presence of laser-generating heat energy, the polymer should be able to effectively absorb energy from the laser at a given wavelength. However, only a fraction of the energy is absorbed due to the laser reflection and refraction at the particle surface and transmission through the particles [182]. Most of the commercial SLS printers use CO_2 lasers. This is because the polymer powders contain a C–H bond which absorbs the energy at the laser wavelength of 10.6 μm [183]. The thermoplastic powders after being exposed to the CO_2 laser is transformed from an entropy elastic state to a viscous state [182]. To avoid warpage and shrinkage the laser power is desired to be on the lower side. With other printing parameters constant, the increase of laser power from 20 W to 35 W results in an increase in shrinkage from 2.34% to 2.60% and warping from 0.16 mm to 0.21 mm [34].

In SLA, the optical properties of the resin such as transmission, absorption, reflection, and scattering influence the curing depth. Detailed studies reporting the influence of these optical parameters on the cure depth (Equation (2)) are still lacking in the literature. It is recommended that the penetration depth is defined as the depth where laser irradiation is reduced by $1/e$ [24,184]. The absorption of light also highly depends on the concentration of the photoinitiator and the molar extinction coefficient of resin at the given light wavelength [185].

3.1.6. Rheological Properties

The rheological properties of resin highly influence the SLA process and the melt rheology of powders to be used play an essential role in SLS.

The resins used in SLA must possess a melting temperature below the room temperature. The viscosity should ideally be around 1 Pas but it can range from 0.1 Pas for low-molecular-weight polymers to 10 Pas for high molecular weight polymers [186]. The lower viscosity allows the resin to be in a liquid state at the processing temperature enabling chain mobility. For the resins with higher viscosity, the resins can be processed at higher temperatures but this is limited only to formulations that are insensitive to heat [187].

In SLS, a powder with lower surface tension (γ) and lower zero viscosity (η_0) is desired. It is because powder with lower surface tension has higher coalescence which can be sintered into parts of higher density and strength. The requirement of lower η_0 is the reason behind the difficulty in sintering amorphous powders as the result leads to brittle and amorphous parts because of η_0. In amorphous powder, the η_0 is higher even above the glass transition temperature and thus a proper coalescence does not take place [79].

The investigation of the rheological properties of FDM filaments is well described in [188], in which applicability of the Filament Flow Index (FFI) is reported for a number of different filaments for FDM and suggested that the FFI technique can be considered to promptly characterize print filament. Elsewhere, to avoid buckling, the ratio of elastic modulus and viscosity of the FDM filament melts less than $3 \times 10^5 \text{ s}^{-1}$ is recommended [166].

Table 3. Melting temperature and glass transition temperature of a few common thermoplastic filaments for FDM.

	Material	T_m (°C)	T_g (°C)	Printing Temperature (°C)	Temperature of Degradation (°C)	Ref.
	ABS	-	105	230–250	380–430	[189,190]
Thermoplastic Filament	PLA	150	-	200–235	300–400	[190,191]
	PET	255	75	160–210	350–480	[190,191]
	PP	165	−10	230–260	300–500	[191,192]
	PA6	215	46	419.8	220–270	[189,193]

Table 4. Melting temperature, onset melting temperature, crystallization temperature, and sintering window of a few common thermoplastic powders for SLS.

	Material	T_m	T_m, Onset	T_c, Onset	Sintering Window	Ref.
	PA12	185.6	178.0	158.6	19.4	[108]
Thermoplastic powder	PA11	202.9	189.2	168.3	20.9	[194]
	TPU	144.4	122.2	123.9	1.7	[194]
	PC	167.1	157.0	121.9	35.1	[195]
	PP	182.3	177.1	151.3	19.5	[196]

4. Properties of Printed Parts

The tensile and flexural mechanical properties of the FDM, SLS, and SLA printed parts are discussed.

For commercially available print materials, the mechanical properties of the 3D printed parts are generally provided in the datasheet by most manufacturers. A pictographic overview of the mechanical properties based on the data provided by the manufacturer is given in Figure 7. In each 3D printing process, a range of variations in mechanical properties with different print materials can be noticed. This overview can be used to quickly screen the material and printing processes on the basis of the requirement of tensile and flexural properties. Modulus and ultimate strength are two key mechanical properties to be understood either in tensile or flexural loading. The data provided in the datasheet, however, might be based on a certain external condition favourable for generating optimum properties, therefore, these data should not be considered as a final property of the printed part for bespoke print conditions. Furthermore, it should be noted that selecting a print method and materials are also related to other properties such as fatigue properties, microstructure, stability, etc. [197–201]. In the following, the mechanical properties (tensile and flexural) of various print materials and printing processes reported by bespoke studies considering various print conditions are discussed. A summary of mechanical properties reported by a few bespoke studies is presented in Table 5 (tensile properties) and Table 6 (flexural properties). Anisotropy in mechanical properties is observed in all FDM, SLA, and SLS printed parts. This is primarily due to the fundamental process of the part production method in 3D printing (i.e., layer by layer addition of material). Typically, the mechanical properties of the 3D printed parts printed with sample build orientation parallel and perpendicular to the bed or print platform are found to be different, which signifies the anisotropy. Therefore, a part/object should be printed at an optimal orientation to achieve the best mechanical characteristics to meet the demand of the targeted application.

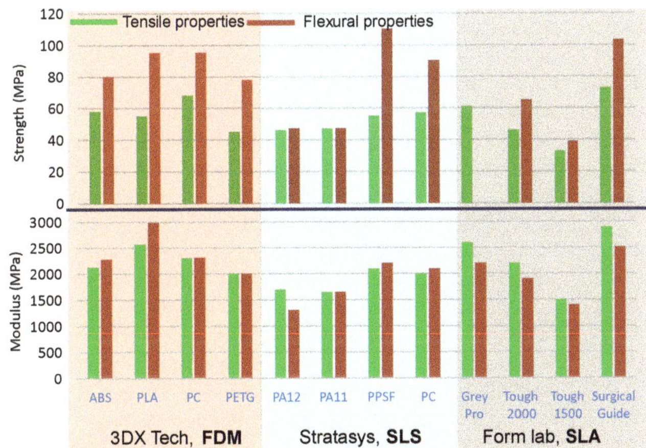

Figure 7. A graphical overview of the mechanical properties of 3D printed parts for a few commercially available materials, data are taken from the respective datasheet available on the supplier's website. Tensile (green coloured) and flexural (red coloured) properties are plotted, the upper graph is strength, and the lower graph is the modulus of corresponding properties of FDM and SLS and SLA printed parts.

The variations in tensile and flexural strength with the change in build orientation and layer thickness in FDM, SLS, and SLA are discussed in the following section.

Please note that, apart from tensile test and flexural tests, nanoindentation is another prominent test for investigating mechanical properties such as modulus, hardness, and elasticity [202–204]. The biggest advantage of the nanoindentation test for the 3D printed part, compared to conventional tests, is that this test method can be used to study the localized anisotropy at various locations on the printed surface with minimum destruction limited to the surface of the material.

4.1. Tensile Properties

Tensile properties are used to study the behaviour of a material under the action of tensile loads. The tensile properties of a few standard materials available for FDM, SLA, and SLS 3D printing are given in Table 5. From the table, it can be concluded that the tensile properties of the polymers highly depend on the material, build orientation, and layer thickness. In FDM, the tensile properties of PLA and ABS have been most prominently studied while the properties of high-performance polymers like PEEK are also available. In SLS, the most widely used polymer powder is PA12 and its mechanical properties are widely studied. On the other hand, in SLA, the material depends on the manufacturer and the application for which it has to be used. The printing parameters such as laser power and bed temperature are varied based on the material being used while materials are printed with various layer thicknesses to alter the printing time which also alters the part strength. In most of the studies, the ISO 527 and ASTM D638 test standards have been used to determine the behaviour of the 3D printed parts under the influence of tensile loading.

The tensile properties of FDM 3D printed parts have been well studied [205–207]. Printing can be performed at various orientations as illustrated in Figure 8 and the mechanical properties, therefore, are influenced by the print orientations. One common finding is that the tensile strength/modulus, when the load is applied in the longitudinal direction, is higher than applying load along the build direction; this is simply due to weak interlayer bonding of the printed parts [208]. Another factor contributing to the tensile strength of the FDM printed part is the raster angle. For instance, the ultimate strength, for PLA, obtained for a raster angle of 45° is higher compared to the raster angles of 0° and 90° [209]. For materials such as PEEK and ABS, the tensile strength of the printed material for raster angles of 0° and 90° was comparable to one another while the raster angle of 45° yielded a considerably lower amount of tensile strength [210–212]. The amount of material that has been deposited on each layer also affects the mechanical properties of FDM printed parts. The tensile strength increases linearly with the layer thickness when specimens are printed in the z-direction [213]. A study by Chacón et al. [213] can be consulted to get a comprehensive summary of the effect of process parameters on mechanical properties of FDM printed PLA and their optimal selection.

There is also a large disparity of mechanical properties for SLS 3D printed parts due to the dependence of various parameters on local process conditions [112]. This leads to properties such as modulus/stiffness and strength being highest along the print direction [214]. The parts built with orientations parallel to the direction of the laser exhibit the highest strength and modulus values while the samples built in the z-axis orientation possess the lowest strength and modulus. For example, the difference of 9.4% in strength and 7% in flexural modulus for these different build orientations is reported [215]. Furthermore, the specimens with the 60° raster angle exhibited the highest tensile strength when compared to the sample printed in 0°, 15°, 30°, 45°, 60°, 75°, and 90° orientations [216], unlike in FDM in which a 45° orientation showed the highest strength [209].

The tensile properties of a commercial photocurable resin (commercially available/manufacturer's grade) have also been widely studied [217,218]. The tensile strength of various build orientations which include flat, and edge are widely reported [217,219,220]. In SLA, flat is similar to x and the edge is similar to y as in FDM printing. Furthermore, each build orientation had sub-orientations of 0°, 45°, and 90°, as in other print methods. Specimens with edge build orientation display higher tensile strength compared to the specimens with flat build orientation. In sub-build orientation, 45° orientations have slightly better properties than the 0° and 90° sub-build orientations in both cases [221]. Build orientation has much less impact on tensile strength when compared to layer thickness [216]. Tensile strength increases when the layer thickness increases while the flexural strength decreases [216]. The increase in tensile strength as layer thickness increases is due to better connection by polymerization of the new layer with the prior layer [74].

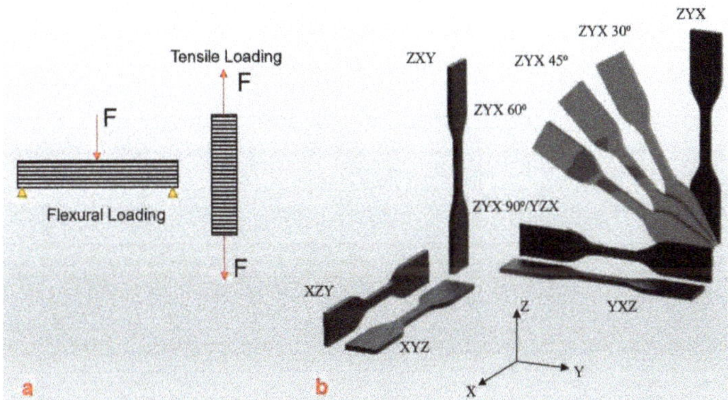

Figure 8. (**a**) Illustration of flexural and tensile loading and (**b**) definition of specimen orientation, redrawn from [16].

4.2. Flexural Properties

Flexural properties are used to determine the behaviour of a material under the action of bending loads. The flexural properties of 3D printed parts have not been studied as widely as the tensile properties. However, there are many studies that have given the flexural properties of some popular polymers of each 3D printing process. Table 6 is the compilation of flexural properties of various polymers fabricated using FDM, SLS, and SLA. The 3-point bending test is used in all of these studies. ASTM D790 is the most commonly used standard test method for the 3-point bending test. Like the tensile properties, the print parameters such as raster angle, part orientation, and layer thickness have a direct impact on the flexural properties of the 3D printed specimen.

In FDM, variation in mechanical properties with varying orientation and layer thickness can be distinctly observed. Layer thickness has the most significant effect on flexural strength. The increase in layer thickness is found to have an increment in flexural strength. For instance, a study focused on flexural strength of the specimen at different layer thicknesses ranging from 0.1 mm to 0.5 mm, reported a maximum flexural strength (59.6 MPa) at 0.5 mm layer thickness and minimum flexural strength (43.6 MPa) at 0.1 mm [222]. The raster angle also has significance on the flexural property [211]. Again, like in tensile loading, the PLA parts with a 90° raster angle showed the least resistance while the 45° orientation showed the highest resistance [209].

In SLS, the flexural strength is again influenced by bed temperature, laser watt power, scan speed, and scan spacing. For instance, the increase in flexural strength with an increase in laser power from 28 W to 36 W and a decrease in flexural strength with an increase in scan speed from 2500 to 4500 mm/s is reported [223]. In the case of the scan spacing, an initial decrease with an increase in scan spacing from 0.25 to 0.35 mm, then a marginal increase from 0.35 to 0.45 mm is reported [223]. Print orientation also has a notable influence on the flexural strength of SLS printed parts. For example, a maximum flexural strength at 0° (59.23 MPa) followed by at 45° (46.25 MPa) and minimum at 90° (19.89 MPa) is reported [224].

Table 5. A summary of the tensile properties of a few common print materials for all three, FDM, SLS, and SLA, printing techniques. Information such as print setting, test standard method, print orientation, tensile modulus, tensile strength, and elongation at break are provided.

Polymer		Supplier	Print Setting	Test Standard	Print Orientation	Tensile Modulus (MPa)	Tensile Strength (MPa)	Elongation (%)	Ref.
Thermoplastic Filament	PLA	3D Systems	100% infill Layer thickness 0.2 mm	ASTM D638	XYZ	1538	38.7	-	[212]
					YXZ	1246	31.1	-	
					XYZ	1350	33.6	-	
	ABS	Qimei Stock, China	100% infill	ASTM D638	XYZ	1200	37	-	[210]
	PEEK	Arevo Labs	100% infill Bed temperature 230 °C	ASTM D638	XYZ	2871	71.36	5.01	[211]
	PC	Stratasys, USA	100% infill	ASTM D638	XYZ	2410	54.6	4.22	[225]
	PP		100% infill Nozzle temperature 165 °C	DIN 53504-S3a	XYZ 0°	1230	34.3	-	[226]
					XYZ 45°	1000	32.0	-	
					XYZ 90°	1050	33.0	-	
ThermoplasticPowder	PA-12	Sinterit	Laser thickness 0.175 mm	ISO 527	ZXY 0°	864 ± 72	42.5 ± 3.1	13.1 ± 2.3	[227]
					ZXY 30°	690 ± 143	28.1 ± 8.4	6.7 ± 1.6	
					ZXY 45°	613 ± 27	16.0 ± 2.3	2.7 ± 0.3	
					ZXY 60°	694 ± 32	25.6 ± 8.9	8.4 ± 5.7	
					ZXY 90°	426 ± 150	17.1 ± 10.0	6.0 ± 3.4	
		Duraform	Layer thickness 0.1 mm Part bed temperature 175 °C Laser power 38 W	ISO 527-1	ZXY	1675 ± 41	47.6 ± 1.5	6.6 ± 0.7	[176]
					YXZ	1610 ± 61	40.6 ± 3.4	3.7 ± 0.6	
		Orgasol IS	Layer thickness 0.1 mm Part bed temperature 164 °C Laser power 48 W	ISO 527-1	ZXY	1700 ± 25	54.7 ± 0.7	12 ± 0.4	[176]
					YXZ	1580 ± 21	29.3 ± 3.6	1.9 ± 0.3	
		EOSINT	Laser power 3.33 W Powder bed temperature 140 °C			205.0 ± 29.3	57.7 ± 10.3	11.5 ± 1.3	[228]
	PA-11		Building chamber temp 157 °C Powder bed temperature 177 °C Layer thickness 0.3 mm	ISO 527-2			7.1 ± 0.5	5.9 ± 0.5	[177]
	PP		Powder bed temperature 150 °C Layer thickness 0.12 mm			-	27.9	-	[196]
		Trial Corporation	Powder bed temperature 150 °C Laser power 13.75 W Layer thickness 0.15 mm	ISO 527-2	ZYX	599.1 ± 14.1	19.9 ± 3.5	122.25	[229]
	PA 6	Mazzafero Tecnopolímeros S.A.	Powder bed temperature 120 °C Laser power 2.34 W			166.6 ± 77.8	62.4 ± 16.0	10.9 ± 3.7	[228]

Table 5. *Cont.*

	Polymer	Supplier	Print Setting	Test Standard	Print Orientation	Tensile Modulus (MPa)	Tensile Strength (MPa)	Elongation (%)	Ref.
	PC	HRPS	Laser power 13.5 W Layer thickness 0.15 mm Bed temperature 100 °C	ISO527-2	ZYX	40.12	1.1	5.05	[230]
	PR 48	Autodesk, USA	Layer thickness 50 μm Print resolution		ZXY 0°	723	-	-	[219]
					ZXY 45°	350	-	-	
					ZXY 90°	376	-	-	
					ZXY 0°	901.4	-	-	
					ZXY 45°	667.1	-	-	
					ZXY 90°	182.2	-	-	
Resin	Clear V4	Formlabs	Layer thickness 50 μm	ISO 527	Mean 0,15°,30°,45°, 60°,75°,90°	2298	60.8	8.05	[217]
	Watershed 11122	DSM Somos	Layer thickness 0.175 mm Laser power 2.5 W Laser scanning speed 3200 mm/s	ASTM D638	XYZ 0°	37.75 ± 1.82	3.45 ± 0.11	11.67 ± 4.97	[221]
					XYZ 45°	43.25 ± 0.98	3.51 ± 0.03	7.60 ± 3.48	
					XYZ 90° /YXZ	38.24 ± 2.22	3.26 ± 0.08	8.53 ± 4.29	
					XZY 0°	46.07 ± 0.99	3.54 ± 0.07	9.27 ± 1.10	
					XZY 45°	47.70 ± 0.52	3.65 ± 0.02	9.00 ± 3.57	
					XZY 90° /YZX	45.72 ± 0.48	3.50 ± 0.05	6.60 ± 0.30	
	Monomer: EGPEA Crosslinker: 1,6-hexanediol diacrylate Photoinitiator: 2-Benzyl-2-(dimethylamino)-4′-morpholinobutyrophenone		Monomer 1, crosslinker 0.4 Monomer 1, crosslinker 1.0	ASTM D638		18.026 ± 0.302	1.861 ± 0.435	0.106 ± 0.025	[220]
						36.586 ± 1.210	2.243 ± 0.709	0.062 ± 0.021	

Table 6. A summary of the flexural properties of a few common print materials for all three, FDM, SLS, and SLA, printing techniques. Information such as print setting, print orientation, flexural modulus, flexural strength, and elongation at break are provided.

Polymer		Supplier	Print Settings	Sample Orientation	Flexural Modulus (MPa)	Flexural Strength (MPa)	Elongation (%)	Ref.
Thermoplastic Filament	ABS		100% infill	XYZ	1750	60	3	[231]
	PLA		100% infill Bed temperature 230 °C Nozzle temperature 210 °C	XYZ 0°	3187	102.203	10.6	[209]
				XYZ 45°	2985	90.649	7.8	
				XYZ 90°	3000	86.136	4.5	
	PEEK	Arevo Labs	100% infill Bed temperature 230 °C Nozzle temperature 340 °C	XYZ 0°	1972	114	10.6	[211]
				XYZ 90°	1954	83.59	5.81	
				XYZ 0° and 90°	2146	88.70	6.58	
ThermoplasticPowder	PA-12	Duraform, 3D Systems			546 ± 28	86 ± 5	11 ± 0.5	[232]
	PC		Laser power 13.5 W Layer thickness 0.15 mm Bed temperature 100 °C		93.83	2.48	-	[230]
	PA 2200	EOS	Laser power 25 W Building chamber temperature 170 °C	XZY 0°	551.24 + −5.6	59.23 + −4.1	4.9 + −0.74 mm	
				XZY 45°	433.05 + −61.4	46.25 + −6.4	4.96 + −0.56 mm	
				XZY 90°	345.39 + −41.5	19.89 + −2.8	3.28 + −1.51 mm	
Resin	Freeprint splint	DETAX GmbH			-	19.5 ± 2.5	-	
	LuxaPrint Ortho Plus	DMG GmbH			-	39.3 ± 2.0	-	[233]
	Nextdent Ortho Clear	Vertex-Dental B.V.			-	91.3 ± 5.9	-	
	Dental SG resin	Formlabs	Layer thickness 0.05 mm	ZXY 0°	1654.35 ± 152.27	117.48 ± 12.39	-	[88]
				ZXY 45°	1467.56 ± 89.36	130.73 ± 5.12	-	
				ZXY 90°	1456.73 ± 149.83	135.69 ± 5.93	-	

In SLA, the build orientation is the main factor influencing flexural strength. The specimen printed with a layer orientation parallel to the axial load is reported to have superior flexural strength and flexural modulus [88]. For example, maximum flexural strength was found at 90° (135.69 MPa), compared to 45° (130.73 MPa) and 0° (117.48 MPa) at ZXY orientation [88].

In Tables 5 and 6, the findings of a number of different studies are presented for the comparison of mechanical properties under the action of tensile and flexural loading of FDM, SLS, and SLA printed parts. The tabular summary can be referred to screen and select a suitable print process and material to meet the requirement of the targeted application in terms of the mechanical properties.

5. Towards Magneto-Active 4D Printing

The 3D printing of smart materials is regarded as 4D printing (4DP) since it is first introduced in 2013 [234]. Smart materials can respond to external stimuli such as heat, pH, magnetic/electric field, etc. [235–240]. As illustrated in Figure 9, 4D printing essentially means the 3D printing of smart materials. The 4D printed structures can change their physical/chemical properties, for example, stiffness, density, etc., and demonstrate various phenomena such as shape memory effects and shape-shifting [235–238,241–244]. The shape memory effect is a phenomenon where a system/structure can remember a certain shape and could be switched from one to another shape (original to programmed shape) in a controlled way in the presence of external stimuli. Shape-shifting is a phenomenon where a system/structure can shift its shape from one to another when triggered by external stimuli. Here, we briefly cover the process and the material requirements to develop polymer-based 4D structures via FDM, SLS, and SLA printing. Our particular focus is magnetic field triggered systems.

The printing materials (resin/powder/filament) must have a magnetic field responsive element(s)/component(s) (typically known as fillers) to be triggered by an external magnetic field to demonstrate the 4D effect. Therefore, the first essential step to developing 4D structures via 3DP is to modify the printing materials by incorporating active components. The widely used magneto-active filler materials are carbonyl iron powders (CIPs), Iron (II, III) oxides, and Fe-Nd-B micro/nanoparticles [245–248]. All these magnetic fillers are yet to be implemented in all 4D printing methods (considered here) mostly due to the filler size. There are, however, other printing methods such as direct ink write (DIW) where nano to micron-sized fillers have been used successfully to a greater extent [236,249–251]. The filler materials should facilitate the re-extrudability of composite filaments for FDM [252,253], production of composite or surface decorated (with nanofillers) micro powders for SLS [254], and high stability in the liquid resin for SLA/DLP [255–257]. Please note that the pure SLA here is modified to its variants, to direct laser processing (DLP) or micro-continuous liquid interface production (μCLIP) or two-photon polymerization (2PP). All variants, however, are based on the light-mediated conversion of liquid resin [24,184,220,258].

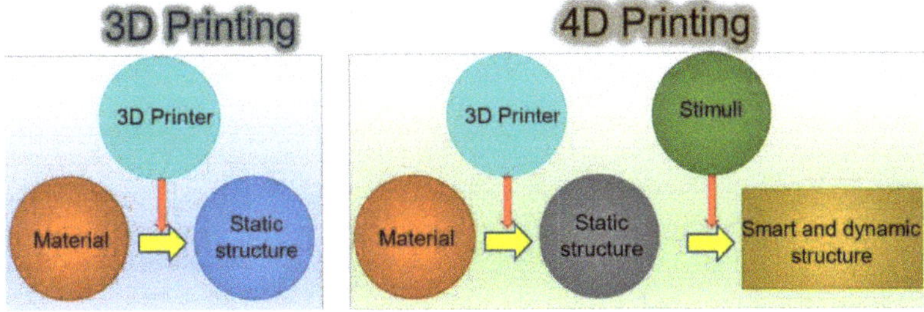

Figure 9. 3D printing and 4D printing and their basic differences, redrawn from [259].

Figure 10 collectively shows material modification methods and key material properties of modified materials to be understood for all three different printing techniques. The main aim is to produce a composite filament with homogenously distributed fillers for FDM, composite powders with homogenously decorated/distributed fillers for SLS, and composite colloidal inks with homogenously dispersed/suspended fillers for SLA variants.

Modifications of filaments for FDM include adding the filler to produce a homogenous mixture of the host material and magnetic fillers and the re-extrusion of composite filaments for printing. For instance, the original filament material (i.e., thermoplastic rubber) was heated to 70 °C to soften the surface then the magnetic fillers (i.e., CIPs) were added and mixed thoroughly. After that, a twin extruder was used to produce a composite filament [253]. In another study, PCL or TPU were mechanically mixed with CIPs first and then hot melted (at 200 °C) within the extruder and the composite filament was extruded [252]. Likewise, a composite filament of Fe_3O_4 nanoparticles and PLA was produced via melt compounding [260]. The processing factors such as melting, mixing, homogenization, granulation of compounds, and viscosity for extrusion play a vital role to produce composite filaments and there are a number of other different studies where the influence and optimization of these parameters are well reported, see Dohmen et al. [261], and others [262–264]. After the extrusion, cross-sectional morphology must be studied to investigate the homogeneity of the filler-matrix system. More importantly, the thermal properties of the composite filament need to be understood in detail for successful printing via FDM. For example, it has been reported that thermal properties (DSC and TGA thermographs) of the modified PLA with Fe_3O_4 nanofillers do not get altered significantly compared to virgin PLA (Figure 10d), hence the printing conditions (nozzle temperature, feed rate, print speed and so on) are altered to only a slighter extent [265], however, optimization of print parameters is still needed.

SLS printing sinters micron-sized powders to produce a part. In order to demonstrate the 4D effect, again the raw material (powders, e.g., PA12) should be modified by incorporating fillers. Compared to FDM and SLA (and its variants), there are very few studies where SLS is implemented to develop 4D structures of magneto-active polymers. To develop composite powders for SLS, the process is well reported in a recent article [254], wherein a novel method called nano-additivation is used. A colloid of magnetic fillers was formed first and then such magnetic particle-based colloid is laser fragmented by irradiating a laser light. The uniformly developed colloid is mixed with PA12 polymer to develop composite polymer powders for SLS (Figure 10b). Thereafter, the thermal material characteristics of nano-additively developed composite PA12 are studied using DSC and TGA methods. To print magnetic parts using such composites, all the process parameters such as temperature, laser output power, scan speed, hatch distance and energy density are studied in detail and optimized. It is reported that the thermal behaviour of the surface functionalized PA12 is similar to that of virgin PA12 powders (Figure 10e). In another very recent study (2021), micron-sized fillers (Nd-Fe-B) are used together with TPU [266]. Therein, composite is prepared just by blending the polymer and fillers, silica nanoparticles are added to improve the flowability. The mixing process is conducted just for 3 min at a rate of 600 rpm. Again, investigation of morphological, thermal, and mechanical properties of TPU/Nd-Fe-B composite is reported. The tensile strength of composites is found to be decreased compared to virgin TPU [266], while thermal properties remained similar to virgin TPU.

In SLA and its variants, which are resin-based systems, photopolymerization governs the conversion of liquid into a solid object in the presence of UV light (Figure 10c). Photopolymerization is a multi-stage and dynamic process therefore researchers have adopted different types of laboratory-based monomers/oligomers or copolymers to produce composite inks as a starting material. Various acrylate-based monomers/oligomers such as urethane acrylate with CIPs [267], with Nd-Fe-B [257], and with Fe oxides [255], and polyethene diacrylate with Fe_2O_3 [268] are reported. The development of composite ink is the mixing of different components (monomer, crosslinker, initiator), magnetic fillers and additives to produce a homogenous and stable ink [269]. Usually, the mixing procedure includes mechanical mixing followed by sonication. For ink, the essential material properties or parameters are resin viscosity, stability/sedimentation of fillers, and cure depth/penetration depth. The ink material properties highly depend on filler concentration and size (micro/nano) [256]. In the process, parameters such as exposer time per projection, layer height, waiting time before exposure, and influence of additives must be investigated and optimized [255]. For instance, viscosity and storage modulus (before and after exposer to UV light) of butyl acrylate-Fe_2O_3 composite inks are required to be very similar to that of virgin butyl acrylate resin (Figure 10f) [255], which defines the limit of the filler loading. Moreover, sedimentation is another major concern if micron-sized fillers are used [256]. There are a number of different studies where both materials and processes for vat-based printing of composite inks are studied in detail, refer to [255,256,269–271].

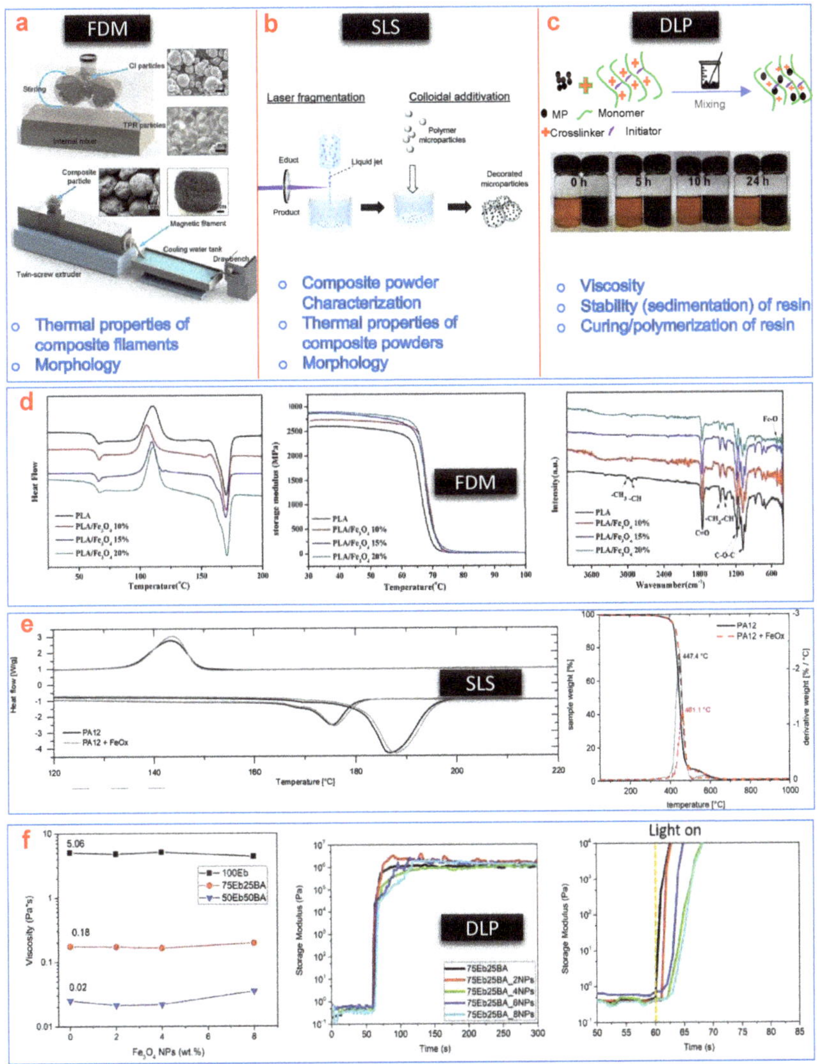

Figure 10. Material modification and requirement to develop magneto-active 4D structures via FDM, SLS, and SLA techniques.

(**a**) Composite filament formation method (adapted from [253]), (**b**) composite powder formation method (adapted from [254]), (**c**) composite ink formation method (adapted from [256]), (**d**) examples of thermal and chemical properties of modified PLA-iron oxide composite filaments for FDM (adapted from [265]), (**e**) examples of thermal properties of modified PA12-iron oxide composite powder for SLS (adapted from [254]), and (**f**) rheological properties of acrylate iron oxide composite ink for DLP (adapted from [255]).

A few typical examples of 4D printed structures of magneto-active polymers using FDM, SLS, and SLA (or its variants) are given in Figure 11. Development of a flower-like structure is one common practice to demonstrate the 4D effect, where such flower-like structures can blossom/open or close in the presence/absence of an external magnetic field. FDM and SLS produce a more rigid structure while resin-based systems (SLA variants) can produce flexible to rigid structures based on the varieties of tailored polymers because of the greater flexibility to control the polymer network. Magneto-active polymeric struc-tures possess a huge potential to exploit in a number of different applications such as in soft robotics and in the biomedical field, where shape-shifting or shape morphing is highly desired [237,238].

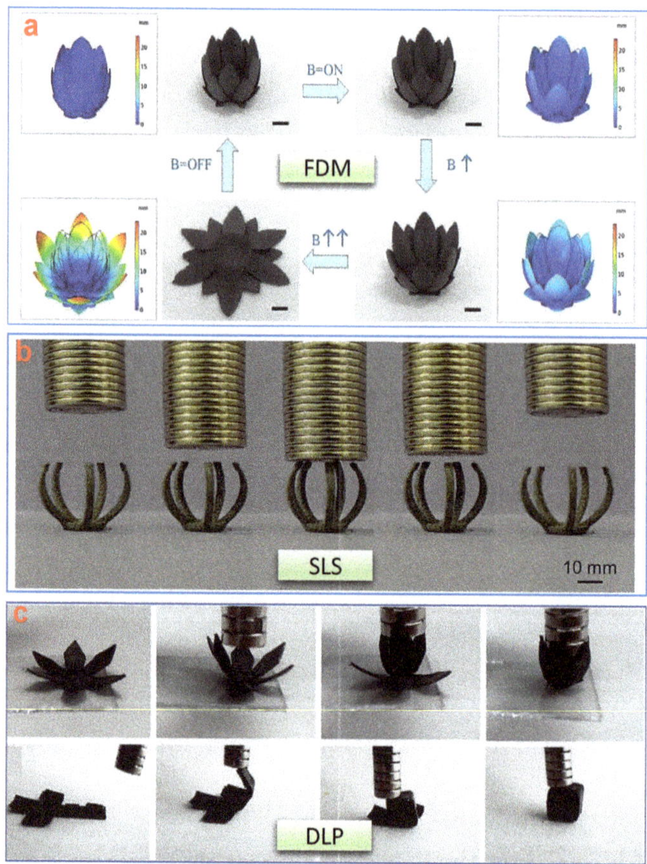

Figure 11. Examples of shape-morphing phenomenon demonstrated by 4D printed magnetic structures. (**a**) 4D effect of the flower-like biomimetic magnetic actuator under an external magnetic field, produced via FDM printing [253], (**b**) 4D effect of a gripper under an external magnetic field, produced via SLS printing [266], and (**c**) 4D effect of flower-like structure and folding of 2D to the 3D structure under an external magnetic field, produced via DLP printing [255].

6. Concluding Remarks

AM is transforming the manufacturing industry with the ability to produce geometrically simple to highly complex and delicate structures. With a variety of 3D printing processes available for a wide range of materials, 3D printing has been extensively adopted in a number of different fields including but are not limited to mechanical engineering, civil engineering, aerospace, electronics, and biomedical. In this review, various aspects of the three most conventional, on the other hand, extensively adopted, 3D printing processes i.e., FDM, SLS, and SLA have been discussed. Correlation of three different aspects, materials, processes, and properties, for these polymer 3D printing techniques is presented. Each of these processes requires materials in a unique form which are filament, powder, and liquid for FDM, SLS, and SLA respectively. Although the fundamental of developing the product layer by layer remains the same, each method has a unique process and parameters of manufacturing to consider. A few process parameters such as CAD design of the model (.stl file), external environmental conditions, and the fundaments of 3D printing are similar in all the processes.

In order to select a 3D printing process for a specific application the requirements of print materials, properties of printed parts, the simplicity of printing, printing time, and layer resolution are essential factors to be considered (Table 7). For instance, SLA has the capability of printing high-resolution parts of up to 10 µm [33], while the minimum layer resolution of the SLS printed part is 20 µm [123]and the FDM has the capability of printing high-resolution parts of up to 40 µm [124]. On the other hand, in terms of the simplicity in printing, FDM is the most suitable technique because the process is as simple as heating the polymer filament to a semi-solid state and depositing it on the print bed. SLS is a comparatively complicated process among others, it requires the movement of two systems: roller and laser light. FDM 3D printers are the most economical and widely available while SLS 3D printers are the expensive ones. According to a recent study [272], a reasonable result in terms of accuracy can be achieved with all print methods; however, the preference should be more based on the print material, the intended application, and the budget. For example, dimensional accuracy and precision of 50 mandibular samples produced from various techniques show that the highest accuracy was found for SLS (0.11 ± 0.016 mm), followed by FDM (0.16 ± 0.009 mm) and SLA (0.45 ± 0.044 mm) [272]. In terms of print time, in their study, SLS (~ 48 min) is the fastest followed by FDM (~2 h and 40 min) and the SLA (~ 5 h 16 min plus post-processing ~15 min). On the other hand, in terms of cost, the SLS printer has the highest purchase price (e.g., EOSINT P 385 ~ USD 150,000), followed by the FDM printer (e.g., Ultimaker 3 Ext. ~USD 4000) and the SLA printer (e.g., Form 2 ~USD 3500), please note that the prices are estimates and may vary by reseller or country [272]. All in all, the FDM, being the simplest printing method, always marks higher in different aspects and thus is a highly reliable option in a number of different applications.

For each process, a wide range of materials are commercially available. For FDM, ABS, and PLA are the most prominently used thermoplastic filaments while, for SLS, PA12 powders are most widely used. For SLA, standard resins with unique formulations for different applications are developed by the manufacturers and there are a number of unique photocurable formulations based on polyester, polycarbonate, and polyether polymers, developed in the research laboratories. Although traditional SLA offers high lateral resolution compared to FDM and SLS, it is inadequately slow for large objects, therefore, a number of variants of vat-based techniques such as digital light processing (DLP), continuous liquid interface production (CLIP), and two-photon polymerization (2PP) have emerged to facilitate the printing time as well as resolution (2PP provides resolutions in the nm range).

Table 7. A comparative overview of FDM, SLS, and SLA printing processes. The representation is created to provide a guideline only.

	Fused Deposition Modelling (FDM)	Selective Laser Sintering (SLS)	Stereolithography (SLA)
Operational principal	Material extrusion	Laser sintering	UV curing
Resolution			
Accuracy			
Surface finish			
Design complexity			
Ease of use			
Printing time			
Advantages	Fast printing process Low part production cost Wide variety of materials are required	Functional parts Design freedom No support structures is required	High-resolution parts can be achieved Wide range of functional applications High accuracy
Limitations	Poor surface finish Support structures are required	Rough surface finish Lengthy printing time	Limited materials High maintenance cost is required

4D printing, on the other hand, offers developing highly innovative and sophisticated devices/structures for various applications such as in soft robotics and the biomedical field. As reviewed in this work, FDM and SLA (mostly its variants) are increasingly adopted 4D printing techniques compared to SLS owing to their facile process and appropriateness for a variety of materials. The fundamental question/challenge of 4DP is how to modify the print materials (ink/resin/powder) as the well-established or commercially available materials cannot directly be used. The addition of various stimuli-active filler materials such as magneto-active, electro-active, and so on are to be incorporated successfully. Tailoring material properties, as well as optimization of the 3D print methods, are different from each other, i.e., for FDM, SLS, and SLA. In order to make printable materials as well as to meet the requirement of the targeted 4D application, a detailed investigation of the print materials and print processes is necessary. 4D printing is a highly interdisciplinary field and the expertise of material modification, process, and characteristics of targeted 4D effect is required for successful printing.

The material-process requirements discussed here in this review for a magneto-active polymer-based system remains similar for other active systems to develop other stimuli-active or even multi-stimuli-active 4D structures. It is believed that the correlation of material-process-properties of conventional polymeric material-based 3D printing together with an example of 4D printing provides a methodical guideline for 3D/4D printing of polymer materials using FDM, SLS, and SLA (or its variants).

Author Contributions: Conceptualization: all authors; A.K. and A.K.B. conceived the idea, designed, and organized the main contents; writing—original draft preparation, writing—review and editing, A.K., E.L., R.S., H.M.P., P.L.S. and A.K.B. All authors have read and agreed to the published version of the manuscript.

Funding: This research received no external funding.

Conflicts of Interest: The authors declare no conflict of interest.

Abbreviations

AM	Additive Manufacturing
CAD	Computer-Aided Design
ASTM	American Society for Testing and Materials
FDM	Fused Deposition Modelling
LOM	Laminated Object Manufacturing
WAAM	Wire and Arc Additive Manufacturing
EBF3	Electron Beam Free Form Fabrication
SLS	Selective Laser Sintering
EBM	Electron Beam Melting
SLM	Selective Laser Melting
LMD	Laser Metal Deposition
SLA	Stereolithography
MJ	Material Jetting
STL	Standard Tessellation Language
FFF	Fused Filament Fabrication
PLA	Polylactic Acid
SLA	Selective Laser Sintering
PP	Polypropylene
TPU	Thermoplastic Polyurethane
TPE	Thermoplastic Elastomer
ABS	Acrylonitrile Butadiene Styrene
TGA	Thermogravimetric Analysis
PA	Polyamide
PEEK	Polyether Ether Ketone
TGA	Thermogravimetric Analysis
PCL	Polycaprolactone
PLGA	Polylactic Glycolic Acid
PEI	Polyethylenimine
PEKK	Polyetherketoneketone
ASA	Acrylonitrile Styrene Acrylate
PMMA	Polymethyl Methacrylate
PS	Polystyrene
PET	polyethylene Terephthalate
PES	Polyethersulfone
PVA	Polyvinyl Alcohol
PPF	Polypropylene Fumarate
PTMC	Polytrimethylene Carbonate
PEG	Polyethylene Glycol
DSC	Differential Scanning Calorimetry
CIPs	Carbonyl Iron Powders
DLP	Direct Laser Processing
µCLIP	Micro-Continuous Liquid Interface Production
DIW	Direct Ink Write

References

1. Goodridge, R.D.; Tuck, C.J.; Hague, R.J.M. Laser sintering of polyamides and other polymers. *Prog. Mater. Sci.* **2012**, *57*, 229–267. [CrossRef]
2. Aboulkhair, N.T.; Simonelli, M.; Parry, L.; Ashcroft, I.; Tuck, C.; Hague, R. 3D printing of Aluminium alloys: Additive Manufacturing of Aluminium alloys using selective laser melting. *Prog. Mater. Sci.* **2019**, *106*, 100578. [CrossRef]
3. Sing, S.L.; An, J.; Yeong, W.Y.; Wiria, F.E. Laser and electron-beam powder-bed additive manufacturing of metallic implants: A review on processes, materials and designs. *J. Orthop. Res.* **2016**, *34*, 369–385. [CrossRef] [PubMed]
4. Khoo, Z.X.; Teoh, J.E.M.; Liu, Y.; Chua, C.K.; Yang, S.; An, J.; Leong, K.F.; Yeong, W.Y. 3D printing of smart materials: A review on recent progresses in 4D printing. *Virtual Phys. Prototyp.* **2015**, *10*, 103–122. [CrossRef]
5. Ngo, T.D.; Kashani, A.; Imbalzano, G.; Nguyen, K.T.Q.; Hui, D. Additive manufacturing (3D printing): A review of materials, methods, applications and challenges. *Compos. Part. B Eng.* **2018**, *143*, 172–196. [CrossRef]
6. Zhu, C.; Li, T.; Mohideen, M.M.; Hu, P.; Gupta, R.; Ramakrishna, S.; Liu, Y. Realization of circular economy of 3D printed plastics: A review. *Polymers* **2021**, *13*, 744. [CrossRef] [PubMed]
7. Tareq, M.S.; Rahman, T.; Hossain, M.; Dorrington, P. Additive manufacturing and the COVID-19 challenges: An in-depth study. *J. Manuf. Syst.* **2021**, *60*, 787–798. [CrossRef]
8. Goh, G.D.; Sing, S.L.; Yeong, W.Y. A review on machine learning in 3D printing: Applications, potential, and challenges. *Artif. Intell. Rev.* **2021**, *54*, 63–94. [CrossRef]
9. Pérez, M.; Carou, D.; Rubio, E.M.; Teti, R. Current advances in additive manufacturing. *Procedia CIRP* **2020**, *88*, 439–444. [CrossRef]
10. Hossain, M.A.; Zhumabekova, A.; Paul, S.C.; Kim, J.R. A review of 3D printing in construction and its impact on the labor market. *Sustainability* **2020**, *12*, 8492. [CrossRef]
11. 3D Concrete Printing Market to Reach $56.4 Million by 2021—3Printr.com. Available online: https://www.3printr.com/3d-concrete-printing-market-reach-56-4-million-2021-1239664/ (accessed on 31 August 2021).
12. Low, Z.X.; Chua, Y.T.; Ray, B.M.; Mattia, D.; Metcalfe, I.S.; Patterson, D.A. Perspective on 3D printing of separation membranes and comparison to related unconventional fabrication techniques. *J. Memb. Sci.* **2017**, *523*, 596–613. [CrossRef]
13. Jiménez, M.; Romero, L.; Domínguez, I.A.; Espinosa, M.D.M.; Domínguez, M. Additive Manufacturing Technologies: An Overview about 3D Printing Methods and Future Prospects. *Complexity* **2019**, *2019*, 9656938. [CrossRef]
14. Espalin, D.; Muse, D.W.; MacDonald, E.; Wicker, R.B. 3D Printing multifunctionality: Structures with electronics. *Int. J. Adv. Manuf. Technol.* **2014**, *72*, 963–978. [CrossRef]
15. Goyanes, A.; Det-Amornrat, U.; Wang, J.; Basit, A.W.; Gaisford, S. 3D scanning and 3D printing as innovative technologies for fabricating personalized topical drug delivery systems. *J. Control. Release* **2016**, *234*, 41–48. [CrossRef] [PubMed]
16. Goh, G.D.; Yap, Y.L.; Tan, H.K.J.; Sing, S.L.; Goh, G.L.; Yeong, W.Y. Process–Structure–Properties in Polymer Additive Manufacturing via Material Extrusion: A Review. *Crit. Rev. Solid State Mater. Sci.* **2020**, *45*, 113–133. [CrossRef]
17. Goh, G.D.; Agarwala, S.; Goh, G.L.; Dikshit, V.; Sing, S.L.; Yeong, W.Y. Additive manufacturing in unmanned aerial vehicles (UAVs): Challenges and potential. *Aerosp. Sci. Technol.* **2017**, *63*, 140–151. [CrossRef]
18. Valvez, S.; Reis, P.N.B.; Susmel, L.; Berto, F. Fused Filament Fabrication-4D-Printed Shape Memory Polymers: A Review. *Polymers* **2021**, *13*, 701. [CrossRef]
19. Jang, T.-S.; Jung, H.-D.; Pan, H.M.; Han, W.T.; Chen, S.; Song, J. 3D printing of hydrogel composite systems: Recent advances in technology for tissue engineering. *Int. J. Bioprint.* **2018**, *4*, 126. [CrossRef]
20. Buchanan, C.; Gardner, L. Metal 3D printing in construction: A review of methods, research, applications, opportunities and challenges. *Eng. Struct.* **2019**, *180*, 332–348. [CrossRef]
21. Gadagi, B.; Lekurwale, R. A review on advances in 3D metal printing. *Mater. Today Proc.* **2021**, *45*, 277–283. [CrossRef]
22. Chahal, V.; Taylor, R.M. A review of geometric sensitivities in laser metal 3D printing. *Virtual Phys. Prototyp.* **2020**, *15*, 227–241. [CrossRef]
23. Murr, L.E. A Metallographic Review of 3D Printing/Additive Manufacturing of Metal and Alloy Products and Components. *Metallogr. Microstruct. Anal.* **2018**, *7*, 103–132. [CrossRef]
24. Tan, L.J.; Zhu, W.; Zhou, K. Recent progress on polymer materials for additive manufacturing. *Adv. Funct. Mater.* **2020**, *30*, 2003062. [CrossRef]
25. Saleh Alghamdi, S.; John, S.; Roy Choudhury, N.; Dutta, N.K. Additive Manufacturing of Polymer Materials: Progress, Promise and Challenges. *Polymers* **2021**, *13*, 753. [CrossRef] [PubMed]
26. Shahbazi, M.; Jäger, H. Current Status in the Utilization of Biobased Polymers for 3D Printing Process: A Systematic Review of the Materials, Processes, and Challenges. *ACS Appl. Bio Mater.* **2021**, *4*, 325–369. [CrossRef]
27. Larrañeta, E.; Dominguez-Robles, J.; Lamprou, D.A. Additive Manufacturing Can Assist in the Fight against COVID-19 and Other Pandemics and Impact on the Global Supply Chain. *3D Print. Addit. Manuf.* **2020**, *7*, 100–103. [CrossRef]
28. Luis, E.; Pan, H.M.; Sing, S.L.; Bastola, A.K.; Goh, G.D.; Goh, G.L.; Tan, H.K.J.; Bajpai, R.; Song, J.; Yeong, W.Y. Silicone 3D Printing: Process Optimization, Product Biocompatibility, and Reliability of Silicone Meniscus Implants. *3D Print. Addit. Manuf.* **2019**, *6*, 319–332. [CrossRef]
29. Luis, E.; Pan, H.M.; Bastola, A.K.; Bajpai, R.; Sing, S.L.; Song, J.; Yeong, W.Y. 3D printed silicone meniscus implants: Influence of the 3D printing process on properties of silicone implants. *Polymers* **2020**, *12*, 2136. [CrossRef]

30. Mazzanti, V.; Malagutti, L.; Mollica, F. FDM 3D printing of polymers containing natural fillers: A review of their mechanical properties. *Polymers* **2019**, *11*, 1094. [CrossRef]
31. Valino, A.D.; Dizon, J.R.C.; Espera, A.H.; Chen, Q.; Messman, J.; Advincula, R.C. Advances in 3D printing of thermoplastic polymer composites and nanocomposites. *Prog. Polym. Sci.* **2019**, *98*, 101162. [CrossRef]
32. Singh, S.; Ramakrishna, S.; Berto, F. 3D Printing of polymer composites: A short review. *Mater. Des. Process. Commun.* **2020**, *2*, 1–13. [CrossRef]
33. Wang, X.; Jiang, M.; Zhou, Z.; Gou, J.; Hui, D. 3D printing of polymer matrix composites: A review and prospective. *Compos. Part. B Eng.* **2017**, *110*, 442–458. [CrossRef]
34. Chen, Z.; Li, Z.; Li, J.; Liu, C.; Lao, C.; Fu, Y.; Liu, C.; Li, Y.; Wang, P.; He, Y. 3D printing of ceramics: A review. *J. Eur. Ceram. Soc.* **2019**, *39*, 661–687. [CrossRef]
35. Hwa, L.C.; Rajoo, S.; Noor, A.M.; Ahmad, N.; Uday, M.B. Recent advances in 3D printing of porous ceramics: A review. *Curr. Opin. Solid State Mater. Sci.* **2017**, *21*, 323–347. [CrossRef]
36. Chen, Z.; Sun, X.; Shang, Y.; Xiong, K.; Xu, Z.; Guo, R.; Cai, S.; Zheng, C. Dense ceramics with complex shape fabricated by 3D printing: A review. *J. Adv. Ceram.* **2021**, *10*, 195–218. [CrossRef]
37. Rasaki, S.A.; Xiong, D.; Xiong, S.; Su, F.; Idrees, M.; Chen, Z. Photopolymerization-based additive manufacturing of ceramics: A systematic review. *J. Adv. Ceram.* **2021**, *10*, 442–471. [CrossRef]
38. Tay, Y.W.D.; Panda, B.; Paul, S.C.; Noor Mohamed, N.A.; Tan, M.J.; Leong, K.F. 3D printing trends in building and construction industry: A review. *Virtual Phys. Prototyp.* **2017**, *12*, 261–276. [CrossRef]
39. Sing, S.L.; Yeong, W.Y.; Wiria, F.E.; Tay, B.Y.; Zhao, Z.; Zhao, L.; Tian, Z.; Yang, S. Direct selective laser sintering and melting of ceramics: A review. *Rapid Prototyp. J.* **2017**, *23*, 611–623. [CrossRef]
40. Ma, G.; Wang, L. A critical review of preparation design and workability measurement of concrete material for largescale 3D printing. *Front. Struct. Civ. Eng.* **2018**, *12*, 382–400. [CrossRef]
41. Shakor, P.; Nejadi, S.; Paul, G.; Malek, S. Review of emerging additive manufacturing technologies in 3d printing of cementitious materials in the construction industry. *Front. Built Environ.* **2019**, *4*, 85. [CrossRef]
42. Mohan, M.K.; Rahul, A.V.; De Schutter, G.; Van Tittelboom, K. Extrusion-based concrete 3D printing from a material perspective: A state-of-the-art review. *Cem. Concr. Compos.* **2021**, *115*, 103855. [CrossRef]
43. Mechtcherine, V.; Nerella, V.N.; Will, F.; Näther, M.; Otto, J.; Krause, M. Large-scale digital concrete construction—CONPrint3D concept for on-site, monolithic 3D-printing. *Autom. Constr.* **2019**, *107*, 102933. [CrossRef]
44. Mekonnen, B.G.; Bright, G.; Walker, A. A study on state of the art technology of laminated object manufacturing (Lom). In *CAD/CAM, Robotics and Factories of the Future*; Springer: New Delhi, India, 2016; pp. 207–216. [CrossRef]
45. Gupta, R.; Dalakoti, M.; Narasimhulu, A. A Critical Review of Process Parameters in Laminated Object Manufacturing Process BT. In *Advances in Materials Engineering and Manufacturing Processes*; Singh, I., Bajpai, P.K., Panwar, K., Eds.; Springer: Singapore, 2020; pp. 31–39.
46. Dermeik, B.; Travitzky, N. Laminated Object Manufacturing of Ceramic-Based Materials. *Adv. Eng. Mater.* **2020**, *22*, 2000256. [CrossRef]
47. Solomon, I.J.; Sevvel, P.; Gunasekaran, J. A review on the various processing parameters in FDM. *Mater. Today Proc.* **2020**, *37*, 509–514. [CrossRef]
48. Popescu, D.; Zapciu, A.; Amza, C.; Baciu, F.; Marinescu, R. FDM process parameters influence over the mechanical properties of polymer specimens: A review. *Polym. Test.* **2018**, *69*, 157–166. [CrossRef]
49. Wickramasinghe, S.; Do, T.; Tran, P. FDM-Based 3D printing of polymer and associated composite: A review on mechanical properties, defects and treatments. *Polymers* **2020**, *12*, 1529. [CrossRef]
50. Liu, Z.; Wang, Y.; Wu, B.; Cui, C.; Guo, Y.; Yan, C. A critical review of fused deposition modeling 3D printing technology in manufacturing polylactic acid parts. *Int. J. Adv. Manuf. Technol.* **2019**, *102*, 2877–2889. [CrossRef]
51. Dey, A.; Roan Eagle, I.N.; Yodo, N. A Review on Filament Materials for Fused Filament Fabrication. *J. Manuf. Mater. Process.* **2021**, *5*, 69.
52. Parulski, C.; Jennotte, O.; Lechanteur, A.; Evrard, B. Challenges of fused deposition modeling 3D printing in pharmaceutical applications: Where are we now? *Adv. Drug Deliv. Rev.* **2021**, *175*, 113810. [CrossRef]
53. Xia, C.; Pan, Z.; Polden, J.; Li, H.; Xu, Y.; Chen, S.; Zhang, Y. A review on wire arc additive manufacturing: Monitoring, control and a framework of automated system. *J. Manuf. Syst.* **2020**, *57*, 31–45. [CrossRef]
54. Raut, L.P.; Taiwade, R.V. Wire Arc Additive Manufacturing: A Comprehensive Review and Research Directions. *J. Mater. Eng. Perform.* **2021**, *30*, 4768–4791. [CrossRef]
55. Xu, J.; Zhu, J.; Fan, J.; Zhou, Q.; Peng, Y.; Guo, S. Microstructure and mechanical properties of Ti–6Al–4V alloy fabricated using electron beam freeform fabrication. *Vacuum* **2019**, *167*, 364–373. [CrossRef]
56. Wei, C.; Li, L. Recent progress and scientific challenges in multi-material additive manufacturing via laser-based powder bed fusion. *Virtual Phys. Prototyp.* **2021**, *16*, 347–371. [CrossRef]
57. Charoo, N.A.; Barakh Ali, S.F.; Mohamed, E.M.; Kuttolamadom, M.A.; Ozkan, T.; Khan, M.A.; Rahman, Z. Selective laser sintering 3D printing–an overview of the technology and pharmaceutical applications. *Drug Dev. Ind. Pharm.* **2020**, *46*, 869–877. [CrossRef] [PubMed]

58. Galati, M.; Iuliano, L. A literature review of powder-based electron beam melting focusing on numerical simulations. *Addit. Manuf.* **2018**, *19*, 1–20. [CrossRef]

59. Gokuldoss, P.K.; Kolla, S.; Eckert, J. Additive manufacturing processes: Selective laser melting, electron beam melting and binder jetting-selection guidelines. *Materials* **2017**, *10*, 672. [CrossRef] [PubMed]

60. Yap, C.Y.; Chua, C.K.; Dong, Z.L.; Liu, Z.H.; Zhang, D.Q.; Loh, L.E.; Sing, S.L. Review of selective laser melting: Materials and applications. *Appl. Phys. Rev.* **2015**, *2*, 041101. [CrossRef]

61. Azarniya, A.; Colera, X.G.; Mirzaali, M.J.; Sovizi, S.; Bartolomeu, F.; St Weglowski, M.K.; Wits, W.W.; Yap, C.Y.; Ahn, J.; Miranda, G.; et al. Additive manufacturing of Ti–6Al–4V parts through laser metal deposition (LMD): Process, microstructure, and mechanical properties. *J. Alloys Compd.* **2019**, *804*, 163–191. [CrossRef]

62. Mazzarisi, M.; Campanelli, S.L.; Angelastro, A.; Dassisti, M. Phenomenological modelling of direct laser metal deposition for single tracks. *Int. J. Adv. Manuf. Technol.* **2020**, *111*, 1955–1970. [CrossRef]

63. Manapat, J.Z.; Chen, Q.; Ye, P.; Advincula, R.C. 3D Printing of Polymer Nanocomposites via Stereolithography. *Macromol. Mater. Eng.* **2017**, *302*, 1–13. [CrossRef]

64. Melchels, F.P.W.; Feijen, J.; Grijpma, D.W. A review on stereolithography and its applications in biomedical engineering. *Biomaterials* **2010**, *31*, 6121–6130. [CrossRef]

65. Huang, J.; Qin, Q.; Wang, J. A review of stereolithography: Processes and systems. *Processes* **2020**, *8*, 1138. [CrossRef]

66. Zhang, J.; Hu, Q.; Wang, S.; Tao, J.; Gou, M. Digital Light Processing Based Three-dimensional Printing for Medical Applications. *Int. J. Bioprint.* **2019**, *6*, 242. [CrossRef]

67. Zhao, Z.; Tian, X.; Song, X. Engineering materials with light: Recent progress in digital light processing based 3D printing. *J. Mater. Chem. C* **2020**, *8*, 13896–13917. [CrossRef]

68. Wiese, M.; Kwauka, A.; Thiede, S.; Herrmann, C. Economic assessment for additive manufacturing of automotive end-use parts through digital light processing (DLP). *CIRP J. Manuf. Sci. Technol.* **2021**, *35*, 268–280. [CrossRef]

69. Tee, Y.L.; Tran, P.; Leary, M.; Pille, P.; Brandt, M. 3D Printing of polymer composites with material jetting: Mechanical and fractographic analysis. *Addit. Manuf.* **2020**, *36*, 101558. [CrossRef]

70. Silva, M.R.; Pereira, A.M.; Sampaio, Á.M.; Pontes, A.J. Assessment of the dimensional and geometric precision of micro-details produced by material jetting. *Materials* **2021**, *14*, 1989. [CrossRef] [PubMed]

71. Kardel, K.; Khoshkhoo, A.; Carrano, A.L. Design guidelines to mitigate distortion in material jetting specimens. *Rapid Prototyp. J.* **2021**, *27*, 1148–1160. [CrossRef]

72. Anderson, T. The Application of 3D Printing for Healthcare. Available online: https://www.itij.com/latest/long-read/application-3d-printing-healthcare (accessed on 2 August 2021).

73. Young, R.J.; Lovell, P.A. *Introduction to Polymers*; CRC Press: Boca Raton, FL, USA, 2011; ISBN 143-9-894-159.

74. Stansbury, J.W.; Idacavage, M.J. 3D printing with polymers: Challenges among expanding options and opportunities. *Dent. Mater.* **2016**, *32*, 54–64. [CrossRef]

75. Ahmed, W.; Siraj, S.; Al-Marzouqi, A.H. Embracing Additive Manufacturing Technology through Fused Filament Fabrication for Antimicrobial with Enhanced Formulated Materials. *Polymers* **2021**, *13*, 1523. [CrossRef] [PubMed]

76. Ilyas, R.A.; Sapuan, S.M.; Harussani, M.M.; Hakimi, M.Y.A.Y.; Haziq, M.Z.M.; Atikah, M.S.N.; Asyraf, M.R.M.; Ishak, M.R.; Razman, M.R.; Nurazzi, N.M.; et al. Polylactic Acid (PLA) Biocomposite: Processing, Additive Manufacturing and Advanced Applications. *Polymers* **2021**, *13*, 1326. [CrossRef] [PubMed]

77. Dickson, A.N.; Abourayana, H.M.; Dowling, D.P. 3D Printing of Fibre-Reinforced Thermoplastic Composites Using Fused Filament Fabrication—A Review. *Polymers* **2020**, *12*, 2188. [CrossRef] [PubMed]

78. Mokrane, A.; Boutaous, M.; Xin, S. Process of selective laser sintering of polymer powders: Modeling, simulation, and validation. *Comptes Rendus Mec.* **2018**, *346*, 1087–1103. [CrossRef]

79. Schmid, M.; Amado, A.; Wegener, K. Polymer powders for selective laser sintering (SLS). *AIP Conf. Proc.* **2015**, *1664*, 160009. [CrossRef]

80. Kruth, J.; Levy, G.; Schindel, R.; Craeghs, T.; Yasa, E. Consolidation of Polymer Powders by Selective Laser Sintering. In Proceedings of the 3rd International Conference on Polymers and Moulds Innovations, Ghent, Belgium, 2008; pp. 15–30. Available online: https://www.semanticscholar.org/paper/Consolidation-of-polymer-powders-by-selective-laser-Kruth-Levy/c6737300dba9cba56b08da81fac647ea0a44e1aa (accessed on 8 September 2021).

81. Gueche, Y.A.; Sanchez-Ballester, N.M.; Cailleaux, S.; Bataille, B.; Soulairol, I. Selective Laser Sintering (SLS), a New Chapter in the Production of Solid Oral Forms (SOFs) by 3D Printing. *Pharmaceutics* **2021**, *13*, 1212. [CrossRef]

82. Pagac, M.; Hajnys, J.; Ma, Q.P.; Jancar, L.; Jansa, J.; Stefek, P.; Mesicek, J. A review of vat photopolymerization technology: Materials, applications, challenges, and future trends of 3d printing. *Polymers* **2021**, *13*, 598. [CrossRef] [PubMed]

83. Revilla-León, M.; Meyers, M.J.; Zandinejad, A.; Özcan, M. A review on chemical composition, mechanical properties, and manufacturing work flow of additively manufactured current polymers for interim dental restorations. *J. Esthet. Restor. Dent.* **2019**, *31*, 51–57. [CrossRef] [PubMed]

84. Andreu, A.; Su, P.-C.; Kim, J.-H.; Siang, C.; Kim, S.; Kim, I.; Lee, J.; Noh, J.; Suriya, A.; Yoon, Y.-J. 4D printing materials for Vat Photopolymerization. *Addit. Manuf.* **2021**, *44*, 102024.

85. Davoudinejad, A. Chapter 5—Vat photopolymerization methods in additive manufacturing. In *Handbooks in Advanced Manufacturing*; Pou, J., Riveiro, A., Davim, J.P.B.T.-A.M., Eds.; Elsevier: Amsterdam, The Netherlands, 2021; pp. 159–181, ISBN 978-0-12-818411-0.
86. Wong, K.V.; Hernandez, A. A Review of Additive Manufacturing. *ISRN Mech. Eng.* **2012**, *2012*, 1–10. [CrossRef]
87. Hu, J. Study on STL-based slicing process for 3D printing. In Proceedings of the 28th Annual International, Solid Freeform Fabrication Symposium–An Additive Manufacturing Conference, Göteborg Sweden, 30 June 2001–4 July 2001; pp. 885–895.
88. Unkovskiy, A.; Bui, P.H.B.; Schille, C.; Geis-Gerstorfer, J.; Huettig, F.; Spintzyk, S. Objects build orientation, positioning, and curing influence dimensional accuracy and flexural properties of stereolithographically printed resin. *Dent. Mater.* **2018**, *34*, e324–e333. [CrossRef]
89. Shaqour, B.; Abuabiah, M.; Abdel-Fattah, S.; Juaidi, A.; Abdallah, R.; Abuzaina, W.; Qarout, M.; Verleije, B.; Cos, P. Gaining a better understanding of the extrusion process in fused filament fabrication 3D printing: A review. *Int. J. Adv. Manuf. Technol.* **2021**, *114*, 1279–1291. [CrossRef]
90. Gautam, R.; Idapalapati, S.; Feih, S. Printing and characterisation of Kagome lattice structures by fused deposition modelling. *Mater. Des.* **2018**, *137*, 266–275. [CrossRef]
91. Garzon-Hernandez, S.; Garcia-Gonzalez, D.; Jérusalem, A.; Arias, A. Design of FDM 3D printed polymers: An experimental-modelling methodology for the prediction of mechanical properties. *Mater. Des.* **2020**, *188*, 108414. [CrossRef]
92. Garzon-Hernandez, S.; Arias, A.; Garcia-Gonzalez, D. A continuum constitutive model for FDM 3D printed thermoplastics. *Compos. Part. B Eng.* **2020**, *201*, 108373. [CrossRef]
93. Andrew, J.J.; Alhashmi, H.; Schiffer, A.; Kumar, S.; Deshpande, V.S. Energy absorption and self-sensing performance of 3D printed CF/PEEK cellular composites. *Mater. Des.* **2021**, *208*, 109863. [CrossRef]
94. Zanjanijam, A.R.; Major, I.; Lyons, J.G.; Lafont, U.; Devine, D.M. Fused Filament Fabrication of PEEK: A Review of Process-Structure-Property Relationships. *Polymers* **2020**, *12*, 1665. [CrossRef]
95. Awasthi, P.; Banerjee, S.S. Fused Deposition Modeling of Thermoplastics Elastomeric Materials: Challenges and Opportunities. *Addit. Manuf.* **2021**, *46*, 102177.
96. Aliheidari, N.; Christ, J.; Tripuraneni, R.; Nadimpalli, S.; Ameli, A. Interlayer adhesion and fracture resistance of polymers printed through melt extrusion additive manufacturing process. *Mater. Des.* **2018**, *156*, 351–361. [CrossRef]
97. Syrlybayev, D.; Zharylkassyn, B.; Seisekulova, A.; Akhmetov, M.; Perveen, A.; Talamona, D. Optimisation of strength properties of FDM printed parts—A critical review. *Polymers* **2021**, *13*, 1587. [CrossRef]
98. Sukindar, N.A.; Ariffin, M.K.A.; Hang Tuah Baharudin, B.T.; Jaafar, C.N.A.; Ismail, M.I.S. Analyzing the effect of nozzle diameter in fused deposition modeling for extruding polylactic acid using open source 3D printing. *J. Teknol.* **2016**, *78*, 7–15. [CrossRef]
99. Triyono, J.; Sukanto, H.; Saputra, R.M.; Smaradhana, D.F. The effect of nozzle hole diameter of 3D printing on porosity and tensile strength parts using polylactic acid material. *Open Eng.* **2020**, *10*, 762–768. [CrossRef]
100. Geng, D.; Zhao, J. Analysis and Optimization of Warpage Deformation in 3D Printing Training Teaching—Taking Jilin University Engineering Training Center as an example. *Int. Workshop Educ. Reform Soc. Sci.* **2019**, *300*, 839–842. [CrossRef]
101. Wu, W.; Ye, W.; Wu, Z.; Geng, P.; Wang, Y.; Zhao, J. Influence of layer thickness, raster angle, deformation temperature and recovery temperature on the shape-memory effect of 3D-printed polylactic acid samples. *Materials* **2017**, *10*, 970. [CrossRef] [PubMed]
102. Rosli, A.A.; Shuib, R.K.; Ishak, K.M.K.; Hamid, Z.A.A.; Abdullah, M.K.; Rusli, A. Influence of bed temperature on warpage, shrinkage and density of various acrylonitrile butadiene styrene (ABS) parts from fused deposition modelling (FDM). *AIP Conf. Proc.* **2020**, *2267*, 1–10. [CrossRef]
103. Ding, S.; Zou, B.; Wang, P.; Ding, H. Effects of nozzle temperature and building orientation on mechanical properties and microstructure of PEEK and PEI printed by 3D-FDM. *Polym. Test.* **2019**, *78*, 105948. [CrossRef]
104. Chung, M.; Radacsi, N.; Robert, C.; McCarthy, E.D.; Callanan, A.; Conlisk, N.; Hoskins, P.R.; Koutsos, V. On the optimization of low-cost FDM 3D printers for accurate replication of patient-specific abdominal aortic aneurysm geometry. *3D Print. Med.* **2018**, *4*, 1–9. [CrossRef]
105. Senthilkumaran, K.; Pandey, P.M.; Rao, P.V.M. Influence of building strategies on the accuracy of parts in selective laser sintering. *Mater. Des.* **2009**, *30*, 2946–2954. [CrossRef]
106. Bourell, D.L. Sintering in Laser Sintering. *JOM* **2016**, *68*, 885–889. [CrossRef]
107. Stoia, D.I.; Linul, E.; Marsavina, L. Influence of manufacturing parameters on mechanical properties of porous materials by selective laser sintering. *Materials* **2019**, *12*, 871. [CrossRef] [PubMed]
108. Martynková, G.S.; Slíva, A.; Kratošová, G.; Barabaszová, K.Č.; Študentová, S.; Klusák, J.; Brožová, S.; Dokoupil, T.; Holešová, S. Polyamide 12 materials study of morpho-structural changes during laser sintering of 3d printing. *Polymers* **2021**, *13*, 810. [CrossRef]
109. Kruth, J.P.; Mercelis, P.; Van Vaerenbergh, J.; Froyen, L.; Rombouts, M. Binding mechanisms in selective laser sintering and selective laser melting. *Rapid Prototyp. J.* **2005**, *11*, 26–36. [CrossRef]
110. Wudy, K.; Lanzl, L.; Drummer, D. Selective laser sintering of filled polymer systems: Bulk properties and laser beam material interaction. *Phys. Procedia* **2016**, *83*, 991–1002. [CrossRef]
111. Ilkgun, O. Effects of Production Parameters on Porosity and Hole. Master's Thesis, Middle East Technical University, Ankara, Iran, 2005; pp. 1–144.

112. Pilipović, A.; Brajlih, T.; Drstvenšek, I. Influence of processing parameters on tensile properties of SLS polymer product. *Polymers* **2018**, *10*, 1208. [CrossRef]

113. Castoro, M. Impact of Laser Power and Build Orientation on the Mechanical Properties of Selectively Laser Sintered Parts. In Proceedings of the National Conference on Undergraduate Research (NCUR), WI, USA, 2013. Available online: http://libjournals.unca.edu/ncur/2013-2/ (accessed on 8 September 2021).

114. Konstantinou, I.; Vosniakos, G.C. Rough-cut fast numerical investigation of temperature fields in selective laser sintering/melting. *Int. J. Adv. Manuf. Technol.* **2018**, *99*, 29–36. [CrossRef]

115. Li, X.; Dong, J. Study on Curve of Pe-heating Temperature Control in Selective Laser Sintering. In Proceedings of the 2009 International Symposium on Web Information Systems and Applications, Maynooth, Ireland, 7–8 December 2009; pp. 156–158.

116. Bagheri, A.; Jin, J. Photopolymerization in 3D Printing. *ACS Appl. Polym. Mater.* **2019**, *1*, 593–611. [CrossRef]

117. Martín-Montal, J.; Pernas-Sánchez, J.; Varas, D. Experimental Characterization Framework for SLA Additive Manufacturing Materials. *Polymers* **2021**, *13*, 1147. [CrossRef] [PubMed]

118. Okolie, O.; Stachurek, I.; Kandasubramanian, B.; Njuguna, J. 3D Printing for Hip Implant Applications: A Review. *Polymers* **2020**, *12*, 2682. [CrossRef] [PubMed]

119. Burke, G.; Devine, D.M.; Major, I. Effect of Stereolithography 3D Printing on the Properties of PEGDMA Hydrogels. *Polymers* **2020**, *12*, 2015. [CrossRef] [PubMed]

120. Mondschein, R.J.; Kanitkar, A.; Williams, C.B.; Verbridge, S.S.; Long, T.E. Polymer structure-property requirements for stereolithographic 3D printing of soft tissue engineering scaffolds. *Biomaterials* **2017**, *140*, 170–188. [CrossRef]

121. Lipson, H.; Kurman, M. *Fabricated: The New World of 3D Printing*; Wiley: Hoboken, NJ, USA, 2013.

122. Kim, D.; Shim, J.; Lee, D.; Shin, S.; Nam, N. Effects of Post-Curing Time on the Mechanical and Color Properties of Three-Dimensional Printed. *Polymers* **2020**, *12*, 2762. [CrossRef]

123. Muzaffar, A.; Ahamed, M.B.; Deshmukh, K.; Kovářík, T.; Křenek, T.; Pasha, S.K.K. *3D and 4D Printing of pH-Responsive and Functional Polymers and Their Composites*; ScienceDirect: Amsterdam, The Netherlands, 2019; ISBN 978-012-816-8-059.

124. Alsoufi, M.S.; Alhazmi, M.W.; Suker, D.K.; Yunus, M.; Malibari, R.O. From 3D models to FDM 3D prints: Experimental study of chemical treatment to reduce stairs-stepping of semi-sphere profile. *AIMS Mater. Sci.* **2019**, *6*, 1086–1106. [CrossRef]

125. 3D ARGO 500 Printer with Technology for High Viscosity Polymers. Available online: https://www.roboze.com/en/3d-printers/argo-500.html (accessed on 6 August 2021).

126. Nexa3D NXV Review—Professional Desktop Resin 3D Printer. Available online: https://www.aniwaa.com/product/3d-printers/nexa3d-nxv/ (accessed on 6 August 2021).

127. PICO2—Products—Asiga. Available online: https://www.asiga.com/products/printers/pico2/ (accessed on 6 August 2021).

128. Stratasys F900 Is the Ultimate Manufacturing-Grade Production 3D Printer. Available online: https://www.javelin-tech.com/3d/stratasys-3d-printer/stratasys-f900/ (accessed on 6 August 2021).

129. HSE 280i HT—Essentium. Available online: https://www.essentium.com/3d-printers/high-speed-extrusion-280/ (accessed on 6 August 2021).

130. CreatBot PEEK-300 3D Printer. Available online: https://www.creatbot.com/en/creatbot-peek-300.html (accessed on 6 August 2021).

131. Industrial Anisoprinting | Anisoprint. Available online: https://anisoprint.com/solutions/industrial/ (accessed on 6 August 2021).

132. INDUSTRY F421—Industrial 3D Printer | 3DGence. Available online: https://3dgence.com/3d-printers/industry-f421/ (accessed on 6 August 2021).

133. Peek 3D Printer | WASP. Available online: https://www.3dwasp.com/en/peek-3d-printer/ (accessed on 6 August 2021).

134. Cincinnati MAAM—Vision Miner. Available online: https://visionminer.com/products/cincinnati-maam?_vsrefdom=adwords&gc_id=1481937543&gclid=Cj0KCQjwu7OIBhCsARIsALxCUaP81y8Nysma5uR6yvTM_fokqlCFO-3m4XdHvodPvjkl7ow8jLFOmToaAkm4EALw_wcB (accessed on 6 August 2021).

135. Tractus3D T850P—High Temperature 3D Printer in Our PRO Series. Available online: https://tractus3d.com/products/t850p/ (accessed on 6 August 2021).

136. AON M2+ High Temp Industrial 3D Printer | AON3D. Available online: https://www.aon3d.com/aon-m2-industrial-3d-printer/ (accessed on 6 August 2021).

137. Kumovis R1 3D Printer for Medical Production | Kumovis. Available online: https://kumovis.com/3d-printer/ (accessed on 6 August 2021).

138. Sintratec Kit—Sintratec AG. Available online: https://sintratec.com/product/sintratec-kit/ (accessed on 6 August 2021).

139. Desktop SLS Printer—Red Rock 3D. Available online: http://www.redrocksls.com/ (accessed on 6 August 2021).

140. Sinterit Lisa PRO—Open SLS 3D Printer. Available online: https://www.sinterit.com/sinterit-lisa-pro/ (accessed on 6 August 2021).

141. Fuse 1: Benchtop Selective Laser Sintering (SLS) 3D Printer. Available online: https://formlabs.com/asia/3d-printers/fuse-1/ (accessed on 6 August 2021).

142. Sintratec S2—Sintratec AG. Available online: https://sintratec.com/product/sintratec-s2/ (accessed on 6 August 2021).

143. The Perfect Professional CO_2 Sintering 3D Printer Fully Configurable by the User. Available online: https://www.sharebot.it/en/sharebot-snowwhite-3d-printer/ (accessed on 6 August 2021).

144. Gravity—SLS 3D-Printer for Office Use—Wematter.se. Available online: https://wematter3d.com/gravity-sls-3d-printer/ (accessed on 6 August 2021).
145. XYZprinting 3D Printers | Mfgpro230_xS. Available online: https://pro.xyzprinting.com/landing-page/en-US/mfgpro230_xs (accessed on 6 August 2021).
146. A Revolutionary Thermoplastic 3D Printer | Nexa3D. Available online: https://nexa3d.com/3d-printers/thermoplastic-printers/ (accessed on 6 August 2021).
147. EP-P3850 Plastic 3D Printer Brochure—SHINING 3D. Available online: https://www.shining3d.com/downloads/ep-p3850-plastic-3d-printer/ (accessed on 6 August 2021).
148. Affordable Industrial SLS®3D Printer—ProMaker P1000 | Prodways FR. Available online: https://www.prodways.com/en/industrial-3d-printers/promaker-p1000/ (accessed on 6 August 2021).
149. SLS Printer FORMIGA P 110 Velocis. Available online: https://www.eos.info/en/additive-manufacturing/3d-printing-plastic/eos-polymer-systems/formiga-p-110-velocis (accessed on 6 August 2021).
150. ProX SLS 6100-3D Printer | 3D Systems. Available online: https://www.3dsystems.com/3d-printers/prox-sls-6100 (accessed on 6 August 2021).
151. Farsoon Technologies—Open for Industry. Available online: http://en.farsoon.com/solution_list01_detail/FrontColumns_navigation01-1497537754897FirstColumnId=2&FrontColumns_navigation01-1497537754897SecondColumnId=58&productId=25.html (accessed on 6 August 2021).
152. Nyomo Minny Review—Compact, Desktop Resin DLP 3D Printer. Available online: https://www.aniwaa.com/product/3d-printers/nyomo-minny/ (accessed on 6 August 2021).
153. XYZprinting Nobel 1.0 Stereolithography 3D Printer 3L10XXUS00C. Available online: https://www.bhphotovideo.com/c/product/1180391-REG/xyzprinting_3l10xxus00c_nobel_printer.html/specs (accessed on 6 August 2021).
154. Form Wash and Form Cure Tech Specs. Available online: https://formlabs.com/asia/post-processing/wash-cure/tech-specs/ (accessed on 6 August 2021).
155. Photocentric Liquid Crystal Pro | Large High Resolution Prints | Dream 3D. Available online: https://www.dream3d.co.uk/product/photocentric-liquid-crystal-pro/ (accessed on 6 August 2021).
156. DWS X Pro S 3D Printer In-Depth Review—Pick 3D Printer. Available online: https://pick3dprinter.com/dws-x-pro-s-review/ (accessed on 6 August 2021).
157. Uniontech Large Volume SLA RSPro Series—Europac3D UK. Available online: https://europac3d.com/3d-printers/sla-uniontech-rspro/ (accessed on 6 August 2021).
158. 3D Systems ProX 950 3D Printer—Reviews, Specs, Price. Available online: https://www.treatstock.com/machines/item/156-prox-950 (accessed on 6 August 2021).
159. Wang, J.; Porter, R.S. On the viscosity-temperature behavior of polymer melts. *Rheol. Acta* **1995**, *34*, 496–503. [CrossRef]
160. Flowers, P.F.; Reyes, C.; Ye, S.; Kim, M.J.; Wiley, B.J. 3D printing electronic components and circuits with conductive thermoplastic filament. *Addit. Manuf.* **2017**, *18*, 156–163. [CrossRef]
161. Hashemi Sanatgar, R.; Campagne, C.; Nierstrasz, V. Investigation of the adhesion properties of direct 3D printing of polymers and nanocomposites on textiles: Effect of FDM printing process parameters. *Appl. Surf. Sci.* **2017**, *403*, 551–563. [CrossRef]
162. Hohimer, C.J.; Petrossian, G.; Ameli, A.; Mo, C.; Pötschke, P. 3D printed conductive thermoplastic polyurethane/carbon nanotube composites for capacitive and piezoresistive sensing in soft pneumatic actuators. *Addit. Manuf.* **2020**, *34*, 101281. [CrossRef]
163. Gibson, I.; Rosen, D.; Stucker, B. *Introduction and Basic Principles*; Higher Education Press: Beijing, China, 2015; ISBN 978-149-392-1-126.
164. Elkins, K.; Nordby, H.; Janak, C.; Gray, R.W.; Bohn, J.H.; Baird, D.G. Soft elastomers for fused deposition modeling. In Proceedings of the 1997 International Solid Freeform Fabrication Symposium, Austin, TX, USA, 1997; pp. 441–448. Available online: https://repositories.lib.utexas.edu/handle/2152/69898 (accessed on 8 September 2021).
165. Pandey, A.; Pradhan, S.K. Investigations into Complete Liquefier Dynamics and Optimization of Process Parameters for Fused Deposition Modeling. *Mater. Today Proc.* **2018**, *5*, 12940–12955. [CrossRef]
166. Turner, B.N.; Gold, S.A. A review of melt extrusion additive manufacturing processes: II. Materials, dimensional accuracy, and surface roughness. *Rapid Prototyp. J.* **2015**, *21*, 250–261. [CrossRef]
167. Spoerk, M.; Savandaiah, C.; Arbeiter, F.; Sapkota, J.; Holzer, C. Optimization of mechanical properties of glass-spheres-filled polypropylene composites for extrusion-based additive manufacturing. *Polym. Compos.* **2019**, *40*, 638–651. [CrossRef]
168. Schmid, M.; Amado, A.; Wegener, K. Materials perspective of polymers for additive manufacturing with selective laser sintering. *J. Mater. Res.* **2014**, *29*, 1824–1832. [CrossRef]
169. Schmidt, J.; Sachs, M.; Blümel, C.; Winzer, B.; Toni, F.; Wirth, K.E.; Peukert, W. A novel process route for the production of spherical LBM polymer powders with small size and good flowability. *Powder Technol.* **2014**, *261*, 78–86. [CrossRef]
170. Manufacturing, A. A Critical Review for Synergic Kinetics and Strategies for Enhanced Photopolymerizations for 3D-Printing and Additive Manufacturing. *Polymers* **2021**, *13*, 2325. [CrossRef]
171. Taormina, G.; Sciancalepore, C.; Messori, M.; Bondioli, F. 3D printing processes for photocurable polymeric materials: Technologies, materials, and future trends. *J. Appl. Biomater. Funct. Mater.* **2018**, *16*, 151–160. [CrossRef]
172. Miedzińska, D.; Gieleta, R.; Popławski, A. Experimental study on influence of curing time on strength behavior of sla-printed samples loaded with different strain rates. *Materials* **2020**, *13*, 5825. [CrossRef]

173. Jiang, F.; Drummer, D. Curing kinetic analysis of acrylate photopolymer for additive manufacturing by photo-DSC. *Polymers* **2020**, *12*, 1080. [CrossRef]

174. Papadakis, L.; Chantzis, D.; Salonitis, K. On the energy efficiency of pre-heating methods in SLM/SLS processes. *Int. J. Adv. Manuf. Technol.* **2018**, *95*, 1325–1338. [CrossRef]

175. Wang, R.J.; Wang, L.; Zhao, L.; Liu, Z. Influence of process parameters on part shrinkage in SLS. *Int. J. Adv. Manuf. Technol.* **2007**, *33*, 498–504. [CrossRef]

176. Schmid, M.; Kleijnen, R.; Vetterli, M.; Wegener, K. Influence of the origin of polyamide 12 powder on the laser sintering process and laser sintered parts. *Appl. Sci.* **2017**, *7*, 462. [CrossRef]

177. Dechet, M.A.; Lanzl, L.; Werner, Y.; Drummer, D.; Bück, A.; Peukert, W.; Schmidt, J. Manufacturing and application of pa11-glass fiber composite particles for selective laser sintering. In Proceedings of the 30th Annual International Solid Freeform Fabrication Symposium, Austin, TX, USA, 12–14 August 2019.

178. Fitzharris, E.R.; Watanabe, N.; Rosen, D.W.; Shofner, M.L. Effects of material properties on warpage in fused deposition modeling parts. *Int. J. Adv. Manuf. Technol.* **2018**, *95*, 2059–2070. [CrossRef]

179. Armillotta, A.; Bellotti, M.; Cavallaro, M. Warpage of FDM parts: Experimental tests and analytic model. *Robot. Comput. Integr. Manuf.* **2018**, *50*, 140–152. [CrossRef]

180. Spoerk, M.; Holzer, C.; Gonzalez-Gutierrez, J. Material extrusion-based additive manufacturing of polypropylene: A review on how to improve dimensional inaccuracy and warpage. *J. Appl. Polym. Sci.* **2020**, *137*, 48545. [CrossRef]

181. Yang, T.C. Effect of extrusion temperature on the physico-mechanical properties of unidirectional wood fiber-reinforced polylactic acid composite (WFRPC) components using fused depositionmodeling. *Polymers* **2018**, *10*, 976. [CrossRef] [PubMed]

182. Osmanlic, F.; Wudy, K.; Laumer, T.; Schmidt, M.; Drummer, D.; Körner, C. Modeling of laser beam absorption in a polymer powder bed. *Polymers* **2018**, *10*, 784. [CrossRef] [PubMed]

183. Suslick, K.S. *Encyclopedia of Physical Science and Technology. Sonoluminescence and Sonochemistry Massachusetts*; Elsevier Science Ltd.: Amsterdam, The Netherlands, 2001; pp. 1–20.

184. Ligon, S.C.; Liska, R.; Stampfl, J.; Gurr, M.; Mülhaupt, R. Polymers for 3D Printing and Customized Additive Manufacturing. *Chem. Rev.* **2017**, *117*, 10212–10290. [CrossRef] [PubMed]

185. Zhang, X.; Jiang, X.N.; Sun, C. Micro-stereolithography of polymeric and ceramic microstructures. *Sens. Actuators A Phys.* **1999**, *77*, 149–156. [CrossRef]

186. Schüller-Ravoo, S.; Teixeira, S.M.; Feijen, J.; Grijpma, D.W.; Poot, A.A. Flexible and elastic scaffolds for cartilage tissue engineering prepared by stereolithography using poly(trimethylene carbonate)-based resins. *Macromol. Biosci.* **2013**, *13*, 1711–1719. [CrossRef] [PubMed]

187. Garay, A.C.; Paese, L.T.; Souza, J.A.; Amico, S.C. Studies on thermal and viscoelastic properties of vinyl ester resin and its composites with glass fiber. *Rev. Mater.* **2015**, *20*, 64–71. [CrossRef]

188. Chen, J.; Smith, D.E. Filament rheological characterization for fused filament fabrication additive manufacturing: A low-cost approach. *Addit. Manuf.* **2021**, *47*, 102208. [CrossRef]

189. Singh, R.; Kumar, R.; Ahuja, I.P.S. Mechanical, thermal and melt flow of aluminum-reinforced PA6/ABS blend feedstock filament for fused deposition modeling. *Rapid Prototyp. J.* **2018**, *24*, 1455–1468. [CrossRef]

190. Wojtyła, S.; Klama, P.; Baran, T. Is 3D printing safe? Analysis of the thermal treatment of thermoplastics: ABS, PLA, PET, and nylon. *J. Occup. Environ. Hyg.* **2017**, *14*, D80–D85. [CrossRef]

191. Trhlíková, L.; Zmeskal, O.; Psencik, P.; Florian, P. Study of the thermal properties of filaments for 3D printing. *AIP Conf. Proc.* **2016**, *1752*, 040027. [CrossRef]

192. Esmizadeh, E.; Tzoganakis, C.; Mekonnen, T.H. Degradation behavior of polypropylene during reprocessing and its biocomposites: Thermal and oxidative degradation kinetics. *Polymers* **2020**, *12*, 1627. [CrossRef]

193. Hao, X.; Guo, Y.; Li, Y.; Yang, Y.; Shen, Y.; Hao, X.; Wang, J. Study on the Structure and Properties of Novel Bio-based Polyamide56 Fiber Compared with Normal Polyamide Fibers BT. In Proceedings of the 2015 International Conference on Materials, Environmental and Biological Engineering, Guilin, China, 28–30 March 2015; Atlantis Press: Amsterdam, The Netherlands, 2015; pp. 147–152.

194. Wei, S.T.; Zhou, K.; Chao, C. A Comprehensive Investigation on 3D Printing of Polyamide 11 and Thermoplastic Polyurethane via Multi Jet Fusion. *Polymers* **2021**, *13*, 2139.

195. Ituarte, I.F.; Wiikinkoski, O.; Jansson, A. Additive manufacturing of polypropylene: A screening design of experiment using laser-based powder bed fusion. *Polymers* **2018**, *10*, 1293. [CrossRef]

196. Fang, L.; Wang, Y.; Xu, Y. Preparation of polypropylene powder by dissolution-precipitation method for selective laser sintering. *Adv. Polym. Technol.* **2019**, *2019*, 5803895. [CrossRef]

197. Shanmugam, V.; Das, O.; Babu, K.; Marimuthu, U.; Veerasimman, A.; Johnson, D.J.; Neisiany, R.E.; Hedenqvist, M.S.; Ramakrishna, S.; Berto, F. Fatigue behaviour of FDM-3D printed polymers, polymeric composites and architected cellular materials. *Int. J. Fatigue* **2021**, *143*, 106007. [CrossRef]

198. Safai, L.; Cuellar, J.S.; Smit, G.; Zadpoor, A.A. A review of the fatigue behavior of 3D printed polymers. *Addit. Manuf.* **2019**, *28*, 87–97. [CrossRef]

199. Yao, T.; Ouyang, H.; Dai, S.; Deng, Z.; Zhang, K. Effects of manufacturing micro-structure on vibration of FFF 3D printing plates: Material characterisation, numerical analysis and experimental study. *Compos. Struct.* **2021**, *268*, 113970. [CrossRef]

200. Erokhin, K.S.; Gordeev, E.G.; Ananikov, V.P. Revealing interactions of layered polymeric materials at solid-liquid interface for building solvent compatibility charts for 3D printing applications. *Sci. Rep.* **2019**, *9*, 1–14. [CrossRef]
201. Liu, T.; Guessasma, S.; Zhu, J.; Zhang, W.; Nouri, H.; Belhabib, S. Microstructural defects induced by stereolithography and related compressive behaviour of polymers. *J. Mater. Process. Technol.* **2018**, *251*, 37–46. [CrossRef]
202. Bano, S.; Iqbal, T.; Ramzan, N.; Farooq, U. Study of surface mechanical characteristics of abs/pc blends using nanoindentation. *Processes* **2021**, *9*, 637. [CrossRef]
203. Batakliev, T.; Georgiev, V.; Ivanov, E.; Kotsilkova, R.; Di Maio, R.; Silvestre, C.; Cimmino, S. Nanoindentation analysis of 3D printed poly(lactic acid)-based composites reinforced with graphene and multiwall carbon nanotubes. *J. Appl. Polym. Sci.* **2019**, *136*, 3–7. [CrossRef]
204. Mansour, M.; Tsongas, K.; Tzetzis, D. Measurement of the mechanical and dynamic properties of 3D printed polylactic acid reinforced with graphene. *Polym. Technol. Mater.* **2019**, *58*, 1234–1244. [CrossRef]
205. Vălean, C.; Marşavina, L.; Mărghitaşl, M.; Linul, E.; Razavi, J.; Berto, F. Effect of manufacturing parameters on tensile properties of FDM printed specimens. *Procedia Struct. Integr.* **2020**, *26*, 313–320. [CrossRef]
206. Ding, S.; Kong, L.; Jin, Y.; Lin, J.; Chang, C.; Li, H.; Liu, E.; Liu, H. Influence of the molding angle on tensile properties of FDM parts with orthogonal layering. *Polym. Adv. Technol.* **2020**, *31*, 873–884. [CrossRef]
207. Fountas, N.A.; Kostazos, P.; Pavlidis, H.; Antoniou, V.; Manolakos, D.E.; Vaxevanidis, N.M. Experimental investigation and statistical modelling for assessing the tensile properties of FDM fabricated parts. *Procedia Struct. Integr.* **2020**, *26*, 139–146. [CrossRef]
208. Heidari-Rarani, M.; Rafiee-Afarani, M.; Zahedi, A.M. Mechanical characterization of FDM 3D printing of continuous carbon fiber reinforced PLA composites. *Compos. Part. B Eng.* **2019**, *175*, 107147. [CrossRef]
209. Letcher, T.; Waytashek, M. Material property testing of 3D-printed specimen in pla on an entry-level 3D printer. In *ASME International Mechanical Engineering Congress and Exposition*; American Society of Mechanical Engineers: New York, NY, USA, 2014. [CrossRef]
210. Weng, Z.; Wang, J.; Senthil, T.; Wu, L. Mechanical and thermal properties of ABS/montmorillonite nanocomposites for fused deposition modeling 3D printing. *Mater. Des.* **2016**, *102*, 276–283. [CrossRef]
211. Rahman, K.M.; Letcher, T.; Reese, R. Mechanical properties of additively manufactured peek components using fused filament fabrication. In *ASME International Mechanical Engineering Congress and Exposition*; American Society of Mechanical Engineers: New York, NY, USA, 2015. [CrossRef]
212. Afrose, M.F.; Masood, S.H.; Iovenitti, P.; Nikzad, M.; Sbarski, I. Effects of part build orientations on fatigue behaviour of FDM-processed PLA material. *Prog. Addit. Manuf.* **2016**, *1*, 21–28. [CrossRef]
213. Chacón, J.M.; Caminero, M.A.; García-Plaza, E.; Núñez, P.J. Additive manufacturing of PLA structures using fused deposition modelling: Effect of process parameters on mechanical properties and their optimal selection. *Mater. Des.* **2017**, *124*, 143–157. [CrossRef]
214. Kazmer, D. *Three-Dimensional Printing of Plastics*, 2nd ed.; Elsevier Inc.: Amsterdam, The Netherlands, 2017; ISBN 978-032-339-0-408.
215. Ajoku, U.; Saleh, N.; Hopkinson, N.; Hague, R.; Erasenthiran, P. Investigating mechanical anisotropy and end-of-vector effect in laser-sintered nylon parts. *Proc. Inst. Mech. Eng. Part. B J. Eng. Manuf.* **2006**, *220*, 1077–1086. [CrossRef]
216. Dizon, J.R.C.; Espera, A.H.; Chen, Q.; Advincula, R.C. Mechanical characterization of 3D-printed polymers. *Addit. Manuf.* **2018**, *20*, 44–67. [CrossRef]
217. Cosmi, F.; Dal Maso, A. A mechanical characterization of SLA 3D-printed specimens for low-budget applications. *Mater. Today Proc.* **2019**, *32*, 194–201. [CrossRef]
218. Yang, Y.; Li, L.; Zhao, J. Mechanical property modeling of photosensitive liquid resin in stereolithography additive manufacturing: Bridging degree of cure with tensile strength and hardness. *Mater. Des.* **2019**, *162*, 418–428. [CrossRef]
219. Ahmad, K.W.H.; Mohamad, Z.; Othman, N.; Man, S.H.C.; Jusoh, M. The mechanical properties of photopolymer prepared via 3d stereolithography printing: The effect of uv curing time and anisotropy. *Chem. Eng. Trans.* **2020**, *78*, 565–570. [CrossRef]
220. Borrello, J.; Nasser, P.; Iatridis, J.C.; Costa, K.D. 3D printing a mechanically-tunable acrylate resin on a commercial DLP-SLA printer. *Addit. Manuf.* **2018**, *23*, 374–380. [CrossRef]
221. Chantarapanich, N.; Puttawibul, P.; Sitthiseripratip, K.; Sucharitpwatskul, S.; Chantaweroad, S. Study of the mechanical properties of photo-cured epoxy resin fabricated by stereolithography process. *Songklanakarin J. Sci. Technol.* **2013**, *35*, 91–98.
222. Ayrilmis, N.; Kariz, M.; Kwon, J.H.; Kitek Kuzman, M. Effect of printing layer thickness on water absorption and mechanical properties of 3D-printed wood/PLA composite materials. *Int. J. Adv. Manuf. Technol.* **2019**, *102*, 2195–2200. [CrossRef]
223. Negi, S.; Sharma, R.K.; Dhiman, S. Experimental investigation of sls process for flexural strength improvement of PA-3200GF parts. *Mater. Manuf. Process.* **2015**, *30*, 644–653. [CrossRef]
224. Marsavina, L.; Stoia, D.I. Flexural properties of selectively sintered polyamide and Alumide. *Mater. Des. Process. Commun.* **2020**, *2*, 1–5. [CrossRef]
225. Domingo-Espin, M.; Puigoriol-Forcada, J.M.; Garcia-Granada, A.A.; Llumà, J.; Borros, S.; Reyes, G. Mechanical property characterization and simulation of fused deposition modeling Polycarbonate parts. *Mater. Des.* **2015**, *83*, 670–677. [CrossRef]
226. Carneiro, O.S.; Silva, A.F.; Gomes, R. Fused deposition modeling with polypropylene. *Mater. Des.* **2015**, *83*, 768–776. [CrossRef]
227. Tomanik, M.; Żmudzińska, M.; Wojtków, M. Mechanical and Structural Evaluation of the PA12 Desktop Selective Laser Sintering Printed Parts Regarding Printing Strategy. *3D Print. Addit. Manuf.* **2021**, *1*, 1–9. [CrossRef]

228. Salmoria, G.V.; Leite, J.L.; Vieira, L.F.; Pires, A.T.N.; Roesler, C.R.M. Mechanical properties of PA6/PA12 blend specimens prepared by selective laser sintering. *Polym. Test.* **2012**, *31*, 411–416. [CrossRef]

229. Zhu, W.; Yan, C.; Shi, Y.; Wen, S.; Liu, J.; Shi, Y. Investigation into mechanical and microstructural properties of polypropylene manufactured by selective laser sintering in comparison with injection molding counterparts. *Mater. Des.* **2015**, *82*, 37–45. [CrossRef]

230. Shi, Y.; Chen, J.; Wang, Y.; Li, Z.; Huang, S. Study of the selective laser sintering of polycarbonate and postprocess for parts reinforcement. *Proc. Inst. Mech. Eng. Part. L J. Mater. Des. Appl.* **2007**, *221*, 37–42. [CrossRef]

231. Ning, F.; Cong, W.; Qiu, J.; Wei, J.; Wang, S. Additive manufacturing of carbon fiber reinforced thermoplastic composites using fused deposition modeling. *Compos. Part. B Eng.* **2015**, *80*, 369–378. [CrossRef]

232. Salmoria, G.V.; Paggi, R.A.; Lago, A.; Beal, V.E. Microstructural and mechanical characterization of PA12/MWCNTs nanocomposite manufactured by selective laser sintering. *Polym. Test.* **2011**, *30*, 611–615. [CrossRef]

233. Berli, C.; Thieringer, F.M.; Sharma, N.; Müller, J.A.; Dedem, P.; Fischer, J.; Rohr, N. Comparing the mechanical properties of pressed, milled, and 3D-printed resins for occlusal devices. *J. Prosthet. Dent.* **2020**, *124*, 780–786. [CrossRef]

234. Skylar Tibbits, The Emergence of 4D Printing, "TED Talk" 2013. Available online: https://www.ted.com/talks/skylar_tibbits_the_emergence_of_4d_printing?language=yi (accessed on 2 August 2021).

235. Bastola, A.K.; Rodriguez, N.; Behl, M.; Soffiatti, P.; Rowe, N.P.; Lendlein, A. Cactus-inspired design principles for soft robotics based on 3D printed hydrogel-elastomer systems. *Mater. Des.* **2021**, *202*, 109515. [CrossRef]

236. Kim, Y.; Yuk, H.; Zhao, R.; Chester, S.A.; Zhao, X. Printing ferromagnetic domains for untethered fast-transforming soft materials. *Nature* **2018**, *558*, 274–279. [CrossRef]

237. Rafiee, M.; Farahani, R.D.; Therriault, D. Multi-material 3D and 4D printing: A survey. *Adv. Sci.* **2020**, *7*, 1902307. [CrossRef]

238. Joshi, S.; Rawat, K.; Karunakaran, C.; Rajamohan, V.; Mathew, A.T.; Koziol, K.; Thakur, V.K.; Balan, A.S.S. 4D printing of materials for the future: Opportunities and challenges. *Appl. Mater. Today* **2020**, *18*, 100490. [CrossRef]

239. Moreno, M.A.; Gonzalez-Rico, J.; Lopez-Donaire, M.L.; Arias, A.; Garcia-Gonzalez, D. New experimental insights into magneto-mechanical rate dependences of magnetorheological elastomers. *Compos. Part. B Eng.* **2021**, *224*, 109148. [CrossRef]

240. Hu, G.F.; Damanpack, A.R.; Bodaghi, M.; Liao, W.H. Increasing dimension of structures by 4D printing shape memory polymers via fused deposition modeling. *Smart Mater. Struct.* **2017**, *26*, 125023. [CrossRef]

241. Tirado-Garcia, I.; Garcia-Gonzalez, D.; Garzon-Hernandez, S.; Rusinek, A.; Robles, G.; Martinez-Tarifa, J.M.; Arias, A. Conductive 3D printed PLA composites: On the interplay of mechanical, electrical and thermal behaviours. *Compos. Struct.* **2021**, *265*, 113744. [CrossRef]

242. Bodaghi, M.; Damanpack, A.R.; Liao, W.H. Triple shape memory polymers by 4D printing. *Smart Mater. Struct.* **2018**, *27*, 65010. [CrossRef]

243. Zolfagharian, A.; Kaynak, A.; Kouzani, A. Closed-loop 4D-printed soft robots. *Mater. Des.* **2020**, *188*, 108411. [CrossRef]

244. Bodaghi, M.; Serjouei, A.; Zolfagharian, A.; Fotouhi, M.; Rahman, H.; Durand, D. Reversible energy absorbing meta-sandwiches by FDM 4D printing. *Int. J. Mech. Sci.* **2020**, *173*, 105451. [CrossRef]

245. Bastola, A.K.; Paudel, M.; Li, L.; Li, W. Recent progress of magnetorheological elastomers: A review. *Smart Mater. Struct.* **2020**, *29*, 123002. [CrossRef]

246. Bastola, A.K.; Hossain, M. A review on magneto-mechanical characterizations of magnetorheological elastomers. *Compos. Part. B Eng.* **2020**, *200*, 108348. [CrossRef]

247. Merazzo, K.M.J.; Pereira, N.; Lima, A.; Iglesias, M.R.; Fernandes, L.; Lanceros-Mendez, S.; Martins, P. Magnetic materials: A journey from finding north to an exciting printed future. *Mater. Horizons* **2021**. [CrossRef]

248. Wu, S.; Hu, W.; Ze, Q.; Sitti, M.; Zhao, R. Multifunctional magnetic soft composites: A review. *Multifunct. Mater.* **2020**, *3*, 042003. [CrossRef]

249. Bastola, A.K.; Hoang, V.T.; Li, L. A novel hybrid magnetorheological elastomer developed by 3D printing. *Mater. Des.* **2017**, *114*, 391–397. [CrossRef]

250. Zhang, Y.; Wang, Q.; Yi, S.; Lin, Z.; Wang, C.; Chen, Z.; Jiang, L. 4D Printing of Magnetoactive Soft Materials for On-Demand Magnetic Actuation Transformation. *ACS Appl. Mater. Interfaces* **2021**, *13*, 4174–4184. [CrossRef]

251. Bastola, A.K.; Paudel, M.; Li, L. Development of hybrid magnetorheological elastomers by 3D printing. *Polymers* **2018**, *149*, 213–228. [CrossRef]

252. Qi, S.; Fu, J.; Xie, Y.; Li, Y.; Gan, R.; Yu, M. Versatile magnetorheological plastomer with 3D printability, switchable mechanics, shape memory, and self-healing capacity. *Compos. Sci. Technol.* **2019**, *183*, 107817. [CrossRef]

253. Cao, X.; Xuan, S.; Sun, S.; Xu, Z.; Li, J.; Gong, X. 3D Printing Magnetic Actuators for Biomimetic Applications. *ACS Appl. Mater. Interfaces* **2021**, *13*, 30127–30136. [CrossRef] [PubMed]

254. Hupfeld, T.; Salamon, S.; Landers, J.; Sommereyns, A.; Doñate-Buendía, C.; Schmidt, J.; Wende, H.; Schmidt, M.; Barcikowski, S.; Gökce, B. 3D printing of magnetic parts by laser powder bed fusion of iron oxide nanoparticle functionalized polyamide powders. *J. Mater. Chem. C* **2020**, *8*, 12204–12217. [CrossRef]

255. Lantean, S.; Barrera, G.; Pirri, C.F.; Tiberto, P.; Sangermano, M.; Roppolo, I.; Rizza, G. 3D Printing of Magnetoresponsive Polymeric Materials with Tunable Mechanical and Magnetic Properties by Digital Light Processing. *Adv. Mater. Technol.* **2019**, *4*, 1–10. [CrossRef]

256. Ji, Z.; Yan, C.; Yu, B.; Wang, X.; Zhou, F. Multimaterials 3D Printing for Free Assembly Manufacturing of Magnetic Driving Soft Actuator. *Adv. Mater. Interfaces* **2017**, *4*, 1–6. [CrossRef]
257. Nagarajan, B.; Mertiny, P.; Qureshi, A.J. Magnetically loaded polymer composites using stereolithography—Material processing and characterization. *Mater. Today Commun.* **2020**, *25*, 101520. [CrossRef]
258. Hossain, M.; Liao, Z. An additively manufactured silicone polymer: Thermo-viscoelastic experimental study and computational modelling. *Addit. Manuf.* **2020**, *35*, 101395. [CrossRef]
259. Bodaghi, M.; Noroozi, R.; Zolfagharian, A.; Fotouhi, M.; Norouzi, S. 4D Printing Self-Morphing Structures. *Materials* **2019**, *12*, 1353. [CrossRef] [PubMed]
260. Zhao, W.; Zhang, F.; Leng, J.; Liu, Y. Personalized 4D printing of bioinspired tracheal scaffold concept based on magnetic stimulated shape memory composites. *Compos. Sci. Technol.* **2019**, *184*, 107866. [CrossRef]
261. Dohmen, E.; Saloum, A.; Abel, J. Field-structured magnetic elastomers based on thermoplastic polyurethane for fused filament fabrication. *Philos. Trans. R. Soc. A Math. Phys. Eng. Sci.* **2020**, *378*, 20190257. [CrossRef]
262. Kumar, S.; Singh, R.; Singh, T.P.; Batish, A. On mechanical characterization of 3-D printed PLA-PVC-wood dust-Fe3O4 composite. *J. Thermoplast. Compos. Mater.* **2019**, 0892705719879195. [CrossRef]
263. Prem, N.; Sindersberger, D.; Monkman, G.J. Mini-Extruder for 3D Magnetoactive Polymer Printing. *Adv. Mater. Sci. Eng.* **2019**, *2019*, 8715718. [CrossRef]
264. Calascione, T.M.; Fischer, N.A.; Lee, T.J.; Thatcher, H.G.; Nelson-Cheeseman, B.B. Controlling magnetic properties of 3D-printed magnetic elastomer structures via fused deposition modeling. *AIP Adv.* **2021**, *11*, 25223. [CrossRef]
265. Zhang, F.; Wang, L.; Zheng, Z.; Liu, Y.; Leng, J. Magnetic programming of 4D printed shape memory composite structures. *Compos. Part. A Appl. Sci. Manuf.* **2019**, *125*, 105571. [CrossRef]
266. Wu, H.; Wang, O.; Tian, Y.; Wang, M.; Su, B.; Yan, C.; Zhou, K.; Shi, Y. Selective Laser Sintering-Based 4D Printing of Magnetism-Responsive Grippers. *ACS Appl. Mater. Interfaces* **2021**, *13*, 12679–12688. [CrossRef]
267. Shinoda, H.; Azukizawa, S.; Maeda, K.; Tsumori, F. Bio-mimic motion of 3D-printed gel structures dispersed with magnetic particles. *J. Electrochem. Soc.* **2019**, *166*, B3235. [CrossRef]
268. Zhu, W.; Li, J.; Leong, Y.J.; Rozen, I.; Qu, X.; Dong, R.; Wu, Z.; Gao, W.; Chung, P.H.; Wang, J.; et al. 3D-Printed Artificial Microfish. *Adv. Mater.* **2015**, *27*, 4411–4417. [CrossRef] [PubMed]
269. Shao, G.; Ware, H.O.T.; Li, L.; Sun, C. Rapid 3D Printing Magnetically Active Microstructures with High Solid Loading. *Adv. Eng. Mater.* **2020**, *22*, 1900911. [CrossRef]
270. Shao, G.; Ware, H.O.T.; Huang, J.; Hai, R.; Li, L.; Sun, C. 3D printed magnetically-actuating micro-gripper operates in air and water. *Addit. Manuf.* **2021**, *38*, 101834. [CrossRef]
271. Domingo-Roca, R.; Jackson, J.C.; Windmill, J.F.C. 3D-printing polymer-based permanent magnets. *Mater. Des.* **2018**, *153*, 120–128. [CrossRef]
272. Msallem, B.; Sharma, N.; Cao, S.; Halbeisen, F.S.; Zeilhofer, H.-F.; Thieringer, F.M. Evaluation of the Dimensional Accuracy of 3D-Printed Anatomical Mandibular Models Using FFF, SLA, SLS, MJ, and BJ Printing Technology. *J. Clin. Med.* **2020**, *9*, 817. [CrossRef]

 *polymers*

Article

Direct Writing Corrugated PVC Gel Artificial Muscle via Multi-Material Printing Processes

Bin Luo [1,2,3,*], Yiding Zhong [2], Hualing Chen [1,2,*], Zicai Zhu [2] and Yanjie Wang [4]

1   State Key Laboratory for Strength and Vibration of Mechanical Structures, Xi'an Jiaotong University, Xi'an 710049, China
2   School of Mechanical Engineering, Xi'an Jiaotong University, Xi'an 710049, China; ydzhong@zju.edu.cn (Y.Z.); zicaizhu@xjtu.edu.cn (Z.Z.)
3   School of Mechanical and Energy Engineering, Shaoyang University, Shaoyang 422000, China
4   Changzhou Campus, School of Mechanical and Electrical Engineering, Hohai University, Changzhou 213022, China; yj.wang1985@gmail.com
*   Correspondence: lb19870911@stu.xjtu.edu.cn (B.L.); hlchen@mail.xjtu.edu.cn (H.C.)

Citation: Luo, B.; Zhong, Y.; Chen, H.; Zhu, Z.; Wang, Y. Direct Writing Corrugated PVC Gel Artificial Muscle via Multi-Material Printing Processes. *Polymers* **2021**, *13*, 2734. https://doi.org/10.3390/polym13162734

Academic Editor: Houwen Matthew Pan

Received: 16 July 2021
Accepted: 13 August 2021
Published: 15 August 2021

Publisher's Note: MDPI stays neutral with regard to jurisdictional claims in published maps and institutional affiliations.

Abstract: Electroactive PVC gel is a new artificial muscle material with good performance that can mimic the movement of biological muscle in an electric field. However, traditional manufacturing methods, such as casting, prevent the broad application of this promising material because they cannot achieve the integration of the PVC gel electrode and core layer, and at the same time, it is difficult to create complex and diverse structures. In this study, a multi-material, integrated direct writing method is proposed to fabricate corrugated PVC gel artificial muscle. Inks with suitable rheological properties were developed for printing four functional layers, including core layers, electrode layers, sacrificial layers, and insulating layers, with different characteristics. The curing conditions of the printed CNT/SMP inks under different applied conditions were also discussed. The structural parameters were optimized to improve the actuating performance of the PVC gel artificial muscle. The corrugated PVC gel with a span of 1.6 mm had the best actuating performance. Finally, we printed three layers of corrugated PVC gel artificial muscle with good actuating performance. The proposed method can help to solve the inherent shortcomings of traditional manufacturing methods of PVC gel actuators. The printed structures have potential applications in many fields, such as soft robotics and flexible electronic devices.

Keywords: direct writing; PVC gel; artificial muscle; rheological behavior; integrated printing

1. Introduction

Artificial muscle is a material or structure that can undergo various forms of deformation under external excitations (i.e., electricity [1–5], heat [6–8], light [9,10], magnetic field [11,12], fluid pressure [13–15], and chemical stimulation [16]) and output strain and stress. Artificial muscle is named for its ability to achieve biological muscle-like function, and it has broad application prospects in fields such as robots [4,13,17–19], flexible electronics [20,21], smart textiles [22], and medical rehabilitation [23,24]. Among the various actuating modes of artificial muscles defined according to the external excitation, the electrical actuating mode has advantages of easy control and a wide application range. Electroactive polymers (EAPs) are a class of electrically actuated artificial muscle materials that are of great interest to researchers. Typical representatives of EAPs include ionic polymer/metal composites (IPMCs), dielectric elastomers (DEs), and electroactive polyvinyl chloride (PVC) gel. Among these materials, IPMC has advantages of low operating voltage (1–3 V) and a response time of seconds; however, its actuating force is limited, the working energy density is less than other EAPs, and its durability is poor in dry environments [25]. DE has the advantage of a fast response and can work in air; however, DEs generally require operating voltages of up to several kV to meet electric field strength requirements, which increases the risk of electrical breakdown. However, PVC gel has advantages of high output stress and strain, quick response, good thermal stability, moderate working voltage, low power consumption, and a long cycle life [5], as well as the advantages of DE and IPMC. Therefore, it has more potential in the application [26].

Deformation principle of PVC-gel actuator is shown in Figure 1; in an applied electric field, PVC molecules migrate to the anode, the gel is polarized, and a Maxwell force is generated, which results in creep deformation of PVC gel near the anode and compression of PVC gel in the thickness direction. The PVC gel can return to its original shape under its own elasticity when the electric field is turned off [27].

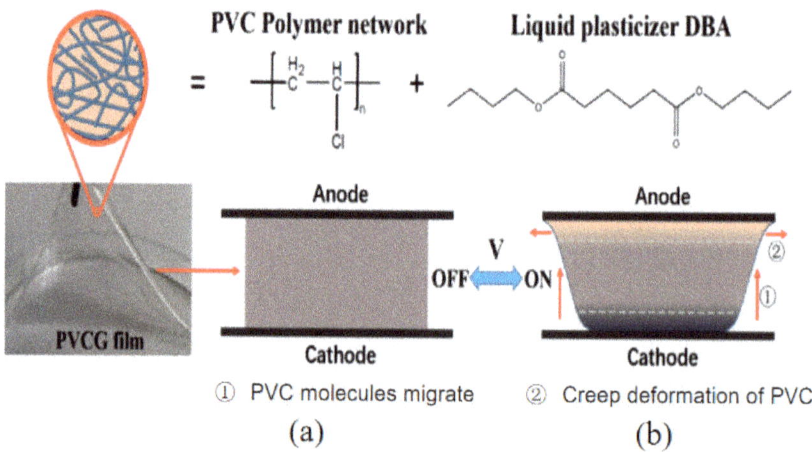

Figure 1. Deformation principle of PVC-gel actuator. (**a**) Discharge (**b**) Charge.

The PVC-gel actuators in application include the traditional mesh anode-planar PVC gel-planar cathode structure (referred to as the planar PVC gel actuator) [5,24] and, more recently, the planar electrode-corrugated PVC gel structure (referred to as the corrugated PVC gel actuator) [20,28]. For any of the structures, the basic actuating unit is the sandwich structure as electrode layer–core layer–electrode layer. A key common feature of the two typical actuators is the existence of pore structures, which are necessary for the actuators to deform effectively; for example, the migration of the peak material to the valley floor or the transfer of planar materials into the mesh pores.

The current manufacturing method of PVC gel is to mix PVC, dibutyl adipate (DBA), and tetrahydrofuran (THF) together to form a precursor material and then cast the material into molds to obtain films with designed shapes [5]. To make a PVC gel actuator, the traditional method also requires the fabricated PVC gel films to be cut and then manually stacked with metal electrodes. Obviously, the casting method is only suitable for manufacturing simple 2D structures, and it is difficult to produce complex, diversified, flexible 3D structures. To overcome this issue, additive manufacturing is a possible choice. Rossiter and co-workers proposed a filament additive manufacturing technique to create complex (3D) PVC gel structures. The 3D printing process was introduced, where a precursor material (PVC precursor powder mix with DIDA) is extruded into a thermoplastic filament for 3D printing [29]. However, the electrode printing process results in many difficulties because most electrode materials are not thermoplastic, which is essential for PVC-gel as actuators.

Recently, multiple reports have shown that a wide variety of materials have been used in 3D direct writing, including hydro-gels [30], nano-particles [31], polyelectrolytes, ceramics [32], shape memory polymers, and carbon materials [33]. Direct writing is a layer-by-layer assembly technique in which shear-thinning inks are extruded through a nozzle in a programmable pattern, upon which the inks rapidly solidify via gelation, evaporation, or temperature-induced phase change [32]. There are few reports with respect to PVC-gel solution printing by direct writing. Meanwhile, electrode printing, when adopting the direct writing process, has been successful in the presence of a conductive solution that can be cured, such as carbon nanotubes (CNTs) doped polymer and ionic gels [33]. In this work, integrated printing process of electrodes and core layers for fabricating PVC gel artificial muscle were proposed. CNTs doped shape memory polymers (CNT/SMP) were designed for PVC-gel electrode. Sacrificial materials (F127) were used for creating different PVC core structure. Inks with suitable rheological properties were developed for printing four functional layers, including core PVC layers, electrode layers, sacrificial layers, and insulating layers, with different characteristics. The corrugated PVC gel structural parameters were designed and optimized to improve the actuating performance of the PVC gel artificial muscle. Several techniques were used to evaluate the feasibility of the process, including conductivity tests, rheology measurements, and electro-mechanical tests.

2. Experimental Section

2.1. Materials Preparation and Structure Design

Polyvinyl chloride (PVC, degree of polymerization 4000) was purchased from Scientific Polymer Products Inc., New York, NY, USA. Dibutyl adipate (DBA) was obtained from Hubei Jusheng Technology Co., Ltd., Tianmen, P.R. China. Tetrahydrofuran (THF) was purchased from Tianjin Fuyu Fine Chemical Co., Ltd., Tianjin, P.R. China. Pluronic F-127 was purchased from Sigma-Aldrich, St. Louis, MO, USA. The multi-walled CNTs (Time Nano China, TSW3, diameter 10–20 nm, length 0.5–2 μm, >98%) were from Chengdu Organic Chemicals Co., Ltd., Chinese Academy of Sciences, Beijing, P.R. China. Dimethylacetamide (DMAC) was obtained from Aladdin Reagent Co., Ltd., Shanghai, P.R. China. Shape memory polymer (SMP, DiAPLEX MM4520) was from Mitsubishi Co., Ltd., Tokyo, Japan. Ecoflex 00-20 was purchased from Smooth-On Inc., Macungie, PA, USA.

As shown in Figure 2a, CNT doped shape memory polymers (CNT/SMP) were chosen for printing PVC-gel electrode because the electrodes' hardness is suitable [31]. The Ecoflex silicone materials, which possess very high insulation performance, were designed for connecting PVC-gel actuator units. To fabricate the flexible artificial muscle structure, it was first necessary to prepare inks suitable for direct writing and to satisfy the functional requirements for core layers, sacrificial layers, electrode layers, and insulating layers. Using a printing device with a pneumatic ink extrusion system (Figure S1), each functional layer was printed in sequence with the prepared inks to produce the composite structure. In particular, pluronic F-127 is a polyethylene oxide (PEO)-polypropylene oxide (PPO)-polyethylene oxide (PEO) triblock copolymer. As shown in Figure 2b, F-127 has thixotropic properties. When the ink was prepared, the molecules cross-linked with each other to form gel networks. When the gel was subjected to shear stress, the networks broke down and the ink became fluid-like so it could be conveniently extruded through a nozzle. After the shear stress was removed, the ink quickly turned to gel with rigidity that could help maintain the shape of the printed structure. In this paper, the F-127 gel inks were printed for constructing a corrugated PVC-gel core structure.

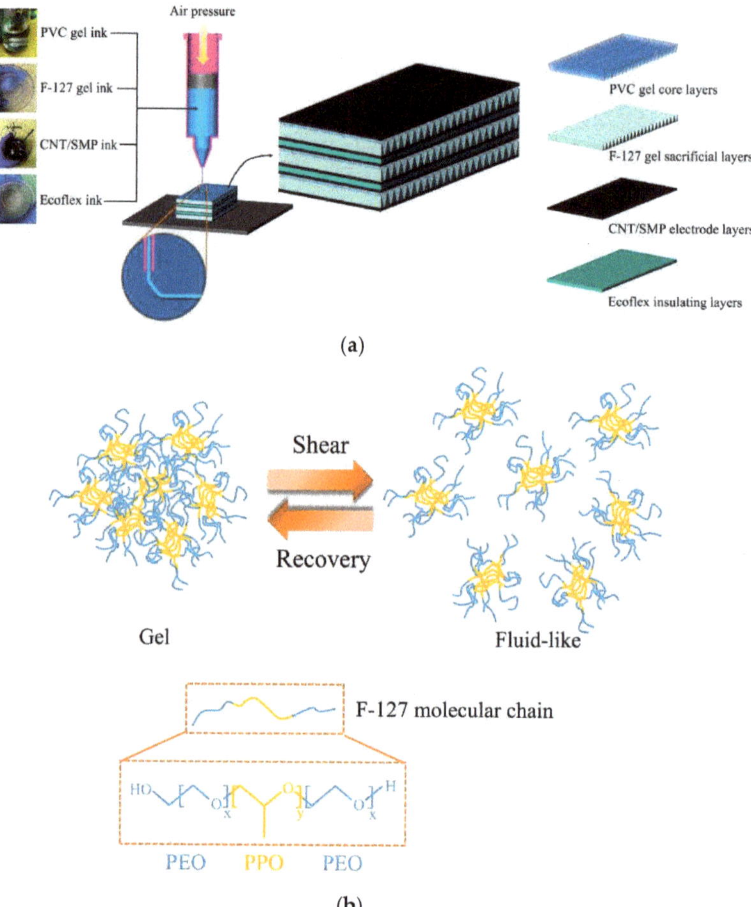

Figure 2. (**a**) Schematic diagram of the printing process of the artificial muscle structure. (**b**) Schematic diagram of the thixotropic property of the F-127 gel ink.

The PVC gel ink consisted of three constituent materials: PVC, DBA, and THF. For preparation of the gel ink, the three constituents were proportionally mixed (referring to research on the relationship between the proportion of PVC gel and its properties in literature [34,35], the mass ratio used in this paper was PVC:DBA:THF = 1:7:12), and then the solution was stirred by a magnetic stirrer for 2–3 days until the gel became homogenous, colorless, and transparent. F-127 gel ink was comprised of F-127 and deionized water. During the preparation process, F-127 was mixed with deionized water in proportion and then the mixture was stirred for 15 min at 2000 rpm by a planetary mixer (HM800) to obtain the colorless, transparent gel. When preparing CNT/SMP composite ink, SMP was first mixed with DMAC, which served as a solvent, and the mixture was heated (60 °C) and magnetically stirred (600 rpm) for approximately 4 h such that the SMP particles were completely dissolved to form a transparent solution. Subsequently, CNTs were added proportionally to the solution and the mixture was dispersed by ultrasound for approximately 3 h until the carbon nanotubes were evenly dispersed in the solution. Finally, using a magnetic stir heater (60 °C, 600 rpm), the solution was concentrated for approximately 5 h to produce ink containing 30 wt.% solid phase that could be used for printing. Ecoflex silicone consisted of two components: Ecoflex-A and Ecoflex-B. To prepare the Ecoflex ink, the A and B components were mixed at a mass ratio of 1:1 and stirred.

2.2. Performance Characterization of the Inks

When analyzing the electrical properties of the CNT/SMP composites, CNT/SMP composites with different CNT contents were made into long strips (40 mm × 5 mm × 1 mm). The resistance of the strips was measured using a source meter (2450 SourceMeter, Keithley, Tektronix, Cleveland, OH, USA) and the conductivity of composites with different proportions was calculated. The rheological behaviors of the prepared inks were measured by a rheometer (MCR 302, Anton Paar, Graz, Austria) fitted with a parallel-plate geometry (PP35Ti, diameter of 35 mm and gap of 0.6 mm) at 25 °C. The apparent viscosity, storage modulus (G'), and loss modulus (G'') of the PVC gel ink, F-127 gel ink, CNT/SMP composite ink, and Ecoflex ink were measured as a function of the angular frequency from 0.628 rad/s to 62.8 rad/s under fixed strain (1%) and shear stress (from 0.1 to 100 Pa) in oscillatory mode at 1 Hz.

2.3. Direct Writing Process

In this paper, the 3D printing used a multi-nozzle direct writing device (Figure S1). In the first step, the 3D models were established in CAD software (SolidWorks 2016) and then imported software to generate G codes. The second step was to install the syringe with ink for printing to the designed position in the direct writing device. In the third step, the designed structure was printed layer-by-layer by controlling the movement of the 3D positioning stage according to G codes and adjusting the output air pressure to control the extrusion speed of the ink.

2.4. Post-Processing of the Printed Structures

After printing the composite structures of the PVC gel actuators, it was necessary to immerse them in water to remove the sacrificial material and obtain the designed corrugated PVC gel. Post-processing by curing was also needed for the printed CNT/SMP composites (for both the curing of the shape and the formation of conductive network characterized by improved conductivity). Therefore, the curing characteristic of the CNT/SMP composite ink and the relationships between the conductivity and time at different curing temperatures were studied. We used 10 wt.% CNT/SMP ink to produce rectangular sheet samples (40 mm × 20 mm × 0.4 mm) that were heated in a vacuum oven or stored at room temperature. At regular intervals, the shape curing was observed, and the resistance was measured to determine the relationship between the conductivity and time. To study integrated printing of PVC gel artificial muscles, we combined the above two post-processing methods, proposing a method for placing the printed structure at room temperature first and then heating in a water bath, and studied the effects of this method on the dissolution of the sacrificial layers and the curing of the CNT/SMP composite ink.

2.5. Performance Characterization of the PVC Gel Artificial Muscle

The actuating performance of the PVC gel artificial muscle was measured by a test system (Figure S2, Supporting Information). The displacement of the artificial muscle with a laser sensor (LK-G500, KEYENCE, Osaka, Japan) was measured using a signal generator (DG4062, RIGOL, Beijing, P.R. China) and voltage amplifier (MODEL 20/20C, TREK, Waterloo, WI, USA) to apply voltage to the electrode layers of the artificial muscle.

3. Results and Discussion

3.1. Printability of Printing Inks

Rheology measurements were conducted for evaluating printability of inks. As shown in Figure 3a,b. The apparent viscosity of the prepared inks decreased with an increase in the angular frequency, proving that they were shear-thinning non-Newtonian fluids appropriate for direct writing. The G' of the PVC gel ink was greater than its G'', and their difference decreased with an increase in angular frequency.

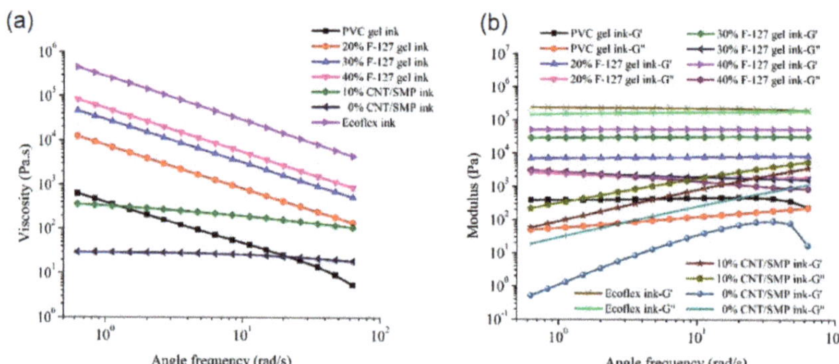

Figure 3. Preparation and performance characterization of the inks. (**a**) Viscosity and (**b**) moduli (storage modulus G' and loss modulus G'') of the PVC gel ink, F-127 gel ink with different concentrations, pure SMP ink, CNT/SMP ink, and Ecoflex ink as a function of the angular frequency.

Therefore, when the shear effect was strong and the angular frequency was high, the PVC gel ink had excellent fluidity and, thus, good filling performance. Alternatively, when the shear effect was weaker and the angular frequency decreased, the difference between G' and G'' increased, which resulted in a decrease in fluidity; therefore, the printed shape was easy to maintain. These characteristics are also demonstrated in Figure 4a. The PVC-gel ink exhibited solid-like gel behavior ($G' > G''$) at stresses (0.1–29.81 Pa) and liquid-like gel behavior ($G' < G''$) at stresses >29.81 Pa, which indicated that this ink is suitable for direct writing [36]. As for F-127 gel ink, at the same angular frequency, with an increase in F-127, the viscosity of the F-127 gel ink increased, and the printability increased. The G' of each experimental group was greater than G'', indicating that the material was in a non-flowing state. With the increase in F-127 content, this behavior was more obvious, and thus, the shape retention ability of the printed structure and the printing accuracy were better. Additionally, 40 wt.% F-127 gel ink storage and loss modulus decreased steadily when the stress was increased from a very low stress of 0.1 Pa up to 10 Pa in Figure 4b, and the 40 wt.% F-127 gel ink exhibited solid-like gel behavior ($G' > G''$) at stresses (0.1–7.41 Pa) and liquid-like gel behavior ($G' < G''$) at stresses >7.41 Pa, which indicated the thixotropic properties. As the main purpose of using this material was to obtain the required pore structure in the core layers with high printing precision, in the following study, 40 wt.% F-127 gel ink was used to print the sacrificial layers.

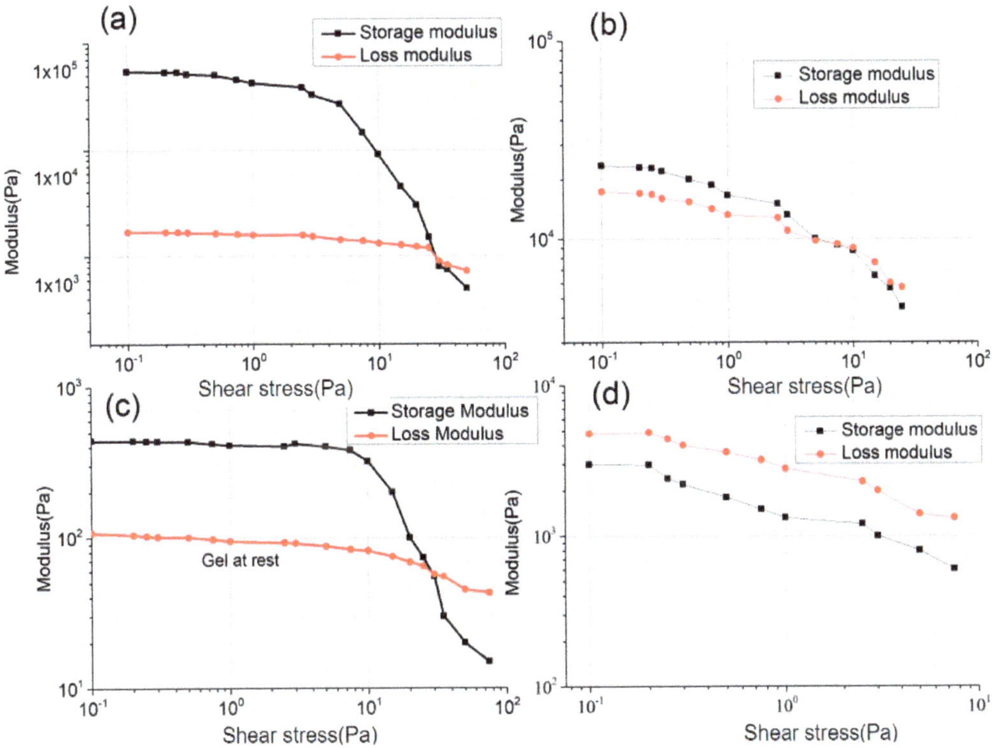

Figure 4. Plot of the storage modulus and loss modulus as a function of shear stress. (**a**) PVC-gel ink, (**b**) F-127 gel ink, (**c**) Ecoflex ink, (**d**) CNT/SMP composite ink.

The rheological properties of the Ecoflex ink were measured after storage at room temperature (25 °C) for 20 min. The ink reached a semi-cured state since the molecular chains of the two components had gradually crosslinked, and the G' of the Ecoflex ink was greater than G''. Therefore, the ink can be used to print stable lines. For both pure SMP ink and CNT/SMP ink with 10 wt% carbon nanotubes, their G'' were greater than their respective G', demonstrating strong fluidity suitable for printing the planar structure in this paper that can also produce good surface quality. Additionally, after CNT loading, the viscosity and moduli (G' and G'') of the ink were greater than that of pure SMP ink, showing that CNT had a good thickening effect. The 10 wt% CNT/SMP composite ink's fluidity is indicated in Figure 4d; its loss modulus was higher than the storage modulus at stresses (0.1–10 Pa). We systematically evaluated the printability of inks by process experiments. The influence of the extrusion pressure (P) and nozzle moving speed (V) on the filament width of the printed inks are listed in Figure 5. At the lowest extrusion pressure and highest nozzle moving speed, a minimum fiber diameter was achieved with the nozzle diameter (0.29 mm) for each ink. The process parameters used in the subsequent printing experiment are as follows: PVC-gel inks (P = 20 KPa, V = 12 mm/s); F127-gel inks (P = 370 KPa, V = 12 mm/s); CNT/SMP inks (P = 260 KPa, V = 6 mm/s); Ecoflex inks (P = 180 KPa, V = 10 mm/s).

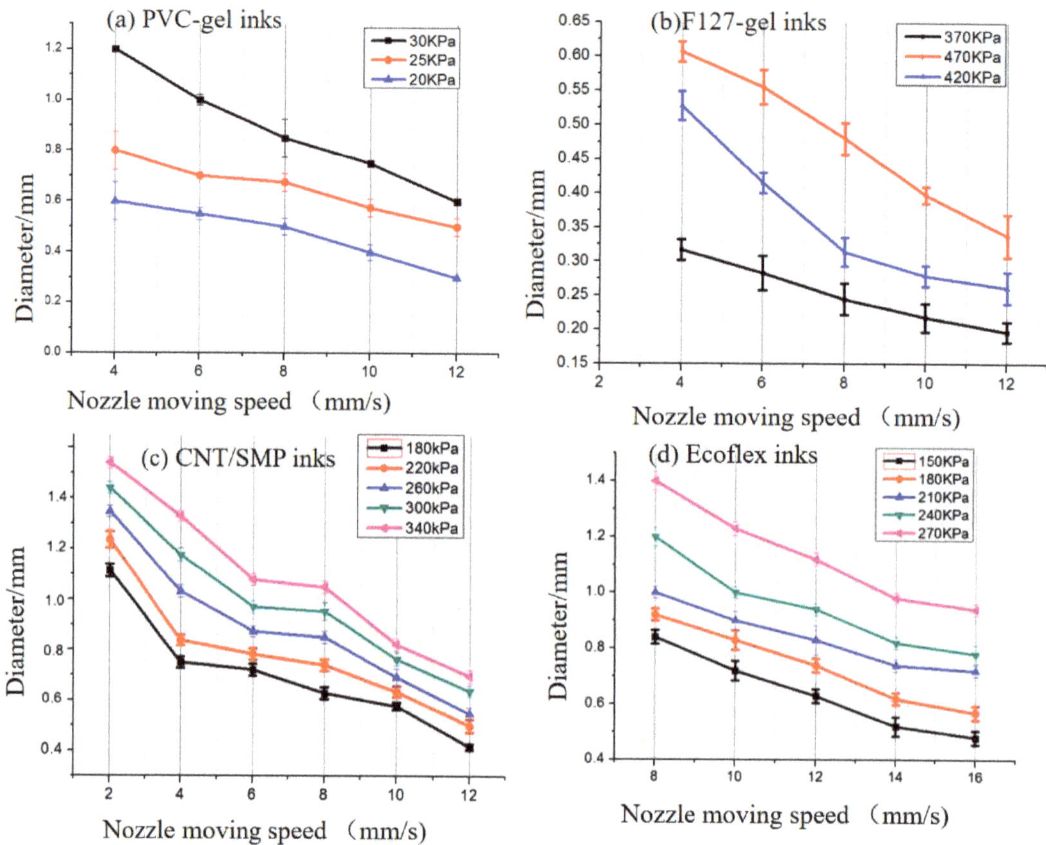

Figure 5. Influence of the extrusion pressure and nozzle moving speed on the printed diameter (**a**) PVC gel ink, (**b**) F-127 gel ink, (**c**) CNT/SMP ink, (**d**) Ecoflex ink.

3.2. Research Results of Curing Methods

The solidification of electrode and the integrated structure shown in Figure 2 is the key to the success of printing PVC-gel actuators. The curing properties of the CNT/SMP ink were analyzed (including both shape curing and conductive network formation characterized by improved conductivity) at different temperatures. After heating at 80 °C for 10 min, 60 °C for 15 min, or storage at room temperature (25 °C) for 48 h, the shape of the composites was completely cured and their surface was uniform and dense. The conductivity–time relationships of the CNT/SMP composite ink at different temperatures are shown in Figure 6a,b. As shown in the figure, higher temperatures resulted in a faster increase in conductivity and a higher final conductivity. After heating at 80 °C for 3 h, the conductivity tended to be stable up to 5.06 S/m. After heating at 60 °C for 4 h, the conductivity reached 3.61 S/m. At room temperature, the curing speed was too slow to meet the requirements of this paper. However, because of compatibility issues of multiple different material systems, the layer-by-layer thermal curing method was not suitable for the integrated printing of multi-material composite structures. For example, the thermal curing of the upper electrode layer may cause the sacrificial layer to lose moisture quickly and melt before the shape of the upper electrode layer is solidified, leading to collapse of the overall structure. Similarly, repeated heating at high temperature for a long time may also affect the performance of PVC gel core layers. Therefore, we proposed a step-wise curing method for the printed electrode layer, which included room temperature storage for 48 h to cure the shape and then uniform heat in a water bath after printing. Uniform heating was used to avoid the effects of repeated heating on the performance of the PVC gel core layers. Water bath heating allowed the sacrificial material to be rapidly dissolved and helped uniformly heat the electrode. Regarding the heating temperature, long-term heating of the structure at 80 °C could greatly attenuate the performance of PVC gel core layers, while heating the structure at 60 °C had little effect. Therefore, 60 °C was selected as the heating temperature. For the printed electrode layers, the first step of incubating at room temperature for 60 h was to ensure that the shapes of the samples were completely cured. As shown in Figure 6c, the conductivity increased slowly at room temperature, and increased rapidly when the samples were heated in a water bath. After the samples were heated in a water bath at 60 °C for 30 min, the conductivity reached 1.87 S/m, which meets the requirements for conductivity of the electrode material in actuators. Therefore, when separately printing the electrode layers, the selected curing condition was 80 °C for 3 h. For integrated printing of PVC gel actuators, the curing method for the electrode layers was first storage at room temperature for 60 h until the shape of the newly printed electrode layer was completely cured, printing of the other layers, and after the shape of the top electrode layer was completely cured, heating of the entire structure in a 60 °C water bath for 30 min.

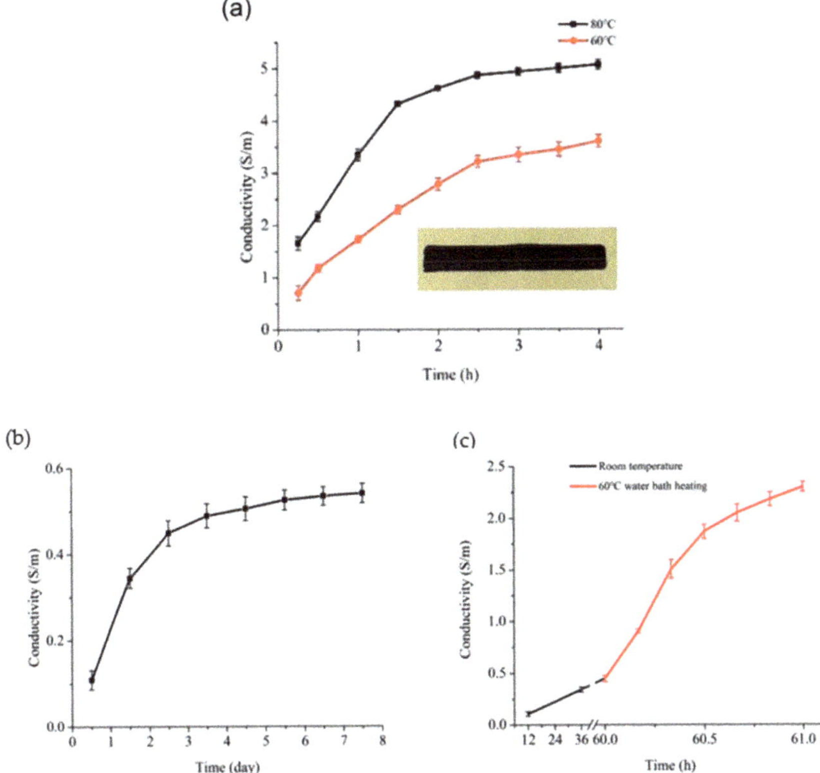

Figure 6. The conductivity of the CNT/SMP composite inks as functions of time (**a**) at 80 °C, 60 °C, (**b**) room temperature (25 °C), and (**c**) stepwise curing conditions.

3.3. 3D Printing Processes of Flexible PVC Gel Artificial Muscle Structures

The printing methods of the core layers, electrode layers, and whole artificial muscles were studied. Initially, the printing method of the corrugated PVC gel core layers was studied. A corrugated F-127 gel sacrificial layer was printed and then the PVC gel was printed on top, filling the pores of the sacrificial layer. The PVC gel was cured at room temperature (25 °C) for approximately 12 h. The resulting composite structure was immersed in water to remove the sacrificial material, and after drying, a corrugated PVC gel core layer was produced, as shown in Figure 7a. The fabricated corrugated PVC gel core layer had good elasticity and transparency with an obvious corrugated shape. The shape of the corrugation in the core layers is shown in Figure 7b. To study the effect of the corrugation scale on the actuating performance of the PVC gel core layers, we designed and printed five groups of corrugated PVC gels with different spans, b. The dimensions of each group were as follows: a = 0.2 mm, h = 2 mm, d = 0.3 mm, and b = 0.8, 1.2, 1.6, 2.0, 2.4 mm. The bottom of the core layer had a planar size of 40 mm × 20 mm.

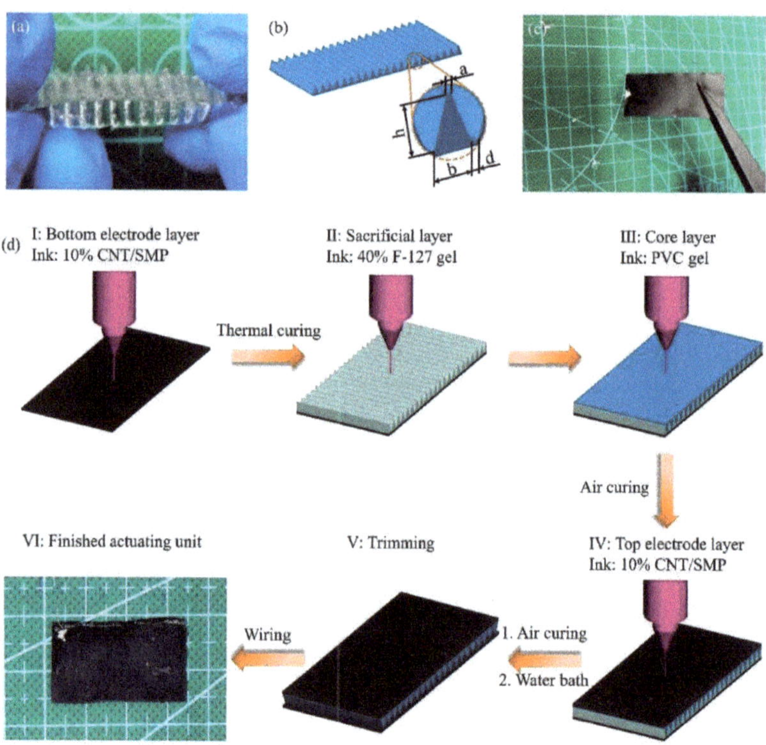

Figure 7. *Cont.*

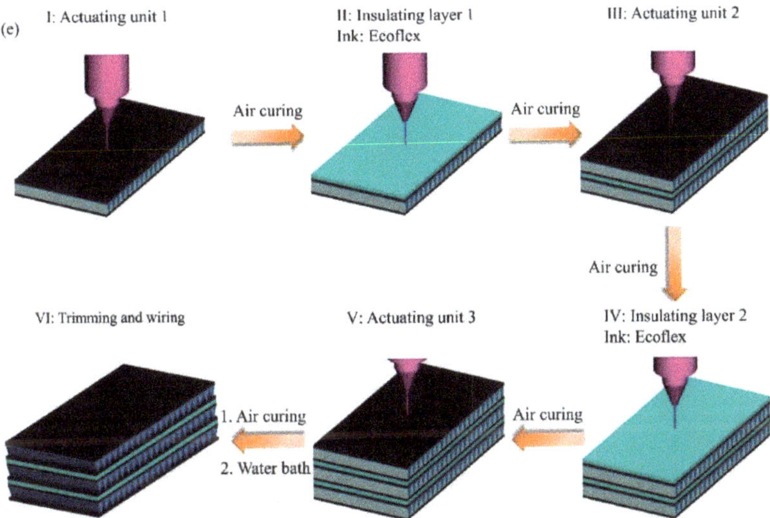

Figure 7. Printing process and printed structures. (**a**) Finished corrugated PVC gel core layer. (**b**) Designed shape of the corrugation in core layers. (**c**) Finished CNT/SMP composite electrode layer. (**d**) Printing process of a single layer PVC gel actuator. (**e**) Printing process of a multilayer PVC gel actuator.

Additionally, we printed CNT/SMP composite electrode layers (40 mm × 20 mm × 0.2 mm) and then cured them at 80 °C for 3 h. Stainless steel wire and conductive silver paste were used to wire them together. As shown in Figure 7c, the finished electrode layer had a dense and smooth surface with suitable rigidity.

Based on the above research, we printed the artificial muscle structure shown in Figure 2a in an integrated way. From the perspective of manufacturing, the functional layers of this structure were integrated printings layer-by-layer at different times to produce a complete structure. Regarding the working principle, all of the electrode layers in contact with the corrugated side of the corrugated PVC gel core layers should be connected to the anode, while all of the electrode layers in contact with the planar side of the corrugated PVC gel core layers should be connected to the cathode. To avoid mutual interference or coupling of different actuating units when powered, the insulating layers of silicone Ecoflex were placed between adjacent actuating units.

As shown in Figure 7d, single layer actuators (actuating units) were first printed. At the beginning, the bottom electrode layer was printed and then heated in a vacuum oven at 80 °C for 3 h to completely cure. Subsequently, the sacrificial layer was printed over the bottom electrode layer. The PVC gel core layer was then printed on the sacrificial layer to fully fill and cover the pores of the sacrificial layer, and the sample was stored at room temperature for approximately 12 h until the THF was sufficiently volatilized and the PVC gel was cured. The top electrode layer was then printed on the core layer, and the sample was stored at room temperature for 60 h until the top electrode layer was cured. After printing, the following post-processing of the sample was performed. First, the composite structure was heated in a water bath at 60 °C for 30 min until the sacrificial layer was completely dissolved. Next, after drying, the edges of the sample were cut to avoid contact between the two electrode layers at the edges, which may result in short circuit of the actuator during testing. Finally, stainless steel wire and conductive silver paste were used to create a lead from the two electrode layers. After heating at 60 °C for approximately 15 min to solidify the conductive silver paste, the integrated printed corrugated PVC gel actuating unit was complete.

On this basis, we studied the printing method of a multilayer PVC gel actuator (two layers and three layers), and Figure 7e illustrates the printing process with a triple layer actuator as an example. Insulating layer 1 was printed on actuating unit 1 and then stored at room temperature for 4 h for complete cure. Similarly, the bottom electrode layer, sacrificial layer, core layer, and top electrode layer were printed in sequence, wherein the electrode layers and core layer were both cured at room temperature before printing the next functional layer, thereby, completing the printing of actuating unit 2. The process continued to superimpose the required number of actuating units. After the overall structure was printed and the electrode layers cured, the entire structure was heated in a water bath at 60 °C for 30 min to remove the sacrificial layer. The sample was then trimmed and leaded to complete fabrication of the multilayer actuators.

3.4. Performance Characterization of the Printed Structures

To test the actuating performance of the printed corrugated PVC gel core layers, three pieces of PVC gel from each of the five experimental groups with varying spans and four pieces of zinc foil electrode (40 mm × 20 mm × 0.07 mm) were superimposed to make a metal electrode-based PVC gel actuator. At different voltages (400 to 800 V, 1 Hz, 50% duty cycle square wave), the strain of the actuators was tested as a function of the corrugated span of core layers, as shown in Figure 8a. The strain of each actuator increased with an increase in voltage. As the corrugated span increased, the strain at each voltage exhibited a single peak form that increased first and then decreased. The corrugated PVC gel with a span of 1.6 mm had the best actuating performance, and the strain of the actuator using this kind of PVC gel core layer reached 9.9% at 800 V. Overall, the PVC gel printed using the additive manufacturing method described in this paper had good electro–deformation performance. Additional experiments in this study used a corrugated PVC gel with a span size 1.6 mm.

To prove the applicability of the printed CNT/SMP composite electrode, we superimposed three pieces of corrugated PVC gel with the 1.6 mm span and four pieces of CNT/SMP composite electrode to form a CNT/SMP composite electrode-based PVC gel actuator and tested the strain versus voltage (400 to 800 V, 1 Hz, 50% duty cycle square wave). Figure 8b shows a comparison of the actuating performance of the CNT/SMP composite electrode-based actuator and a metal electrode-based actuator with the same kind of core layer. The strain of the two actuators at the same voltage were similar. The strain of the CNT/SMP composite electrode-based actuator reached 10.3% at 800 V, proving that the performance of the printed CNT/SMP composite electrode was similar to that of the metal electrode, meeting the requirements of this study.

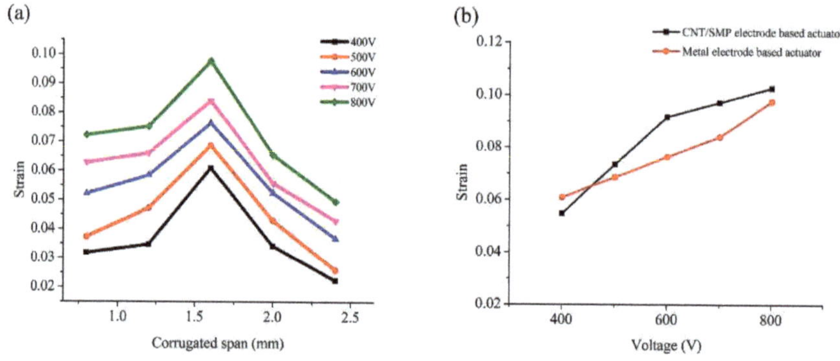

Figure 8. *Cont.*

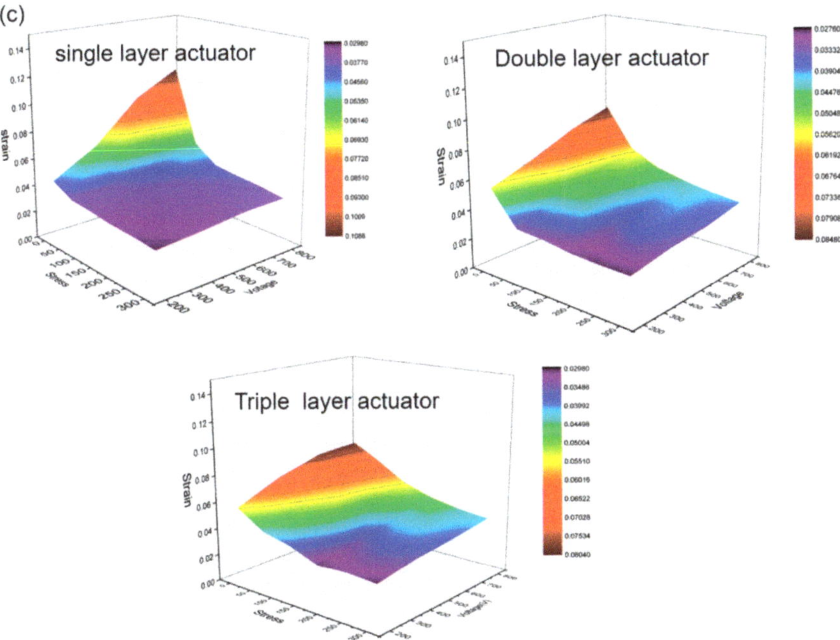

Figure 8. Performance of the printed structures. (**a**) The strain–span curves of metal electrode-based PVC gel actuators at different voltages. (**b**) The strain–voltage curves of the CNT/SMP composite electrode–based actuator and a metal electrode-based actuator with the same kind of core layer. (**c**) The strain–stress curves of the PVC gel actuators with different numbers of layers under voltage from 200 V to 800 V.

For testing the actuating performance of the corrugated PVC gel actuators obtained by integrated printing, the strain–voltage (400 to 800 V, 1 Hz, 50% duty cycle square wave) relationship without load and the strain–load (0 to 70 g) relationship (constant application of 800 V, 1 Hz, 50% duty cycle square wave) of the actuators with varied number of layers (1 to 3) were measured. As shown in Figure 8c, at the same voltage and as the number of layers increased, the strain gradually decreased: at 800 V, the strain of the single layer actuator was as high as 10.9%, the strain of the double layer actuator was up to 8.5%, and the strain of the triple layer actuator reached 8.0%. Since the insulating layers in the multilayer structure increased the total thickness, the strain of the actuators typically decreased to some extent as the number of layers increased. As the load increased, the strain gradually decreased, and the speed gradually slowed. The fewer number of layers, the faster the decreasing speed. When subjected to the same stress (0 to 300 kPa) at the variable voltage (200 to 800 V), the strain increased with the number of layers. As the number of superimposed layers increased, the total output force of the actuator increased, so the load carrying capacity increased and, thus, the negative effect of load on the actuating strain decreased accordingly. The performance of our printed actuators is better than that of the actuators made by casting reported in the literature [37].

4. Conclusions

In summary, we used direct writing to print a series of flexible PVC gel artificial muscles with good actuating performance in the integrated way. Inks with tailored rheological properties met the corresponding functional requirements. PVC gel ink, F-127 gel ink, CNT/SMP composite ink, and Ecoflex ink were used for printing core layers, sacrificial layers, electrode layers, and insulating layers, respectively. The inks were prepared, characterized, and evaluated for curing. Using these inks, multi-material composite artificial muscle structures with complex geometries were fabricated. Performance testing of these structures demonstrated the feasibility of the proposed manufacturing method. Our flexible PVC gel artificial muscles with good actuating performance prepared by integrated printing have potential application in soft robotics, medical rehabilitation, and wearable electronic devices. The proposed method of integrated printing artificial muscles introduces a new route to solve the inherent problems of the traditional fabricating methods of PVC gel, which lays a foundation for the broad application of this new artificial muscle material in various fields.

Supplementary Materials: The following are available online at https://www.mdpi.com/article/10.3390/polym13162734/s1, Figure S1. Experimental setup for the direct writing process; Figure S2. Composition of 3D printing system; Figure S3. Schematic illustration of the system used to test the PVC gel artificial muscle actuating performance. Video S1: Actuating performance of the triple layer printed structures.

Author Contributions: Writing—original draft preparation, B.L.; writing—review and editing, Y.Z.; supervision, H.C.; formal analysis, Z.Z.; investigation, Y.W. All authors have read and agreed to the published version of the manuscript.

Funding: This research was funded byNational Natural Science Foundation of China, grant number (91648110); National Natural Science Foundation of Hunan Province, grant number (2020jj5520, 2020jj4526); Postdoctoral Science Foundation of China, grant number (2021M690871); Excellent youth project of Hunan Provincial Department of Education, grant number (19B515, 19B516). The APC was funded by National Natural Science Foundation of Hunan Province.

Acknowledgments: We thank Ashleigh Cooper for editing the English text of a draft of this manuscript.

Conflicts of Interest: The authors declare no conflict of interest.

References

1. Yan, Y.; Santaniello, T.; Bettini, L.G.; Minnai, C.; Bellacicca, A.; Porotti, R.; Denti, I.; Faraone, G.; Merlini, M.; Lenardi, C. Electroactive ionic soft actuators with monolithically integrated gold nanocomposite electrodes. *Adv. Mater.* **2017**, *29*, 1606109. [CrossRef]
2. Kong, L.; Chen, W. Carbon nanotube and graphene-based bioinspired electrochemical actuators. *Adv. Mater.* **2014**, *26*, 1025–1043. [CrossRef]
3. Zhao, H.; Hussain, A.M.; Duduta, M.; Vogt, D.M.; Wood, R.J.; Clarke, D.R. Compact dielectric elastomer linear actuators. *Adv. Funct. Mater.* **2018**, *28*, 1804328. [CrossRef]
4. Nguyen, C.T.; Phung, H.; Nguyen, T.D.; Lee, C.; Kim, U.; Lee, D.; Moon, H.; Koo, J.; Choi, H.R. A small biomimetic quadruped robot driven by multistacked dielectric elastomer actuators. *Smart Mater. Struct.* **2014**, *23*, 065005. [CrossRef]
5. Li, Y.; Hashimoto, M. PVC gel based artificial muscles: Characterizations and actuation modular constructions. *Sens. Actuators A* **2015**, *233*, 246–258. [CrossRef]
6. Ongaro, F.; Scheggi, S.; Yoon, C.; Van den Brink, F.; Oh, S.H.; Gracias, D.H.; Misra, S. Autonomous planning and control of soft untethered grippers in unstructured environments. *J. Micro-Bio Robot.* **2017**, *12*, 45–52. [CrossRef] [PubMed]
7. Wu, J.; Yuan, C.; Ding, Z.; Isakov, M.; Mao, Y.; Wang, T.; Dunn, M.L.; Qi, H.J. Multi-shape active composites by 3D printing of digital shape memory polymers. *Sci. Rep.* **2016**, *6*, 24224. [CrossRef] [PubMed]
8. Mirvakili, S.M.; Hunter, I.W. Multidirectional artificial muscles from nylon. *Adv. Mater.* **2017**, *29*, 1604734. [CrossRef]
9. Yang, H.; Leow, W.R.; Wang, T.; Wang, J.; Yu, J.; He, K.; Qi, D.; Wan, C.; Chen, X. 3D Printed photoresponsive devices based on shape memory composites. *Adv. Mater.* **2017**, *29*, 1701627. [CrossRef]
10. Bushuyev, O.S.; Aizawa, M.; Shishido, A.; Barrett, C.J. Shape-shifting azo dye polymers: Towards sunlight-driven molecular devices. *Macromol. Rapid Commun.* **2018**, *39*, 1700253. [CrossRef]
11. Zhu, P.; Yang, W.; Wang, R.; Gao, S.; Li, B.; Li, Q. 4D Printing of complex structures with a fast response time to magnetic stimulus. *ACS Appl. Mater. Interfaces* **2018**, *10*, 36435–36442. [CrossRef] [PubMed]
12. Liu, J.; Jiang, C.; Xu, H. Giant magnetostrictive materials. *Sci. China Technol. Sci.* **2012**, *55*, 1319–1326. [CrossRef]
13. Yuk, H.; Lin, S.; Ma, C.; Takaffoli, M.; Fang, N.X.; Zhao, X. Hydraulic hydrogel actuators and robots optically and sonically camouflaged in water. *Nat. Commun.* **2017**, *8*, 14230. [CrossRef]
14. Villegas, D.; Van Damme, M.; Vanderborght, B.; Beyl, P.; Lefeber, D. Third-generation pleated pneumatic artificial muscles for robotic applications: Development and comparison with mckibben muscle. *Adv. Rob.* **2012**, *26*, 1205–1227. [CrossRef]
15. Hawkes, E.W.; Christensen, D.L.; Okamura, A.M. Design and implementation of a 300% strain soft artificial muscle. In Proceedings of the IEEE International Conference on Robotics and Automation (ICRA), 20 May–5 June 2016, Xi'an, China; pp. 4022–4029.

16. Duan, J.; Liang, X.; Zhu, K.; Guo, J.; Zhang, L. Bilayer hydrogel actuators with tight interfacial adhesion fully constructed from natural polysaccharides. *Soft Matter* **2017**, *13*, 345–354. [CrossRef]
17. Umedachi, T.; Vikas, V.; Trimmer, B.A. Softworms: The design and control of non-pneumatic, 3D-printed, deformable robots. *Bioinspir. Biomim.* **2016**, *11*, 025001. [CrossRef]
18. Li, T.; Li, G.; Liang, Y.; Cheng, T.; Dai, J.; Yang, X.; Liu, B.; Zeng, Z.; Huang, Z.; Luo, Y. Fast-moving soft electronic fish. *Sci. Adv.* **2017**, *3*, e1602045. [CrossRef] [PubMed]
19. Gu, G.; Zou, J.; Zhao, R.; Zhao, X.; Zhu, X. Soft wall-climbing robots. *Sci. Robot.* **2018**, *3*, 2874. [CrossRef]
20. Park, W.H.; Bae, J.W.; Shin, E.J.; Kim, S.Y. Development of a flexible and bendable vibrotactile actuator based on wave-shaped poly (vinyl chloride)/acetyl tributyl citrate gels for wearable electronic devices. *Smart Mater. Struct.* **2016**, *25*, 115020. [CrossRef]
21. Zarek, M.; Layani, M.; Cooperstein, I.; Sachyani, E.; Cohn, D.; Magdassi, S. 3D printing of shape memory polymers for flexible electronic devices. *Adv. Mater.* **2016**, *28*, 4449–4454. [CrossRef]
22. Xia, H.; Hirai, T. New shedding motion, based on electroactuation force, for micro-and nanoweaving. *Adv. Eng. Mater.* **2013**, *15*, 962–965. [CrossRef]
23. Li, Y.; Hashimoto, M. Design and prototyping of a novel lightweight walking assist wear using PVC gel soft actuators. *Sens. Actuators A* **2016**, *239*, 26–44. [CrossRef]
24. Li, Y.; Hashimoto, M. PVC gel soft actuator-based wearable assist wear for hip joint support during walking. *Smart Mater. Struct.* **2017**, *26*, 125003. [CrossRef]
25. Must, I.; Kaasik, F.; Põldsalu, I.; Mihkels, L.; Johanson, U.; Punning, A.; Aabloo, A. Ionic and capacitive artificial muscle for biomimetic soft robotics. *Adv. Eng. Mater.* **2015**, *17*, 84–94. [CrossRef]
26. Kim, S.Y.; Yeo, M.; Shin, E.J.; Park, W.H.; Jang, J.S.; Nam, B.U.; Bae, J.W. Fabrication and evaluation of variable focus and large deformation plano-convex microlens based on non-ionic poly (vinyl chloride)/dibutyl adipate gels. *Smart Mater. Struct.* **2015**, *24*, 115006. [CrossRef]
27. Ali, M.; Ueki, T.; Tsurumi, D.; Hirai, T. Influence of plasticizer content on the transition of electromechanical behavior of PVC gel actuator. *Langmuir* **2011**, *27*, 7902–7908. [CrossRef]
28. Park, W.H.; Shin, E.J.; Kim, S.Y. Enhanced Design of a soft thin-film vibrotactile actuator based on PVC gel. *Appl. Sci.* **2017**, *7*, 972. [CrossRef]
29. Helps, T.; Taghavi, M.; Rossiter, J. Thermoplastic electroactive gels for 3D-printable artificial muscles. *Smart Mater. Struct.* **2019**, *28*, 085001. [CrossRef]
30. Guo, S.Z.; Qiu, K.; Meng, F.; Park, S.H.; McAlpine, M.C. 3D printed stretchable tactile sensors. *Adv. Mater.* **2017**, *29*, 1701218. [CrossRef]
31. Wang, R.; Yang, W.Y.; Zhu, P.F.; Gao, S.; Li, B.; Li, Q. Creation of 3D terahertz photonic crystals by the direct writing technique with a TiO2 sol-gel ink. *J. Am. Ceram. Soc.* **2018**, *101*, 1967–1973. [CrossRef]
32. Walton, R.L.; Brova, M.J.; Watson, B.H.; Kupp, E.R.; Fanton, M.A.; Meyer, R.J.; Messing, G.L. Direct writing of textured ceramics using anisotropic nozzles. *J. Eur. Ceram. Soc.* **2021**, *41*, 1945–1953. [CrossRef]
33. Wan, X.; Luo, L.; Liu, Y.J.; Leng, J.S. Direct ink writing based 4D printing of materials and their applications. *Adv. Sci.* **2020**, *7*, 2001000. [CrossRef] [PubMed]
34. Shin, E.J.; Park, W.H.; Kim, S.Y. Fabrication of a high-performance bending actuator made with a PVC gel. *Appl. Sci.* **2018**, *8*, 1284. [CrossRef]
35. Xia, H.; Ueki, T.; Hirai, T. Direct observation by laser scanning confocal microscopy of microstructure and phase migration of PVC gels in an applied electric field. *Langmuir* **2011**, *27*, 1207–1211. [CrossRef] [PubMed]
36. Smay, J.E.; Gratson, G.M.; Shepherd, R.F.; Cesarano, J.; Lewis, J.A. Directed colloidal assembly of 3D periodic structures. *Adv. Mater.* **2002**, *14*, 1279–1283. [CrossRef]
37. Helps, T.; Taghavi, M.; Rossiter, J. Towards electroactive gel artificial muscle structures. *Adv. Mater.* **2002**, *14*, 1279–1283.

Article

Experimental Evaluation of Polyphenylsulfone (PPSF) Powders as Fire-Retardant Materials for Processing by Selective Laser Sintering

Yifan Lv [1], Wayne Thomas [2], Rodger Chalk [2], Andrew Hewitt [2] and Sarat Singamneni [1,*]

[1] Additive Manufacturing Research Centre, Auckland University of Technology, WS116, 34 Saint Paul Street, Auckland CBD, Auckland 1010, New Zealand; yifan.lv@aut.ac.nz

[2] Air New Zealand, 5 Geoffrey Roberts Road, Mangere, Auckland 2022, New Zealand; wayne.thomas@airnz.co.nz (W.T.); rodger.chalk@airnz.co.nz (R.C.); andrew.hewitt@airnz.co.nz (A.H.)

* Correspondence: sarat.singamneni@aut.ac.nz; Tel.: +64-9-921-9999

Abstract: Additive manufacturing has progressed rapidly, and the unique attributes of the layer-wise material consolidation are attracting ever increasing application potentials in critical sectors such as medical and aerospace industries. A lack of materials options has been the main bottleneck for the much wider uptake of these promising new technologies. Inventing new material alternatives has been central to most of the research attention in additive manufacturing in recent times. The current research is focused on evaluating the polyphenylsulfone polymer powders for the first time as fire-resistant candidate materials for processing by selective laser sintering, the most promising additive processing method for polymeric material systems. Experimental evaluations were undertaken based on a selective laser sintering test bed. Single layer and multi-layer samples were produced for microstructural and mechanical characterisations. The microstructural evaluations and the mechanical property results indicate sufficient intra- and inter-layer consolidation together with reasonable tensile property responses. The lower viscosity and thermal conductivity characteristics rendered lower tensile strengths, which will require some further attention in the future, for better consolidation and mechanical properties.

Keywords: polyphenylsulfone; PPSF; fire-resistant; additive manufacturing; aircraft interior; selective laser sintering

Citation: Lv, Y.; Thomas, W.; Chalk, R.; Hewitt, A.; Singamneni, S. Experimental Evaluation of Polyphenylsulfone (PPSF) Powders as Fire-Retardant Materials for Processing by Selective Laser Sintering. *Polymers* **2021**, *13*, 2704. https://doi.org/10.3390/polym13162704

Academic Editors: Houwen Matthew Pan and Eric Luis

Received: 2 July 2021
Accepted: 9 August 2021
Published: 13 August 2021

Publisher's Note: MDPI stays neutral with regard to jurisdictional claims in published maps and institutional affiliations.

1. Introduction

Considering their highly inflammable nature and toxicity, fire-safe qualities are mandatory for polymer materials used in aero-space applications, which are normally achieved by chemical or physical treatments [1–5]. The additives often may adversely affect the environment or the properties of the fire-resistant polymer [6,7] but intrinsically fire-resistant options such as poly ether ketone (PEEK) have proved to be effective [8]. Several flame-retardant polymers are effectively in use for aircraft interior parts, confirming to the FAR 25.855 standards [9–11]. However, the-e has been a never-ending urge to identify new materials and processes for more efficient building of fire-retardant interior components of aircrafts [12].

Injection moulding [13], hand lay-up, spray-up, compression moulding, filament winding, pultrusion, resin transfer moulding, vacuum-assisted resin transfer moulding, infusion, and continuous panel processing [1] are common methods traditionally used for making fire-resistant polymer parts for the aircraft interior. While these are well-developed manufacturing solutions, the production lead times are often high and the supply-chains quite intricate. Additive manufacturing methods have recently evolved from the erstwhile rapid prototyping stages, offering potential new material processing solutions with better freedom to achieve more complex and optimum design forms. In particular, with the possibility to simplify the supply-chain and inventory constraints, these new methods are attracting significant attention in the aircraft manufacturing and maintenance tasks [14].

Two of the most common AM techniques for processing polymers are selective laser sintering (SLS) and fused deposition modeling (FDM) based on powder bed fusion and material extrusion techniques, respectively. It is well known that the components processed by SLS or by FDM show properties, which heavily depend on the building orientation and processing parameters. In SLS, the main parameters to be optimised are the laser power, the laser scanning pattern, and speed and the hatch distance [15]. For FDM technique, on the other hand, the extrusion speed, deposition temperature of filament, infill percentage, and raster angle and strategies play critical roles [16]. In both techniques, the parts produced can show different mechanical properties depending on those processing orientation and parameters, as they affect the intra- and inter-layer conditions. This is due to the very nature of additive technologies, producing components in a point-wise material consolidation manner. The advantages and disadvantages, and a general comparison between the two techniques are concluded in [15,17,18]. Summarising, FDM is relatively cheap, allows multi-colour solutions, bigger build sizes, and low post-processing costs. On the other hand, SLS needs no support structures, gives better surface quality and part definitions, allowing to produce movable joints with high resolution and close tolerances. However, FDM is inferior to SLS in terms of part quality, anisotropy of the material consolidation, stair-step problems, higher porosity levels, and the need for support structures. The main drawback with SLS is the limited materials options currently available and highly proprietary nature of the materials that are already in use. Further, SLS parts generally have superior mechanical strength and less anisotropy [17]. For example, with Poly-Ether-Ether-Ketone (PEEK), the tensile strengths are 40 to 55 MPa [19,20], and 80 to 90MPa [21] when processed by FDM and SLS, respectively.

Despite the numerous benefits additive manufacturing can bring to the polymer-based manufacturing of aircraft parts, the progress so far has been limited, and mostly confined to the fused deposition modelling (FDM) methods [22]. Proprietary materials and the black-box-type commercial systems have rendered obstacles for the wide range exploration of different materials and processes in applying the additive methods to the needs of the aircraft industry [23]. In particular, the stairstep effects, fibre discontinuity, and the meso-structural limitations [16] typical of FDM render inherent weaknesses in the consolidated material structures. Selective laser sintering (SLS), on the other hand, is a powder bed fusion process for polymers and the point-by-point laser induced energy allows to achieve a better material consolidation mechanics through controlled inter-particle coalescence [24].

There is also a dearth of materials options for processing aircraft parts using fire-retardant polymeric materials. In particular, the laser sintering route lacks a variety of materials options to choose from [25]. Polyphenylsulfone (PPSF or PPSU) is an amorphous thermoplastic, which is also intrinsically flame retardant, with a decent mechanical strength, and can be a potential competitor to PEEK and Ultem. Polyphenylsulfone is mostly researched for membrane material solutions [26], while some studies show that it can also be processed by injection moulding [13]. The tensile strength of polyphenylsulfone is around 70 MPa. There have been some reports, indicating that it can be processed by FDM [27], achieving mechanical properties comparable with Ultem and polycarbonate. However, there has been no evidence of this material being researched for processing by selective laser sintering. The current paper addresses this research gap through experimental investigations, leading to the understanding of how polyphenylsulfone powders respond to consolidation by laser sintering with varying energy inputs. The results indicate the material to be responding positively to consolidation by the continuously moving laser energy input, though the time and temperature conditions lead to specific challenges in controlling the resulting meso-structures and the mechanical properties.

2. Material and Methods

Polyphenylsulfone (PPSF/PPSU) is an amorphous polymer with a repeating unit of molecular form (-C_6H_4-4-$SO_2C_6H_4$-4-OC_6H_4-4-C_6H_4-4-O-)$_n$. The molecular weight is 1600 g mol^{-1}. The powder was specially ordered from Galaxy Chemical Technology Co Ltd., Shenzhen, Guangdong, China. This material is not available in powder form in the commercial sales market. Since this was the first time the powder was ordered, Galaxy Chemical Technology used the cryo-grinding method to mechanically prepare the powder samples. As a result, the particle morphologies were too complex and also quite varied. However, the average powder particle size is around 42 microns. The powder samples were coated with platinum and examined under the Hitachi SU-70 Scanning electron microscope (SEM) system, and the scanning voltage was set as 15 KV to avoid charging effects. SEM images of the PPSF powder particles shown in Figure 1 clearly indicate the irregular forms and variations in sizes. These irregular shapes of the particles caused by poor cryogenic grinding will hinder the flowability of the powder on the build platform of the laser sintering test bed. In addition, the varying sizes can cause problems to the consolidation of layers based on the heat input from a fast-moving laser energy [28–35].

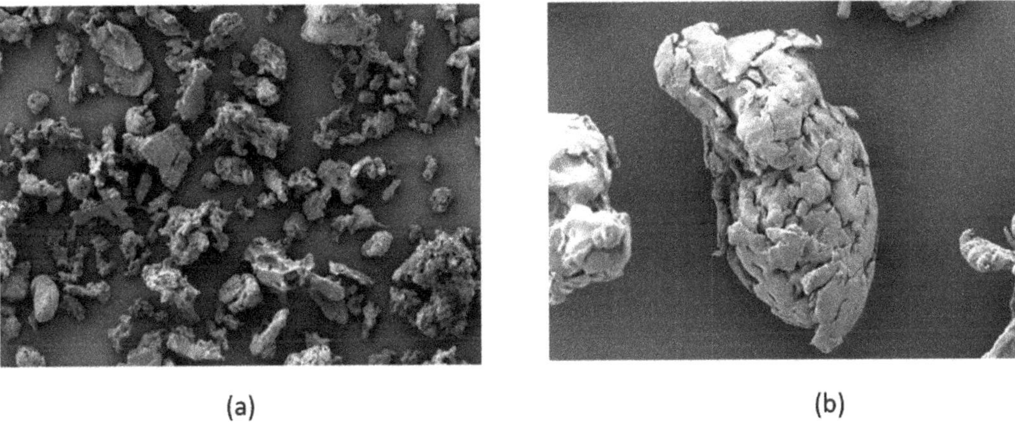

(a)   (b)

Figure 1. Scanning electron microscope (SEM) images of PPSF powder particles, (**a**): 100× zoom, (**b**): 500× zoom.

Differential scanning calorimetry (DSC) is used to measure the heat into or out of the test polymer powder, generating a thermal profile that can be used for establishing the sintering window of these powder materials, before undertaking the laser sintering trials. The DSC system NETZSCH STA 449F5 STA449F5A-0062-M was used to identify the thermal profile of the PPSF powders and to establish the promising ranges of the critical process parameters. A mass of 4.9 mg of PPSF powder was heated from 20 °C to 400 °C with 10.0 °C/min heating gradient during the heating cycle. The DSC curve obtained is shown in Figure 2. Based on the exothermic reaction up as the convention, the small step observed at around 220 °C was taken as the glass transition temperature. The information from the DSC graph was used to establish the critical process parameters for the laser sintering trials.

All the laser sintering trials were performed on a homemade selective laser sintering test bed available at the Additive Manufacturing Research Centre of the Auckland university of Technology. The experimental setup constituted of a 60 W CO_2 laser with sufficient control on the pulse rate and the raster scanning on the powder bed. The powder bed was also tailor-made for semi-automatic dispersal of powders in small quantities for experimental study of the feasibility of laser sintering single and multi-layer specimens with varying process conditions. The complete details of the experimental setup may be obtained from the reference by Velu et al. [28].

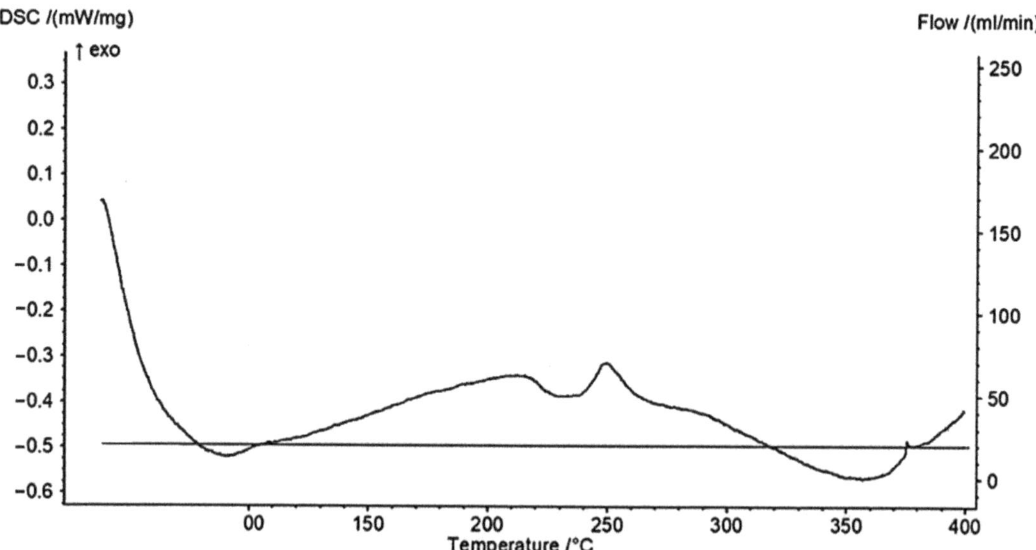

Figure 2. Differential scanning calorimetry (DSC) results based on the PPSF powder.

The theoretical energy density (ED) necessary to sinter a given quantity of the PPSF powder was calculated based on the procedure used by Berretta et al. [29]. Laser powder (p), scan speed (v), and laser beam diameter (D) are the process parameters. The powder bed temperature (T_b) was set at 200 °C, which is just below the glass transition temperature of PPSF. Single-layer samples were made with varying process parameters in order to evaluate the particle coalescence. Based on the literature, the enthalpy of relaxation (h_r) of PPSF is around 0.73 J g^{-1} [30], which is very small due to the amorphous nature of the material. The bulk density (Q) and the packing factor (φ) were estimated as 0.350 g cm^{-3} and 0.8, respectively. The layer thickness (z) is 0.20 mm. T_o is the onset temperature during laser sintering. The specific heat capacity, (C_{pb}) of PPSF varies with temperature and at the powder bed temperature is around 0.44 cal g^{-1} °C^{-1}. The specific heat capacity at the onset sintering temperature (C_{po}) is assumed based on a linear approximation. Based on this assumption, Equation (1) can be used to evaluate the theoretical energy required for sintering. Equation (2) is the theoretical energy density for single layer sintering, which can be used to derive the expression for the sintering energy density for materials with specific heat capacity varying as a function of temperature. Based on a linear approximation within the sintering temperature window, the specific heat capacity can be expressed as in Equation (4). Combining Equations (1)–(4), the semi-empirical Equation (5) can be derived, which can be used to calculate the experimental energy density:

$$Energy\ required\ for\ sintering = [C_p(T_o - T_b) + h_r]Q\varphi \tag{1}$$

$$ED = Energy\ required\ for\ sintering\ \times z \tag{2}$$

$$ED = \left\{ \int_{T_b}^{T_o} [\frac{dC_p}{dT}(T_o - T_b) + h_r]Q\varphi \right\}z \tag{3}$$

$$C_p = \frac{(T - T_o)\left(C_{pb} - C_{po}\right)}{(T_b - T_o)} + C_{po} \tag{4}$$

$$ED = \left\{ h_r Q\varphi + Q\varphi \int_{T_b}^{T_o} \left[\frac{(T - T_o)\left(C_{pb} - C_{po}\right)}{(T_b - T_o)} + C_{po} \right] dT \right\}z \tag{5}$$

Considering that the polymer particles are irregular, the size of the particle and the variation of the viscosity with temperature will affect the sintering time, which in turn will influence the inter-particle coalescence during sintering [31]. According to the Frenkel's model, the higher the product of the viscosity and the particle radius, the longer the sintering time required [31]. The geometrical parameters were measured based on the SEM images, which indicate the original radius, sintered radius, and distance from the centre to the sintering plane of two adjacent particles. Additionally, other properties such as surface tension and viscosity were related to the time (t) of sintering as well, using the information reported by Sedlacek et al. and Mohan [32,33]. The classical sintering equation proposed by Frankel is only suitable for a 2D simplification. The modified Frenkel model proposed by Sun et al. [34] can be used for a 3D situation, which allowed to establish the working energy density of the current powders as discussed next.

Once the sintering across a single layer was thus established, a series of specimens were produced with the energy densities varied at 0.046, 0.054, 0.062, 0.07, and 0.078 J/mm^2; the images of the printed samples are presented in Figure 3. Three levels of laser power settings were used for each energy density—9 W, 13 W, and 20 W—together with the corresponding scan speeds, as stated in Figure 3. A total of 45 single layer samples were produced considering three single layer sample replicas for each process parameter combination. The photographs of the best of the three printed samples with each process parameter combination are shown in Figure 3. These samples are subsequently examined using SEM and the images obtained are presented in Figure 4. Single layer or the first layer is sintered on a solid substrate and not on loose powder. The solid substrate is often made with one material or a material with similar characteristics. Hence, there is relatively lesser variation from single to multi-layers. However, thermal stability issues arise in multi-layer samples as the heat from the substrate cannot easily cross the multi-layers. The external heat source in the form of a heating lamp will compensate and keep the conditions almost similar.

Porosity of the printed samples was established based on the analysis of the SEM images of the single layer sintered samples using the ImageJ software, which is done by the same authors in another study [35]. Three SEM images were taken using samples selected from the single layer samples printed with each of the process parameter combinations. All the images are then used to process by the ImageJ software to analyse the porosity levels. The total specimen areas and number and average sizes of the lack of fusion areas, established from the images processed by ImageJ software, were used to calculate the percent porosity levels. Samples for tensile testing were printed as per the standard ASTM D638. All the tensile tests were performed using the Tinius Olsen H50KS. A S-Beam load cell was used for the test, loading at the rate one millimetre per minute. Each tensile test specimen was printed processing five layers of the PPSF powder, with the process conditions, as discussed in the results section.

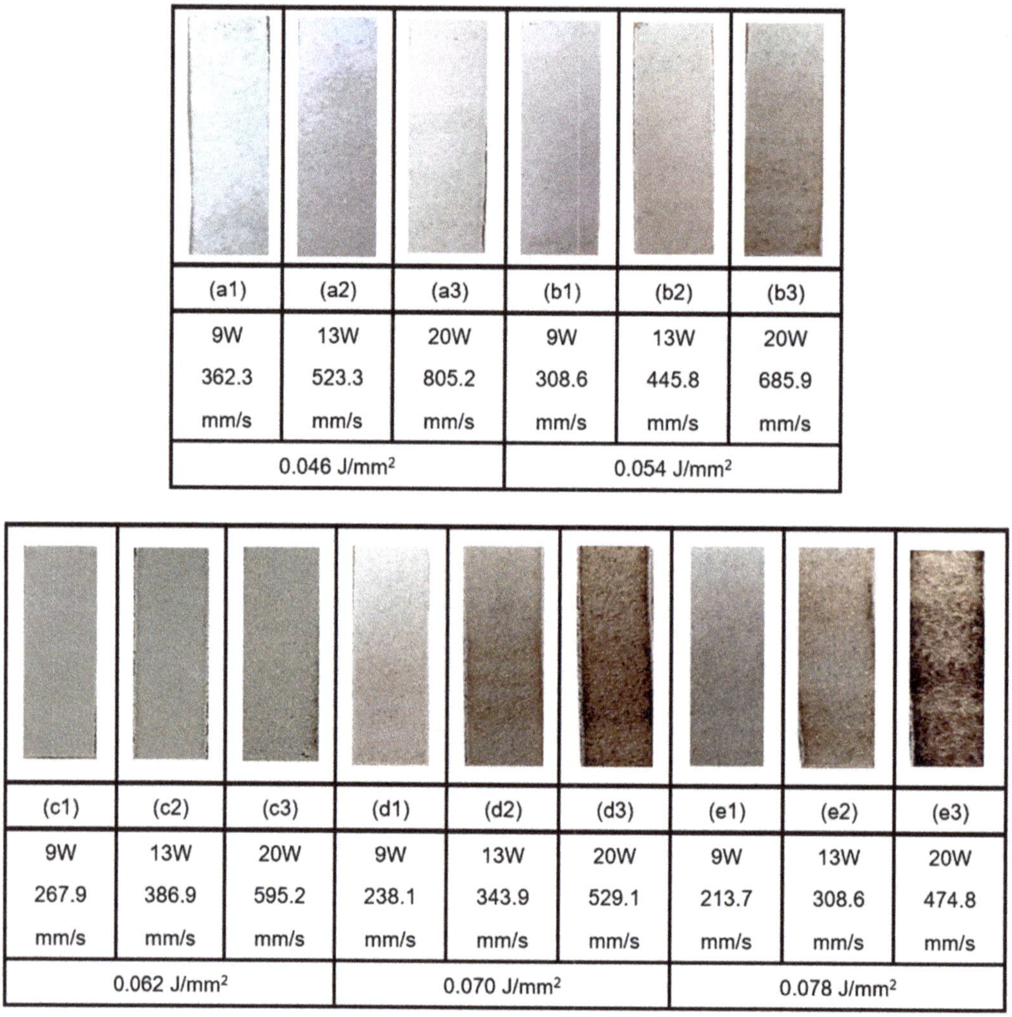

Figure 3. Single layer samples of PPSF produced by SLS; the energy density of 0.046 J/mm², 0.054 J/mm², 0.062 J/mm², 0.070 J/mm², and 0.078 J/mm² correspond with group (**a–e**), respectively. Laser power of 9 W, 13 W, and 20 W are used for the first, second, and third sample in the group, respectively, with varied scanning speed. All samples are of 30×100 mm size.

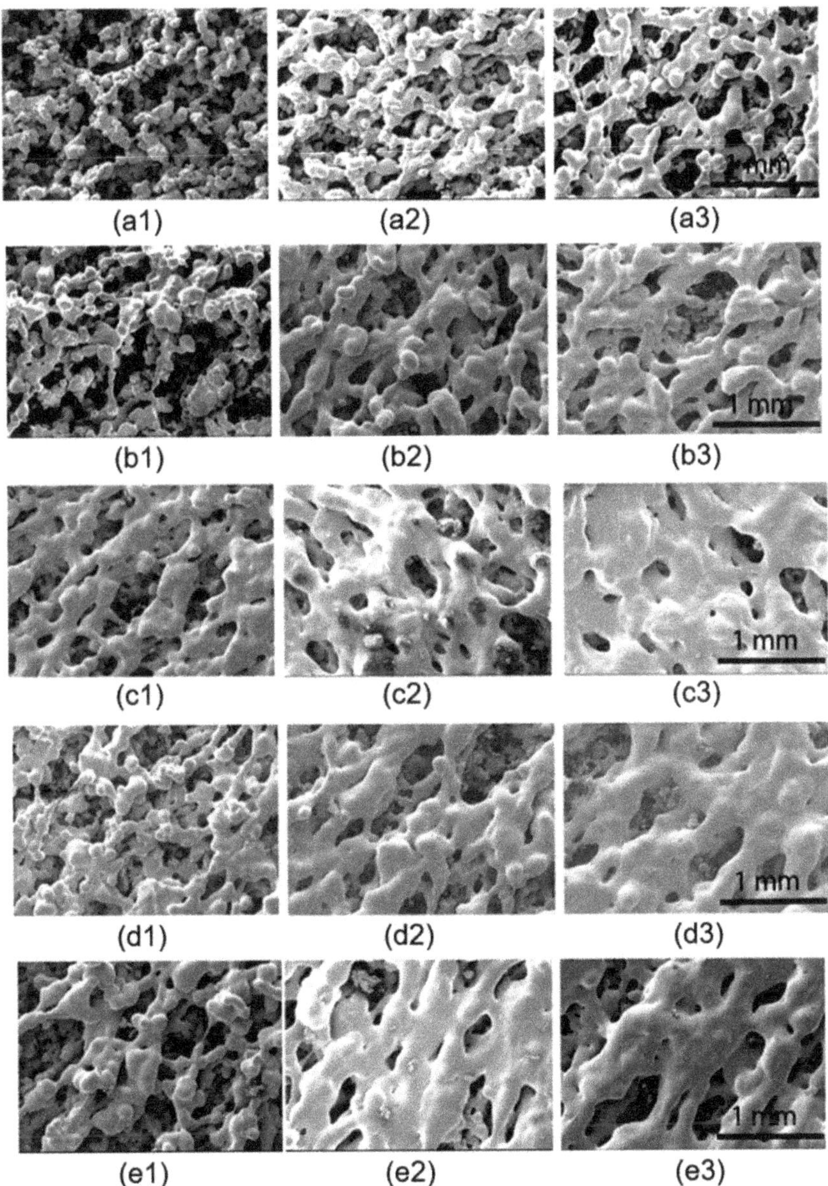

Figure 4. SEM images of single layer samples of PPSF produced by SLS. The energy density of 0.046 J/mm², 0.054 J/mm², 0.062 J/mm², 0.070 J/mm², and 0.078 J/mm² correspond with group (**a–e**), respectively. Laser power of 9 W, 13 W, and 20 W are used for the first, second, and third sample in the group, respectively, with varied scanning speed.

3. Results and Discussions

From the calculations based on the theoretical modelling explained in the previous section, it was established that the sintering rate reaches 2.277×10^{-4} μm/s, when the onset temperature to be 360 °C. This is theoretically sufficient to achieve sufficient consolidation through selective laser sintering. Based on this number, the theoretical energy density necessary to achieve sufficient consolidation was established to be 0.022 J/mm² using

Equation (5). It is common for the theoretical energy density to be lower than the actual energy density required. Starting with this as the first step, the actual energy density necessary was established based on trial-and-error methods. After a few iterations, the first successful single layer specimen with significant inter-particle coalescence was achieved with an energy density of 0.046 J/mm^2, which is twice that of the theoretically estimated energy density. While there are differences in the real thermal conditions and the theoretical estimates, there is also the problem of heat loss from the powder bed, which is not strictly considered in the theoretical calculations

The images of the single layer samples presented in Figure 3 indicate very smooth surface textures and continuous formation of the layers based on powder consolidation achieved by the moving laser beam. There is a physical variation in the form of gradually changing colour from the first set to the last. Evidently, as the energy density increases, the polymer composite is gradually degraded and gets darker possibly due to decomposition and charring. However, a physical examination also revealed that the strength of the sintered samples increases with increasing energy densities, as the excess energy was able to consolidate the particles better. The samples became brittle and charred when the energy density is increased to 0.070 J/mm^2. Based on a compromise between the physically observed strength and consolidation against the degradation, the best energy density for this polymer composite was estimated to be at around 0.062 J/mm^2.

According to a study by Ramgobin et al. under thermo-oxidative atmosphere, the onset of the degradation temperature of PPSF is around 500 °C. With enough activation energy, a two-step decomposition will be activated, first the cleavage of phenyl–sulfone (Ph–SO2) linkages, and then the cleavage of Phenyl–Oxygen or Phenyl–Phenyl bonds. Through thermogravimetric analysis, it was shown that the residual mass of PPSF at 538 °C was at 95 wt% and as low as 40 wt% at 800 °C. This indicates that the loss of material with PPSF is insignificant during thermal degradation. However, charring is probably due to the residual elements of the products of pyrolysis of PPSF [36]. It was also stated that, when PPSU was heated under nitrogen, the degradation temperature is at 600 °C. It may be inferred that a N$_2$ atmosphere in the build chamber would allow to avoid the decomposition and charring of the PPSF powders during selective laser sintering. Arnold et al. suggested that photothermal degradation of polymers caused by laser light may lead to ablation and loss of materials [37]. The ablation responses were correlated to the laser velocity and pulse rates. In the present case, the discolouration of the sintered samples could be due to the higher scan speeds used with the higher laser power settings.

Evidence of discolouration and curling of the specimens may also be noted based on the images of the sintered specimens shown in Figure 3. The discolouration of the edges could be due to the excessive heating that takes place at the end of each raster scan line. The can strategy is zig-zag lines across the width of the specimens. This means the laser will instantaneously stop at the end of each raster path, while the energy is flowing into the specimen at the same rate. This will lead to flow of excessive energy and the consequent discolouration on the longer edges of the specimens, as observed. Evidently there is no such discolouration on the top and the bottom edges of the samples, which is because of the fact that the laser scan lines are oriented parallel to these edges.

The images of the sintered specimens in Figure 3 also show indications of curling of the specimens over the edges. Apparently, the curling phenomenon is more prominent in the cases of the specimens produced at higher power and higher scan velocity settings within the ranges of each energy density, though to a lesser extent in the lower energy density cases. Curling takes place firstly due to lack of sticking of the sintered layer to the base plate. This can be easily fixed by suitably preparing the surface conditions. The uneven heat dispersion in the sintered layer also could cause the curling problem. The central areas are relatively at higher temperature compared to the edges, which may lead to contraction at the edge and consequent curling of the specimens. It is also possible that the excessive heating due to the instantaneous stopping and reversal of the laser beam at the end of each of the scan strokes could lead to embrittlement and loss of elastic nature at the edges, which can cause the curling effects. All these aspects can be controlled by adjusting the process conditions. The surfaces can be better prepared to promote more stronger sticking and the laser energy input can be adjusted to reduce to the necessary extent, as the beam gets closer to the ends of the raster scan lines.

Scanning electron microscopy images taken based on the single layer sintered samples shown in Figure 3 are presented in the same order in Figure 4. The energy density levels increase moving down the rows of images presented. Marked differences may be observed in the inter particle and intra-layer coalescence as the energy density is increased. Both inter-particle and intra layer consolidation is scarce in samples sintered with the lowest energy density 0.046 J/mm^2, as evident from the images of Figure 4(a1–a3). The inter particle coalescence improved significantly showing considerable evidence of the formation of the consolidated polymer strands along the laser scan lines at the energy density 0.054 J/mm^2, as may be seen in Figure 4(b1–b3). However, the inter-strand coalescence is still limited, and the layer formation is significantly restricted in terms of the resulting meso-structures, indicating a lack of energy to achieve sufficient powder consolidation. Both inter-particle and inter-strand coalescence improved with still higher energy densities above 0.062 J/mm^2. Considering the overall dispersion of the consolidated materials and lack of fusion cavities, the energy density at 0.062 J/mm^2 appears to be the optimum setting for effective laser sintering of this powder polymer.

Significant improvements in consolidation may also be visualised moving from the left image to the right, along any row of Figure 4. Though the energy density is the same for all the samples in a given row, the process parameter combination changes from low power and low velocity to high power and high velocity from the left to the right. Evidently, the powder consolidates better with a high power and high velocity setting for any given energy density level. This is often the case with many polymers as the power and velocity settings interact even at the same energy density levels. The high power and high velocity setting are preferred if the viscosity and thermal conductivity properties are lower. The higher power allows to heat the strands more. The lower viscosity leads to increased plasticisation and running of the plasticised strand, but the high velocity of scanning brings the laser back quickly along the next strand, allowing the softened adjacent polymer strands to fuse better. With polymers having lower thermal conductivities, the higher scan velocity helps to quickly move around and revisit the common points along adjacent strands and keep the thermal conditions elevated to the necessary levels promoting inter-strand coalescence [28,35].

Based on these observations, the best layer consolidation corresponds to Figure 4(c3,d3). In both cases, the laser power is the highest of the range used, at 20 W, with the laser scan velocity also at the highest levels corresponding to the energy density values used, 0.062 J/mm^2 and 0.070 J/mm^2, respectively. However, as the energy density goes beyond 0.062 J/mm^2, the samples began to discolour significantly at the higher power setting 20 W. Considering both consolidation and deterioration responses, the energy density level at 0.062 J/mm^2 with the higher power and velocity settings again appears to be the best possible process parameter conditions for the laser sintering of polyphenysulfone polymer powders.

The results of the porosity analysis are presented as the bar charts with the percent porosity levels plotted against the laser power for each energy density, as depicted in Figure 5. It may be consistently observed that the porosity level decreases with increasing laser power, at any given laser energy density level. This is consistent with the observations made based on the examinations of the physical samples and SEM images, as discussed earlier. The lowest porosity level is based on the sample produced with laser power 20 W and energy density 0.062 J/mm^2, which is in complete agreement with the inferences drawn from the other observations. It may also be noted that the porosity levels increase as the energy density level is increased to 0.070 J/mm^2. There may be excessive heating happening when the energy density used is beyond the threshold value suitable for the polymer material investigated. The slightly reduced porosity levels at 0.078 J/mm^2 could be misleading as the sintered layer appears to be overheated, decomposed, and charred and possibly blocking the meso-structural details. The porosity results corresponding to the energy density and laser power setting combinations 0.062 J/mm^2 and 13 W, 0.062 J/mm^2 and 20 W, and 0.078 J/mm^2 and 20 W are considered to be the most optimum in terms of achieving the least porosity or the best sintered layer corresponding to the shortest bar heights in Figure 5.

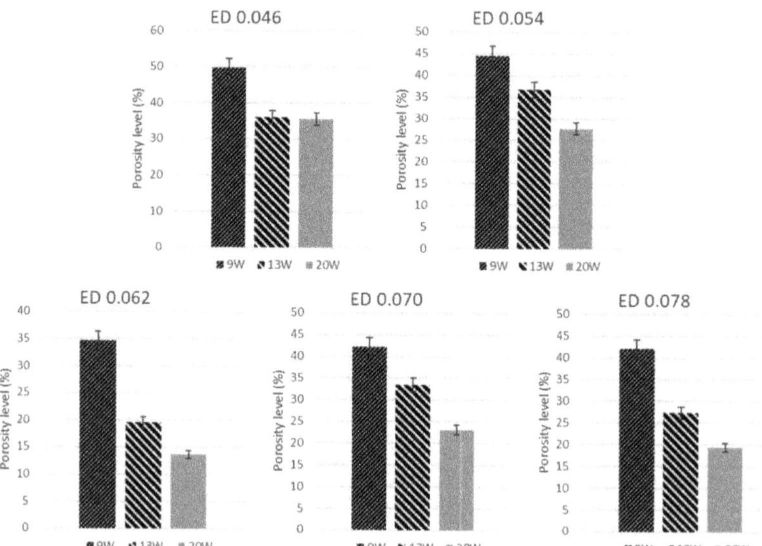

Figure 5. Porosity levels results of each processing parameter.

Tensile test specimens were produced with the three process parameter combinations identified as the optimum settings from the porosity analysis. The process parameter settings and the tensile testing responses are listed in Table 1. Each tensile test specimen is printed as per the ASTM standards mentioned using five layers sintered one upon the other. The images of the printed test specimens are presented in Figure 6. Consistent dimensional and structural qualities could be observed in all the multi-layer tensile test specimens printed. There is some evidence of discolouration as may be noted with the high energy density and high-power samples in Figure 6(C1–C3). From a physical point of view, the specimens produced with 0.062 J/mm^2 and 20 W corresponding to Figure 6(B1–B3) are the best possible outcomes based on selective laser sintering of the current polymer powders.

Table 1. Process conditions and tensile strength results of the multi-layered samples.

	Power (W)	Speed (mm/s)	Energy Density (J/mm^2)	UTS (MPa)
A1				2.280
A2	13	386.9	0.062	1.666
A3				2.810
B1				6.561
B2	20	595.0	0.062	8.049
B3				7.102
C1				7.702
C2	20	474.8	0.78	8.994
C3				9.259

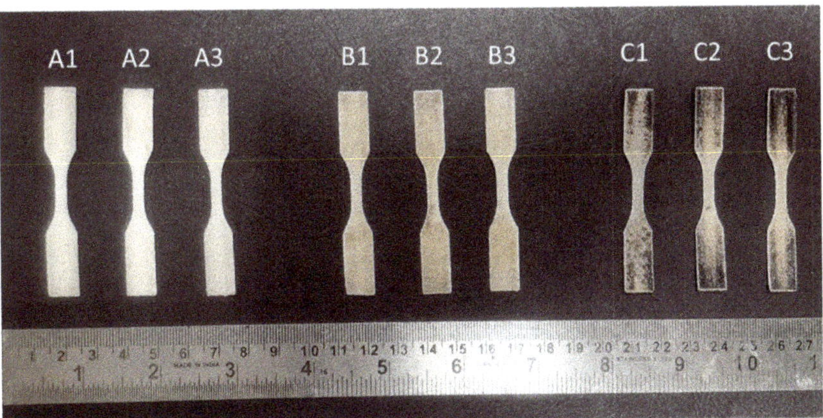

Figure 6. Multi-layered tensile test specimens of PPSF with three sets of laser power, scan speed and energy density combinations, which referring to Table 1.

Comparing the tensile test results presented in Table 1, the first batch corresponding to the A-series scored the least in terms of the ultimate tensile strength, averaging at around 2.252 MPa. Both the B- and C- series samples are far better, averaging at 7.237 MPa and 8.652 MPa, respectively. This drastic loss of strength at the lower power setting 13 W is clearly indicative of the significant role of laser power in controlling the consolidation rates in the sintered polymer layers. The minimum threshold power is 20 W for the scan velocity ranges used in the current experimental work with laser sintering of polyphenylsulfone powders. The C-series produced at the highest energy density exhibit better ultimate tensile strengths than the C-series. However, there is discolouration and deterioration of the polymer, as evident from the images of the physical samples in Figure 6(C1–C3). Considering both ultimate tensile strength and the polymer degradation aspects, the B-series, corresponding to laser energy density 0.062 J/mm^2 and power 20 W appear to be the best specimens produced within the range of factors used in the current experimental investigation. This is also in accordance with the observations made from the physical examination and SEM images and porosity analysis of the single layer samples. The experimental results show a clear trend and relationship between single layer porosity and multilayer properties. Thus, the single layer porosity can be correlated to the multilayer porosity and strength. The reported mechanical strength of PPSF via using fused deposition modelled techniques can achieve roughly 55 MPa [16]. However, the ultimate tensile strength results obtained from the laser sintered samples are much lower compared to the fused deposition modelled counter parts, as reported by Afrose et al. [38]. It may be pertinent to point out that the current results are only based on initial investigations targeted at evaluating the suitability of the polymer for processing by selective laser sintering. Form that viewpoint, the consolidation mechanisms observed clearly prove that polyphenylsulfone powders can be successfully processed by selective laser sintering. Further improvements in tensile properties can be achieved by a more scientific design of experiments considering all the critical process parameters. A careful consideration should be given to achieve uniform powder particle sizes and also closely controlling the atmosphere and temperature conditions inside the build chamber. A careful examination of the fractured surfaces of the tensile specimens clearly indicated brittle fracture modes. The results are satisfactory in terms of proving the PPSF powders for consolidation by laser sintering. However, the SLS test bed used is a make-shift laboratory experimental system. The mechanical properties will be much improved if a commercial system is used at later stages, where commercial applications are targeted.

4. Conclusions

Polyphenylsulfone powders are investigated for processing by selective laser sintering, based on experimental investigations with varying laser process parameter settings. Overall, the polymer powders responded well for the laser sintering process. Single and multilayer sintered samples were evaluated based on physical, meso-structural, porosity, and mechanical property examinations. Upon careful consideration of the results, the energy density level 0.062 J/mm^2, achieved with laser power 20 W and scan speed 595.0 mm/s settings, were identified to be the most optimum process conditions. The best average ultimate tensile strength achieved in the multi-layer sintered samples was at around 8.652 MPa.

- Application of the Frenkel theory of sintering in a modified form resulted in an optimum ED of 0.062 J/mm^2 for laser sintering PPSF powders.
- Experimental observations allowed to establish the best process conditions for single-layer sintering as ED 0.062 J/mm^2, obtained with laser power 20 W and scan speed 595.0 mm/s.
- The consolidation mechanics varied with both geometry and the number of layers printed and for multilayer printing, the optimum conditions slightly shifted to 0.078 J/mm^2, obtained with a laser power of 20 W and a scan speed of 474.8 mm/s.
- The maximum tensile strength of the laser sintered PPSF is 9.259 MPa, obtained with an ED of 0.078 J/mm^2.

Author Contributions: Y.L.: Execution of all the experimental work collecting results and writing the draft of the manuscript and review responses (Y.L.), Manuscript editing and advising (W.T., R.C. and A.H.), Supervision, monitoring the progress, manuscript writing, editing, and review responses (S.S.). All authors have read and agreed to the published version of the manuscript.

Funding: The research leading to this paper was supported by a Callaghan Innovation grant No: PROP-53521-FELLOW-AIRNZ in collaboration with Air New Zealand. The authors wish to thank both Callaghan Innovation and Air New Zealand for making this study happen.

Institutional Review Board Statement: Not applicable.

Informed Consent Statement: Not applicable.

Data Availability Statement: The raw/processed data required to reproduce these findings cannot be shared at this time as the data also forms part of an ongoing study.

Conflicts of Interest: All authors certify that they have NO affiliations with or involvement in any organization or entity with any financial interest (such as honoraria; educational grants; participation in speakers' bureaus; membership, employment, consultancies, stock ownership, or other equity interest; and expert testimony or patent-licensing arrangements), or non-financial interest (such as personal or professional relationships, affiliations, knowledge or beliefs) in the subject matter or materials discussed in this manuscript.

References

1. Horrocks, A.R.; Price, D.; Price, D. *Fire Retardant Materials*; Woodhead Publishing: Sawston, UK, 2001.
2. Njuguna, J.; Pielichowski, K. Polymer nanocomposites for aerospace applications: Properties. *Adv. Eng. Mater.* **2003**, *5*, 769–778. [CrossRef]
3. Wilkie, C.A.; Morgan, A.B. *Fire Retardancy of Polymeric Materials*; CRC Press: Boca Raton, FL, USA, 2009.
4. Hu, Y.; Wang, X. *Flame Retardant Polymeric Materials: A Handbook*; CRC Press: Boca Raton, FL, USA, 2019.
5. Irving, P.E.; Soutis, C. *Polymer Composites in the Aerospace Industry*; Woodhead Publishing: Sawston, UK, 2019.
6. Yang, B.; Mao, Y.; Zhang, Y.; Wei, Y.; Liu, W.; Qiu, Y. Fast-curing halogen-free flame-retardant epoxy resins and their application in glass fiber-reinforced composites. *Text. Res. J.* **2019**, *89*, 3700–3707. [CrossRef]
7. Yin, H.; Sypaseuth, F.D.; Schubert, M.; Schoch, R.; Bastian, M.; Schartel, B. Routes to halogen-free flame-retardant polypropylene wood plastic composites. *Polym. Adv. Technol.* **2019**, *30*, 187–202. [CrossRef]
8. LI, M.; Zhang, T. Flame-retardant properties of peek fabric. *J. Text. Res.* **2010**, *1*, 19.
9. Council, N.R. *Improved Fire-And Smoke-Resistant Materials for Commercial Aircraft Interiors: A Proceedings*; National Academies Press: Washington, DC, USA, 1995; Volume 477.

10. Lyon, R.E. *Fire-Resistant Materials: Research Overview*; Federal Aviation Administration Atlantic City Nj Airport And Aircraft Safety: Washington, DC, USA, 1997.
11. Santos, C.V.d.; Leiva, D.R.; Costa, F.R.; Gregolin, J.A.R. Materials selection for sustainable executive aircraft interiors. *Mater. Res.* **2016**, *19*, 339–352. [CrossRef]
12. Kausar, A.; Rafique, I.; Muhammad, B. Aerospace application of polymer nanocomposite with carbon nanotube, graphite, graphene oxide, and nanoclay. *Polym. Plast. Technol. Eng.* **2017**, *56*, 1438–1456. [CrossRef]
13. Crump, S.S.; Priedeman, W.R., Jr.; Hanson, J.J. Rapid Prototype Injection Molding. Google Patents EP1497093B1, 21 December 2011.
14. Ghadge, A.; Karantoni, G.; Chaudhuri, A.; Srinivasan, A. Impact of additive manufacturing on aircraft supply chain performance. *J. Manuf. Technol. Manag.* **2018**, *29*, 846–865. [CrossRef]
15. Wong, K.V.; Hernandez, A. A review of additive manufacturing. *Int. Sch. Res. Not.* **2012**, *2012*. [CrossRef]
16. Popescu, D.; Zapciu, A.; Amza, C.; Baciu, F.; Marinescu, R. FDM process parameters influence over the mechanical properties of polymer specimens: A review. *Polym. Test.* **2018**, *69*, 157–166. [CrossRef]
17. Badini, C.; Padovano, E.; De Camillis, R.; Lambertini, V.G.; Pietroluongo, M. Preferred orientation of chopped fibers in polymer-based composites processed by selective laser sintering and fused deposition modeling: Effects on mechanical properties. *J. Appl. Polym. Sci.* **2020**, *137*, 49152. [CrossRef]
18. Parandoush, P.; Lin, D. A review on additive manufacturing of polymer-fiber composites. *Compos. Struct.* **2017**, *182*, 36–53. [CrossRef]
19. Deng, X.; Zeng, Z.; Peng, B.; Yan, S.; Ke, W. Mechanical properties optimization of poly-ether-ether-ketone via fused deposition modeling. *Materials* **2018**, *11*, 216. [CrossRef]
20. Berretta, S.; Davies, R.; Shyng, Y.; Wang, Y.; Ghita, O. Fused Deposition Modelling of high temperature polymers: Exploring CNT PEEK composites. *Polym. Test.* **2017**, *63*, 251–262. [CrossRef]
21. Hoskins, T.; Dearn, K.; Kukureka, S. Mechanical performance of PEEK produced by additive manufacturing. *Polym. Test.* **2018**, *70*, 511–519. [CrossRef]
22. Mohamed, O.A.; Masood, S.H.; Bhowmik, J.L. Optimization of fused deposition modeling process parameters: A review of current research and future prospects. *Adv. Manuf.* **2015**, *3*, 42–53. [CrossRef]
23. Fetisov, K.; Maksimov, P. Topology optimization and laser additive manufacturing in design process of efficiency lightweight aerospace parts. *J. Phys. Conf. Ser.* **2018**, *1015*, 052006. [CrossRef]
24. Kruth, J.-P.; Wang, X.; Laoui, T.; Froyen, L. Lasers and materials in selective laser sintering. *Assem. Autom.* **2003**, *23*, 357–371. [CrossRef]
25. Goodridge, R.; Tuck, C.; Hague, R. Laser sintering of polyamides and other polymers. *Prog. Mater. Sci.* **2012**, *57*, 229–267. [CrossRef]
26. Kiani, S.; Mousavi, S.M.; Shahtahmassebi, N.; Saljoughi, E. Hydrophilicity improvement in polyphenylsulfone nanofibrous filtration membranes through addition of polyethylene glycol. *Appl. Surf. Sci.* **2015**, *359*, 252–258. [CrossRef]
27. Parker, M.E.; Arbegast, W.; Boysen, A.; West, M. *Eliminating Voids in FDM Processed Polyphenylsulfone, Polycarbonate, and ULTEM 9085 by Hot Isostatic Pressing*; South Dakota School of Mines and Technology: Rapid City, SD, USA, 2009.
28. Velu, R.; Singamneni, S. Selective laser sintering of polymer biocomposites based on polymethyl methacrylate. *J. Mater. Res.* **2014**, *29*, 1883. [CrossRef]
29. Berretta, S.; Evans, K.; Ghita, O. Predicting processing parameters in high temperature laser sintering (HT-LS) from powder properties. *Mater. Des.* **2016**, *105*, 301–314. [CrossRef]
30. Gindt, B.P.; Abebe, D.G.; Tang, Z.J.; Lindsey, M.B.; Chen, J.; Elgammal, R.A.; Zawodzinski, T.A.; Fujiwara, T. Nanoporous polysulfone membranes via a degradable block copolymer precursor for redox flow batteries. *J. Mater. Chem. A* **2016**, *4*, 4288–4295. [CrossRef]
31. Bellehumeur, C.T.; Kontopoulou, M.; Vlachopoulos, J. The role of viscoelasticity in polymer sintering. *Rheol. Acta* **1998**, *37*, 270–278. [CrossRef]
32. Sedlacek, T.; Hausnerova, B.; Filip, P. An evaluation of the pressure-dependent melt viscosity of polyphenylsulfone. *Polym. Eng. Sci.* **2014**, *54*, 711–715. [CrossRef]
33. Mohan, D. *Studies on Fouling Resistant Hybrid Polyphenylsulfone PPSU Ultrafiltration Membranes for Water Pollution Abatement*; Anna University: Tamil Nadu, India, 2016.
34. Sun, M.-S.; Nelson, C.; Beaman, J.J.; Barlow, J.J. A model for partial viscous sintering. In Proceedings of the 1991 International Solid Freeform Fabrication Symposium, Austin, TX, USA, 12–14 August 1991.
35. Lv, Y.; Thomas, W.; Chalk, R.; Hewitt, A.; Singamneni, S. Polyetherimide powders as material alternatives for selective laser-sintering components for aerospace applications. *J. Mater. Res.* **2020**, *35*, 3222–3234. [CrossRef]
36. Ramgobin, A.; Fontaine, G.; Bourbigot, S. Thermal degradation and fire behavior of high performance polymers. *Polym. Rev.* **2019**, *59*, 55–123. [CrossRef]

37. Arnold, N.; Bityurin, N. Model for laser-induced thermal degradation and ablation of polymers. *Appl. Phys. A* **1999**, *68*, 615–625. [CrossRef]
38. Afrose, M.F. *Mechanical and Viscoelastic Properties of Polylactic Acid (PLA) Materials Processed Through Fused Deposition Modelling (FDM)*; School of Engineering Faculty of Science, Engineering and Technology Swinburne University of Technology: Hawthorn, Australia, 2016.

polymers

Article

Manufacturing a First Upper Molar Dental Forceps Using Continuous Fiber Reinforcement (CFR) Additive Manufacturing Technology with Carbon-Reinforced Polyamide

Roland Told [1], Gyula Marada [2], Szilard Rendeki [3,4], Attila Pentek [1], Balint Nagy [4], Ferenc Jozsef Molnar [3] and Peter Maroti [1,2,*]

1   3D Printing and Visualization Centre, University of Pecs, Boszorkány Street 2, 7624 Pécs, Hungary; told.rland@pte.hu (R.T.); pentek.attila@pte.hu (A.P.)
2   Clinical Centre, Department of Dentistry, Oral and Maxillofacial Surgery, University of Pecs, Dischka Győző Street 5, 7621 Pécs, Hungary; marada.gyula@pte.hu
3   Medical Simulation Education Centre, Medical School, University of Pecs, Szigeti Road 12, 7624 Pécs, Hungary; rendeki.szilard@pte.hu (S.R.); fecni1990@gmail.com (F.J.M.)
4   Clinical Centre, Department of Anesthesiology and Intensive Therapy, University of Pecs, Ifjúság Roud 13, 7624 Pécs, Hungary; balintjanosnagy@gmail.com
*   Correspondence: peter.maroti@aok.pte.hu; Tel.: +36-30-442-68-41

Abstract: 3D printing is an emerging and disruptive technology, supporting the field of medicine over the past decades. In the recent years, the use of additive manufacturing (AM) has had a strong impact on everyday dental applications. Despite remarkable previous results from interdisciplinary research teams, there is no evidence or recommendation about the proper fabrication of handheld medical devices using desktop 3D printers. The aim of this study was to critically examine and compare the mechanical behavior of materials printed with FFF (fused filament fabrication) and CFR (continuous fiber reinforcement) additive manufacturing technologies, and to create and evaluate a massive and practically usable right upper molar forceps. Flexural and torsion fatigue tests, as well as Shore D measurements, were performed. The tensile strength was also measured in the case of the composite material. The flexural tests revealed the measured force values to have a linear correlation with the bending between the 10 mm (17.06 N at 5000th cycle) and 30 mm (37.99 N at 5000th cycle) deflection range. The findings were supported by scanning electron microscopy (SEM) images. Based on the results of the mechanical and structural tests, a dental forceps was designed, 3D printed using CFR technology, and validated by five dentists using a Likert scale. In addition, the vertical force of extraction was measured using a unique molar tooth model, where the reference test was carried out using a standard metal right upper molar forceps. Surprisingly, the tests revealed there to be no significant differences between the standard (84.80 N $\pm$ 16.96 N) and 3D-printed devices (70.30 N $\pm$ 4.41 N) in terms of extraction force in the tested range. The results also highlighted that desktop CFR technology is potentially suitable for the production of handheld medical devices that have to withstand high forces and perform load-bearing functions.

Keywords: 3D printing; additive manufacturing; dental forceps; CFR (continuous fiber reinforcement); fatigue test; mechanical testing; composite; carbon; scanning electron microscopy

Citation: Told, R.; Marada, G.; Rendeki, S.; Pentek, A.; Nagy, B.; Molnar, F.J.; Maroti, P. Manufacturing a First Upper Molar Dental Forceps Using Continuous Fiber Reinforcement (CFR) Additive Manufacturing Technology with Carbon-Reinforced Polyamide. *Polymers* **2021**, *13*, 2647. https://doi.org/10.3390/polym13162647

Academic Editors: Pan Houwen Matthew and Eric Luis

Received: 18 July 2021
Accepted: 4 August 2021
Published: 9 August 2021

Publisher's Note: MDPI stays neutral with regard to jurisdictional claims in published maps and institutional affiliations.

1. Introduction

In recent years, additive manufacturing (AM) technology has become an undoubtedly decisive tool in the healthcare industry. It is widely used in model creation and visualization [1], simulator development [2], preoperative planning [3], prototyping, and the small-series production of medical devices such as implants [4], prosthetics and orthotics [5,6], and laboratory equipment, and can also support tissue engineering processes [7]. Furthermore, it can strongly enhance and support patient–doctor communication, and serve patient education purposes as well [8].

3D printers are used on a daily basis in dental care. Several dentistry-related applications are well described and deeply studied, since the technology provides a relatively cost-effective, fast, easy-to-use, and customizable solution for healthcare professionals. Metal dental implants, screws, and abutments can be produced using DMLS (direct metal laser sintering) technology, mainly from metal alloys such as Co-Cr alloy steel or titanium [9,10]. PolyJet™ and FFF (fused filament fabrication) technologies are mainly used for creating replica teeth, skulls, and mandibular or maxillary models in order to perform preoperative planning [11]. The most commonly used devices are desktop SLA (stereolithography) 3D printers, and their applications include model fabrication, surgical guide or mold printing, and complex prototyping processes [9,10]. Additive manufacturing is also essential in orthodontics, restorative dentistry, and endodontics [12]. Despite the fact that dental applications are strongly rely on handheld devices, only a few previous studies are available concerning the use of 3D printing in instrument development and production. Recently, surgical devices printed for long-duration space missions have been reported using FFF technology and ABS (acrylonitrile butadiene styrene) material [13–15], but dentistry-related applications have not yet been reported. This lack of studies potentially indicates that desktop FFF 3D printing technology has serious disadvantages in terms of mechanical and structural stability, due to the anisotropic material characteristics of the end-products, which mainly occur through the use of PLA (polylactic acid) and ABS.

CFR (continuous fiber reinforcement) additive manufacturing has outstanding mechanical properties compared to FFF. The addition of carbon fibers can increase both tensile strength and the modulus of elasticity [16]. It has been observed that parts and models created using carbon-reinforced polyamide can be created with high precision, and are comparable with molded or even metal parts; however, further investigations are essential in order to better understand their structural characteristics, and to avoid weak interlayer connections [17,18].

The aim of this study was to explore the possibilities CFR AM technology can provide in the field of handheld medical device development and production. The mechanical characterization focused on fatigue testing, in order to determine the usability in load-bearing applications. Structural analysis was performed to explore the mechanisms behind the observed results. The results were compared with neat PLA models printed with an FFF 3D printer. Moreover, the goal was to design, fabricate, and test a functional handheld device—in this case, a first upper molar dental extractor, since this medical instrument must be characterized with excellent mechanical properties and the proper level of durability, in order to withstand outstandingly high forces.

2. Materials and Methods

2.1. 3D Printing Parameters and Fabrication of Test Specimens

In order to fabricate the PLA control test specimens, a Craftbot Plus 3 (Craftbot Ltd., Salgótarjáni rd. 12–14, Budapest, Hungary) 3D printer was used, with a 0.4-mm nozzle, 0.2-mm layer height, and 100% infill density. A 245 °C primary extruder temperature and 110 °C heated-bed temperature provided the necessary amount of heat for fabrication. Neat PLA was utilized, (Herz Hungária Kft., Pesti rd. 284, Üllő, Hungary), with a filament diameter of 1.75 mm and a white color. The specimens were sliced using CraftWare™ software (Craftbot Ltd., Salgótarjáni rd. 12–14, Budapest, Hungary).

To produce the composite-based test specimens and the dental extractor, a Markforged X7 continuous fiber reinforcement (CFR)) 3D printer (Markforged, 480 Pleasant St, Watertown, MA, USA) was applied using Onyx carbon fiber composite (Markforged Co.). The specimens were sliced on the www.eiger.io, accessed on 30 July 2021 (Markforged 480 Pleasant, St Watertown, MA, USA) webpage. Onyx—which is polyamide filled with micro carbon fiber with isotropic direction properties—served as a base material. The printing was conducted in accordance with the manufacturing settings with maximum carbon fiber filling. The layer height was 0.125 mm and had a triangular fill pattern. The infill density of the fibers was 37%, and the number of wall layers was set to 2. The sample specimen printing orientation was "X" in case of both FFF and CFR printing.

To create the custom tooth model for testing the forceps, a Prusa I3 MK3 (Prusa Research a.s., Partyzánská 188/7a, Praha 7, Czech Republic) desktop FFF 3D printer was used with polyethylene terephthalate glycol (PETG) (Devil Design Sp. J. Zwirki I Wigury, Poland). The diameter of the filament was 1.75 mm. The layer height was 0.1 mm with 50% infill density, and in accordance with the melting temperature of PETG, the extruder was set to 230 °C and the bed heating was 80 °C. The slicer program was the default PrusaSlicer 2.3.

2.2. Material Testing

2.2.1. Flexural Fatigue Test

In case of the PLA and the carbon composite test specimens, firstly, flexural fatigue tests were carried out using a ZwickRoell e/m actuator material tester (ZwickRoell, 89079, August-Nagel-Straße 11, Ulm, Germany) with a 5-kN load cell. The type of specimen was A1 from the ISO 527–2:2012 standard. To perform the tests, a special gripper was designed that aimed to correct the increment in the cross-section due to deflection (Figure 1, Supplementary Materials). For the other end of the gripper of the test bar, a holder was fabricated and secured in horizontal position throughout the entire test procedure. Using the test specimens, a pulsating pressure test was performed, where the starting force was 0 N. The deflection was 10 mm, 15 mm, 20 mm, 25 mm, 30 mm, 35 mm, 40 mm, 45 mm, and 50 mm, respectively, and it was performed until fracture, with all specimens being measured once. The testing frequency was 2 Hz, and the move was sinusoidal. In the case of the Onyx–carbon composite the tests were carried out until 20,000 cycles, and the PLA was tested until fracture in all cases. The permanent deformation was also measured. Deflection was determined with a Mitutoyo CD-15APX digimatic caliper (Mitutoyo, Sakado, Takatsu-ku, Kawasaki, Kanagawa, Japan), and the cross-section change was measured with a Mitutoyo 102–707 micrometer. The measurement was carried out on all fatigue specimens.

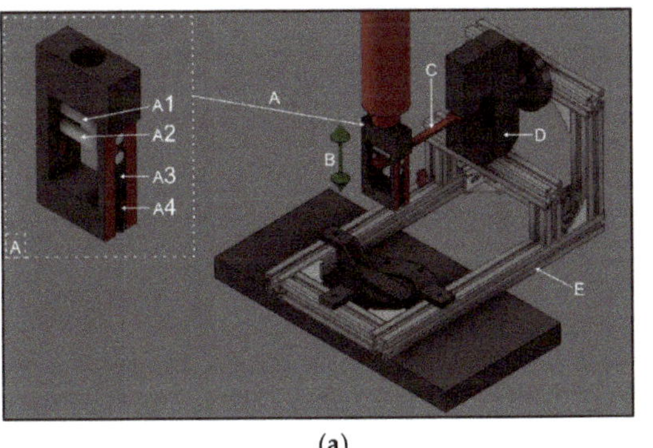

(a)   (b)

Figure 1. (**a**) The models used in the setup of the flexural fatigue test measurement. (**A**) Special gripper; (**A1**) fixed roller; (**A2**) moving roller; (**A3**) spring; (**A4**) adjusting screw; (**B**) moving direction; (**C**) specimen; (**D**) screw grip; (**E**) support frame. (**b**) The assembled device for the flexural fatigue test measurement.

2.2.2. Torsion Fatigue Test

The torsion fatigue tests were performed with a ZwickRoell Z5.0 biaxial material tester (ZwickRoell, 89079, August-Nagel-Straße 11, Ulm, Germany). The diameter of the probe was 6 mm, the parallel length was 50 mm, and the ends of the specimen were modified in order to fit to the biaxial clamping [19] (Figure 2). The axial preload was set to 0.1 N at the start of the test, and then the axial position was fixed. Torsion preload was not applied;

the testing speed was 720°/min, and the testing frequency was 2 Hz. The specimens went under sinusoidal rotation, and were tested every 10°, between 20° and 90°, for up to 20,000 cycles in the case of the composite, or until the specimens broke in the case of PLA specimens. All specimens were measured once. During the tests, the specimens' temperatures were monitored with a Guide B160V infrared camera (Wuhan Guide Infrared Co., Ltd., No. 6, Huanglongshan South Road, East Lake Development Zone, Wuhan, China) to avoid overheating.

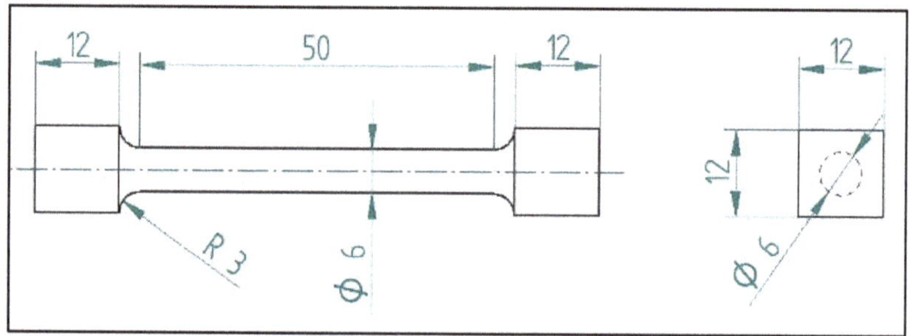

Figure 2. Schematic of torsion fatigue test specimens.

2.2.3. Shore D Measurements

Shore D measurements were performed on both materials, using a ZwickRoell 3131 Shore D testing device (ZwickRoell, 89079, August-Nagel-Straße 11, Ulm, Germany) with the MSZ EN ISO 868:2003 (A, D) standard. The thickness of the test specimens was 4 mm, the room temperature was 25.0 °C, and the relative humidity was 47%. Both the PLA and composite test specimens were measured 5 times.

2.2.4. Tensile Test

In order to investigate the effect of fatigue tests on the flexural and torsional carbon composite test specimens, tensile tests were carried out using a ZwickRoell Z100THW universal material tester (ZwickRoell, 89079, August-Nagel-Straße 11, Ulm, Germany) with a 10-kN load cell. The preload was 0.1 MPa and the test speed was 50 mm/min, according to the ISO 527 standard.

2.2.5. Scanning Electron Microscopy (SEM)

To better understand the mechanical behavior of the composite material, SEM imaging was carried out with 30× and 350× magnification using a JEOL JSM-IT500HR device (3-1-2 Musashino, Akishima, Tokyo, Japan). PLA specimens were not investigated, since previous international studies have described them in detail [20]. The first group of the SEM samples was taken from the specimens with 50-mm bending distance after 20,000 cycles of flexural fatigue testing, while the other group was produced from test specimens that broke in one testing phase. The outer layers of the objects were notched before fracture, in order to prevent further damage to their structure. Before imaging, the composite samples were coated with gold using a JEOL JFC-1300 auto fine coater.

2.2.6. Upper Right First Molar Forceps Scanning, Design, and Printing

To test the CFR technology with the carbon-reinforced polyamide material, a handheld dental extractor was used as a model. The reverse engineering of an upper right first molar forceps was performed suing an Artec Space Spider industrial handheld scanner (Artec, 20 rue des Peupliers, L-2328, Luxembourg, Luxembourg). After disassembling the forceps, the object went through a surface treatment to make the glittering surfaces completely visible during the scanning. This was achieved with a Centrochem (Centrotool Kft., 1102 Budapest, Halom u. 1. Hungary) cracking examination aerosol agent. The digitization of the treated components was finished separately. From the scanning results, the point-cloud processing and mesh generation were carried out in Artec Studio 15 (Artec, 20 rue des Peupliers, L-2328, Luxembourg, Luxembourg). Repair of minor errors in the generated model and the export of the 3D-printable files were done using Blender 3D modeling software (Blender Institute B.V. Buikslotermeerplein 161, ET Amsterdam, The Netherlands).

Due to the applied CFR printing technology, replicated objects will have some flexibility. In order to enable the forceps to exert appropriate strength and improve their usability, minor modifications were made to the geometric structure. An additional 3 mm of thickness was added to the area of the fulcrum in order to make it more resistant to load stress. The jaws are in contact at the closed state; in this way, sufficient force can be applied to carry out the extraction. Additionally, a spacer pair was placed on the inside of the forceps handles to keep the user's finger safe when closing the instrument (Figure 3).

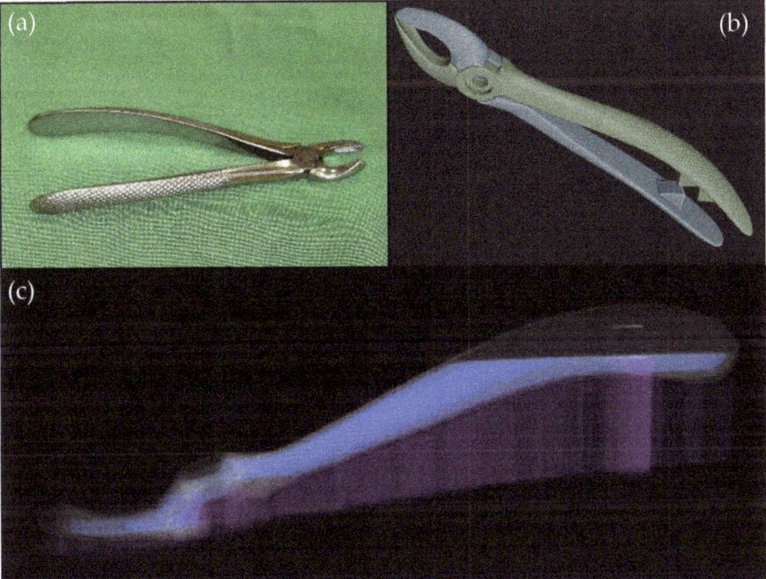

Figure 3. The mains steps of reverse engineering and 3D design of the forceps: (**a**) the original metal dental extractor; (**b**) the modified model in .stl format; (**c**) the sliced model ready for printing.

2.2.7. Clamp Test with Biaxial Tester and Likert Scale

A real-size (1:1) tooth model was designed using Autodesk Fusion 360 software (111 McInnis Parkway, San Rafael, CA, USA) merged to a Zwick Z5.0 biaxial-compatible platform and printed from PETG using a Prusa I3 3D printer, (Figure 4). Five dentists with 2–20 years of experience individually tested the metal and the unique composite forceps using the 3D-printed tooth model that was applied to the testing machine (Supplementary Materials). During the test, the dentists moved their hand holding the extractor, just as they would in case of removing a real tooth, and the device measured the values of the axial force and the torque. They anonymously evaluated and compared the instruments using a Likert scale (Table A1 in Appendix A).

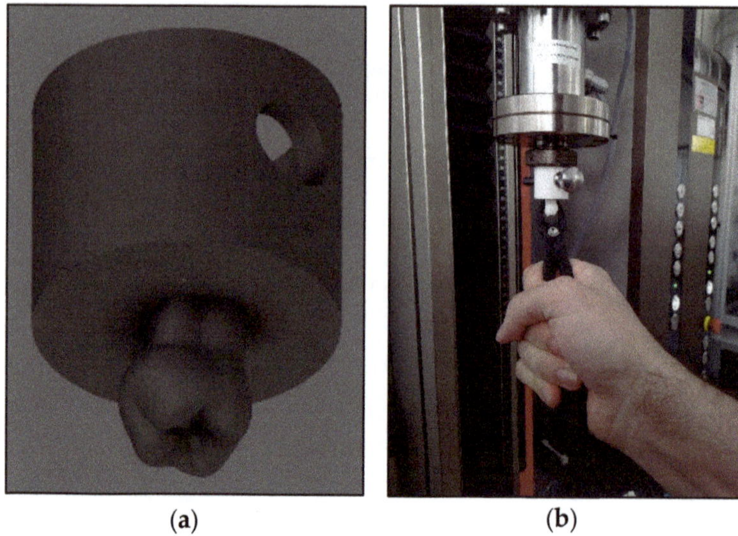

(a)   **(b)**

Figure 4. (**a**) The tooth model in .stl format; (**b**) the printed tooth model applied to the biaxial tester.

2.2.8. Statistics and Analysis

To the measured data of the flexural and torsion tests, an upper envelope curve was fitted. The results of extraction tests were compared via two-sample *t*-test ($p < 0.05$). The statistical analysis, and curve fitting were carried out using Origin 2018 software (OriginLab Corporation, One Roundhouse Plaza, Northampton, MA, USA). The curves in case of the fatigue tests of PLA were fitted using the following function:

$$y = y_0 + A_1 e^{-\frac{x}{t_1}} + A_2 e^{-\frac{x}{t_2}}$$

where y_0, A_1, A_2, t_1, and t_2 are constant values, and x represents the number of cycles.

3. Results

3.1. Material Testing

3.1.1. Flexural Fatigue Test

Flexural fatigue tests were performed on both materials, in order to determine their resistance against forces that occur when the medical instrument is under tilting, bending, or flexing loads. For the PLA material, the fatigue curve corresponded to previous studies, where the characterization was carried out in detail, including the effects of different layer heights, printing speeds, and infill densities and patterns. Rotation and bending tests were also performed [20,21]. The specimen did not break during the fatigue flexural test, since it was sufficiently flexible; therefore, its flexural strength was not measured. Surprisingly, the test specimen broke only after 125,000 cycles with 15-mm deflection. (Figure 5).

The Onyx composite material showed nearly constant force values between 10-mm and 30-mm deflection, with a continuous, slight decrease, measured up to 20,000 cycles. Further analysis highlights that the correlation between the bending and the measured force values was linear between 10 mm and 40 mm, but after 5000 cycles, the linearity was only observable between 10 mm and 30 mm deflection rates. At 10-mm flexion, 15.37 N was measured at the first cycle and 17.06 N at the 5000th cycle, while in the case of 30-mm deflection, 34.22 N was measured at the first cycle and 37.99 N after 5000 cycles (Figure 6). When the measurement was carried out until fracture within one test phase, the flexural strength was measured as 51.16 N and the deflection was 50.56 mm.

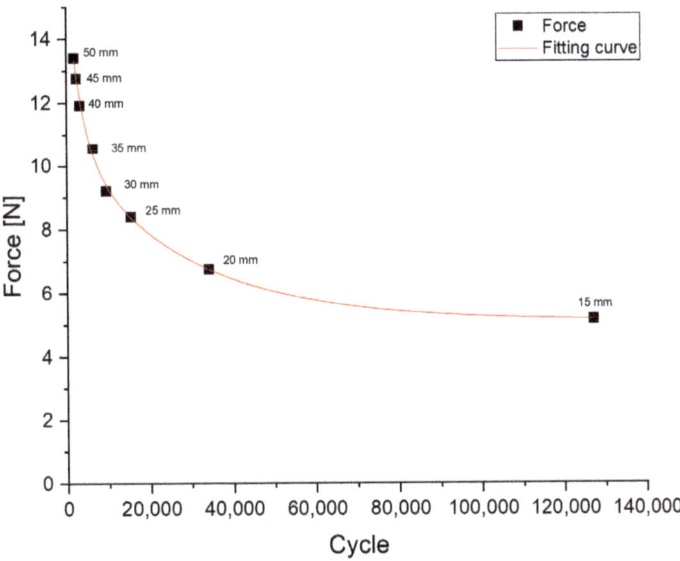

Figure 5. Results of the flexural fatigue test for PLA. The figure indicates the force values measured at the fracture point of each test specimen. The black squares indicate the force values (N) measured at each deflection rate (mm), with the number of cycles needed to break the test specimens.

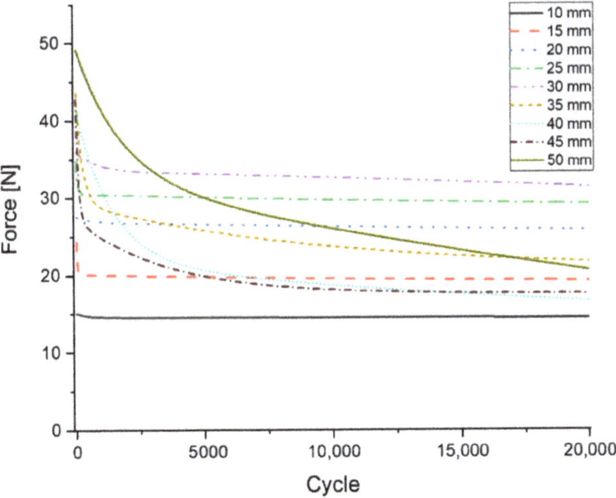

Figure 6. Results of the flexural fatigue test of the composite material. Force values measured at each deflection rate. The colored lines indicate the change in the force values (N), measured up to 200,000 cycles.

After performing the fatigue tests, it was observed that in the test specimens with 35-mm or higher deflection, a protrusion occurred at the point of flexion, along with a phenomenon of permanent deflection, where the highest value was 12.78 mm in the case of 45-mm deflection. The measured parameters are presented in Table 1.

Table 1. The table indicates the location of protrusion points compared to the middle of the test specimens (difference in mm), with the values of permanent deflection.

Bending (mm)	Thickness at Middle Point (mm)	Thickness at Flexural Point (mm)	Difference (mm)	Permanent Deflection (mm)
10	4.182	4.163	−0.019	0
15	4.174	4.18	0.006	0
20	4.154	4.093	−0.061	0
25	4.152	4.133	−0.019	0
30	4.073	4.046	−0.027	0
35	4.16	4.392	0.232	8.16
40	4.153	4.474	0.321	9.20
45	4.121	4.537	0.416	12.78
50	4.059	4.397	0.338	10.78

3.1.2. Torsion Fatigue Test

The torsion fatigue test can reveal the resistance against forces occurring when the healthcare specialists perform twisting, screwing, or scrolling movements with the hand-held device. In case of the PLA test specimens, it was observed that under 35° torsion, fracture occurred within 800 cycles, but at 30° it withstood more than 5500 cycles. At 25° torsion, fracture took place at 11,700 cycles (Figure 7).

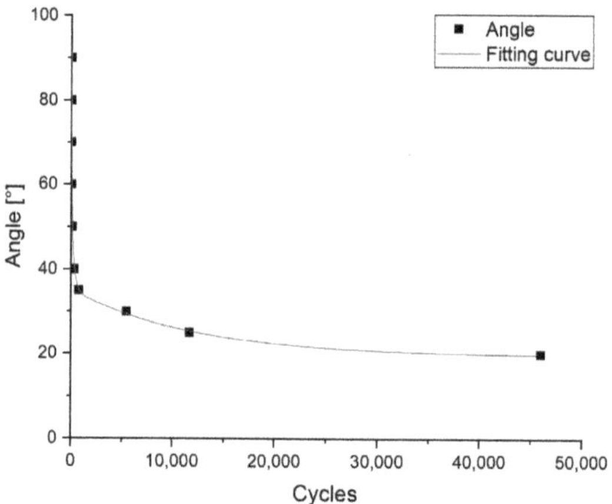

Figure 7. Results of torsion fatigue testing of PLA. Fracture points of test specimens with different rotation values. The black squares indicate the angle (°) and the number of cycles within which the fracture took place.

The composite material at the torsion fatigue test showed similar results to bending; therefore, a classical fatigue pattern was not detected. For the test specimens after 200–300 cycles, the value of torque required to rotate the test bar was set to an approximately constant value, with a minimal decrease (Figure 8). After the test had been performed for 20,000 cycles, the test specimen's torsional flexibility was observed, which was proportional to the maximum angle of rotation. Despite this phenomenon, the mechanical resistance of the specimens in the radial and axial directions was unchanged. In the torsion test specimens, the permanent deformations began after 40° rotation.

3.1.3. Shore D Hardness Measurements

Interestingly, the Shore D measurements did not reveal significant differences ($p < 0.05$) between the PLA and the composite material. The average in the case of PLA was 79.08 ± 0.60, while it was 74.54 ± 0.59 in the case of the composite material.

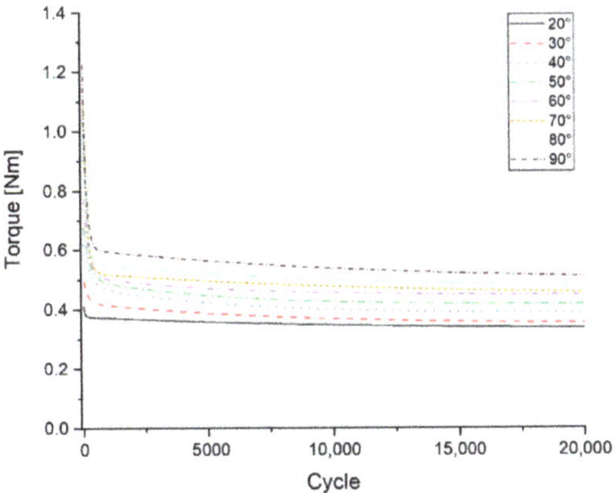

Figure 8. Results of torsion fatigue testing of the composite material. The colored curves show the decrease in Torque (Nm) measured at different rotation values.

3.1.4. Tensile Test

Tensile tests were performed on the composite test specimens after flexion and torsion fatigue experiments. In accordance with the previous measurements, a significant difference was observed drawn from the results of the flexural test (two-sample *t*-test, $p < 0.05$). The average value of tensile strength was 258.30 MPa $\pm$ 10.79 MPa, and the fracture had a V shape at the rounding (Figure 9). Above 35-mm deflection, the average was 71.00 MPa $\pm$ 19.92 MPa, the fractures had a straight pattern, and the fracture locations were observed at the protrusion points (Figure 9). Moreover, the tensile test was carried out on the test specimens used in the torsion fatigue test. The results indicate that there is no major alternation between these values; the average tensile strength was 118 MPa $\pm$ 2 MPa, and the fracture had a V shape at the rounding, just like the flexural specimens (Figure 9).

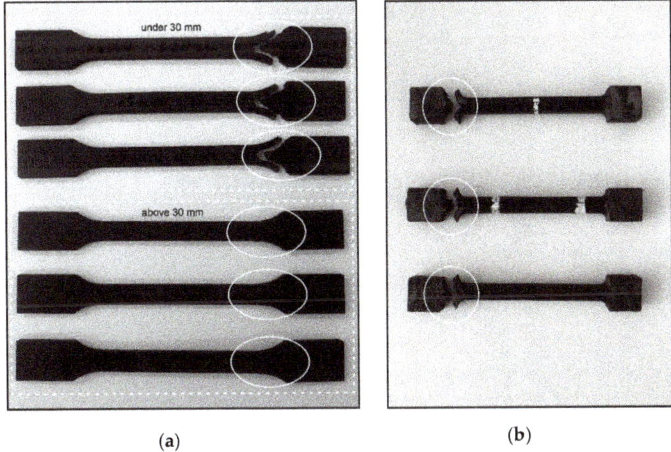

(a)　　　　(b)

Figure 9. (**a**) Composite test specimens used in flexural fatigue testing, after performing the tensile strength test. The white circles indicate a "V"-like fracture pattern in the case of test specimens under 30-mm deflection; the white dotted frame represents the straight fracture line above a 30-mm deflection rate. (**b**) Composite test specimens used in torsion fatigue testing, after performing the tensile strength test. The white circles indicate the "V"-like fracture pattern.

3.2. Results of Scanning Electron Microscopy Imaging

Figure 10 was captured from a flexural test specimen's extension zone, which refers to the horizontally upper side of the test specimen, aligned in "X" orientation. Figure 10a shows the fractured cross-section after breaking the test bar in one cycle, while Figure 10b presents the fractured cross-section after 20,000 cycles. The magnification was 30×. In the case of Figure 10a, the carbon fibers are of equal length on the fracture surface, but Figure 10b clearly demonstrates that the fibers are longer and have different lengths, varying between 500 and 2000 μm. Additionally, in this image, cracks and hollows can be observed.

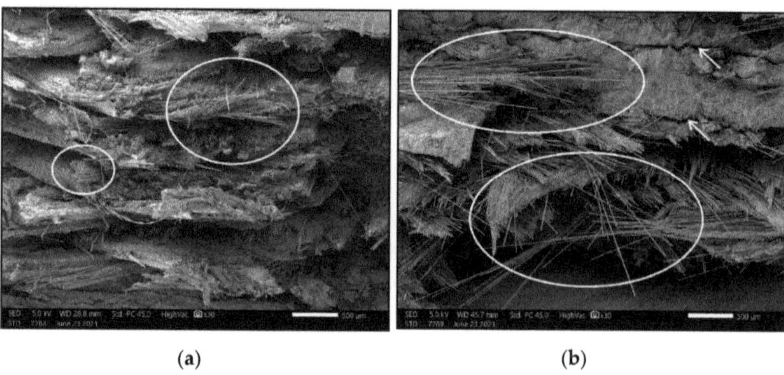

(**a**)   (**b**)

Figure 10. (**a**) Image demonstrating carbon fibers of nearly identical lengths (white circles); the layers are intact. (**b**) Image taken from the specimens used in the flexural fatigue test; the broken surface has cracks and hollows on it (white arrows), and the carbon fibers are elongated. The magnification was 30× in both cases.

Figure 11 was taken from a flexural test specimen's compression zone. This area refers to the horizontally lower half of the test specimen, aligned in "X" orientation. Interestingly, Figure 11a shows orientated and intact carbon fibers of the equal lengths across the entire broken surface of the test specimen broken in one testing phase; meanwhile, Figure 11b presents smaller, fragmented carbon fibers after 20,000 cycles. It is notable that the carbon fiber strings do not have the same orientation—they are facing towards random directions. Both images were captured with 350× magnification.

(**a**)   (**b**)

Figure 11. (**a**) Image showing the carbon fibers in the test specimen broken in one step; the fibers have the same length and orientation (white circle). (**b**) Image taken from the broken surface of the specimen after 20,000 flexural cycles. In this case, the carbon fibers have random orientation, and they are fragmented into small pieces (white arrows). The magnification was 350× in both cases.

3.3. Evaluation of 3D-Printed Molar Extruder

Testing was performed by professional dentists, with up to 20 years of clinical experience since their graduation. As a first step, the specialists performed the tests using a traditional metal forceps. They simulated the force of extraction on the model tooth mounted on the biaxial tester, mimicking a conventional extraction process, and after that, they also pulled the 3D-printed tooth with higher force, simulating a tougher and harder process. The measured data served as controls for the evaluation of the 3D-printed medical instrument. Following the trials with the metal extractor, the composite forceps was tested, and the dentists were asked to apply the highest possible extraction force. The results are summarized in Table 2.

Table 2. Summary of the results of the simulation of tooth extraction.

	Metal Forceps		3D-Printed Composite Forceps
Dentist	Force of Normal Extraction of Tooth (N)	Force of Strong Extraction of Tooth, (N)	Force Performed with Composite Forceps (N)
Dentist 1.	50	80	69.0
Dentist 2.	46	84	71.5
Dentist 3.	79	95	72.8
Dentist 4.	41	60	74.8
Dentist 5.	73	105	63.4
Average (N)	57.80	84.80	70.30
SD (N)	17.05	16.96	4.41

Two-sample *t*-tests were performed to compare the values of the forces measured with the metal forceps and the values of forces measured with the 3D-printed composite forceps. In the case of the simulation of normal extraction, $p = 0.15$, while $p = 0.10$ when strong extraction was mimicked; therefore, there was no significant difference between the compared groups. The average force performed with the 3D-printed device was 70.30 ± 4.41 N. Surprisingly, the standard deviation was the smallest within this group. After the test, all of the dentists filled out a Likert scale. Each question was answered on a scale of 1–5, where 1 signifies "strongly disagree" or "least amount", and 5 signifies "strongly agree" or "at the highest possible amount". Two questions could be answered by "yes" or "no". Furthermore, the respondents had the possibility to offer amendments and suggestions for further improvements. Overall, all specialists gave answers of 3 or 4 for the question *"How firmly could you place the forceps on the model?"*, as well as to the question *"How could you compare the similarity of the axial extraction force you used to what is necessary in a clinical environment?"*. An average of 4.4 points was given to the question that asked about the convenience of using the 3D-printed device. Only two dentists out of five would use the device in its present form for actual treatment, but with some minor modifications, all of them would use it in patient care. The detailed questions and the given answers can be found in Table A1.

4. Discussion

3D printing technologies are reshaping our everyday lives, including the healthcare sector. Numerous international studies have discussed in detail how additive manufacturing can support prevention, diagnosis, and medical intervention. The availability of 3D printers is continuously increasing; therefore, £d printing can be implemented in everyday clinical care. Despite the intensive research and development in the field, handheld medical devices have not been fabricated with CFR technology, nor have fatigue properties been considered previously in terms of these instruments. Similar studies have highlighted that FFF/FDM 3D printing may be suitable for the production of handheld medical devices, but detailed structural and mechanical characterization has not been carried out, and fatigue properties have not been previously investigated [13,14].

Compared to PLA, the examined carbon composite material has no classical fatigue properties; after the 100 cycles, the measured force values been remained nearly constant. Moreover, it was highlighted that CFR technology can provide sturdier and more force-resistant products compared to FFF technology. This is consistent with a previous study, where an external fixator was fabricated using this technology and the Onyx material [22]. In the case of flexural fatigue tests, 30-mm deflection can be considered to be a limit, since a significant decrease in tensile strength was measured above this value, as well as a change from "V"-shaped fracture lines to a straight pattern. This observation was also supported by the SEM images, which revealed that after a long-term fatigue test the carbon fibers broke into smaller pieces. The torsion fatigue test showed a permanent deformation beyond a 40° rotation rate, but the carbon fibers presumably remained fully intact; therefore, the axial and radial resistance did not drop during tests up to 90°—in this case, only the polyamide base material slides away. The tensile test revealed that there is no distribution between the measured values. In terms of Shore D hardness, the values correlated with previous studies [23,24], and underlined that both materials are hard enough to fabricate medical devices, but the flexibility of PLA can be a disadvantage when stability and shape retention are required.

Tooth extraction is a complex surgical process that contains gripping, twisting, and traction in order to expand the bony socket and rupture the periodontal ligament fibers. For this reason, properly geometrically shaped pliers able to reduce the applied strength needed are required. Within the limitations of this study, we measured the applied force of traction.

Based on the initial results, the first upper molar dental forceps was designed using reverse engineering methods, and printed out with a CFR desktop printer. Dental specialists critically evaluated the extractor and compared it to a traditional, metal instrument. Using a unique, real-sized tooth model, it was proven that using the forceps fabricated from the carbon composite material can provide enough force to potentially extract a real tooth in a clinical environment. Furthermore, it was highlighted that this device can be reused several times. The results were consistent with previous findings, which measured vertical tooth extraction forces in real patients, and found that the amount of force required strongly depends on root anatomy and dental status [25,26]. Based on the reports of the dentists, after some minor modifications, our device could be potentially used in real-life conditions. Their reports suggest that the elasticity should be decreased more, the jaws of the device should be more stable, and the internal profile should be more ribbed.

In accordance with previous studies, it was confirmed that additive manufacturing technologies can be useful in the manufacture of medical devices. The main target groups include facilities or missions where continuous supply is not provided—for example, medical humanitarian missions [27], remote medical sites, or even space missions [13,14]. Moreover, recently, several studies highlighted that 3D printing can be a solution in times of global crisis, such as the COVID-19 pandemic [28–30]. Our findings reveal that instruments printed with CFR technology can be potentially used several times, and their mechanical properties provide excellent durability. The instrument is lightweight, easy to customize, and can be fabricated without any special infrastructural needs.

Despite the mechanical characterization being carried out in detail, in the future, it will be extremely important to evaluate the potential disinfection protocols that can be used for medical instruments fabricated with 3D printers without causing major change to their material properties. This aspect is crucial in terms of reusability, and precise, critical evaluations are still needed in terms of the sterilization of these medical instruments [22,31]. Furthermore, cost-effectivity analysis should be considered; however, the availability of 3D printing technologies is continuously increasing [32]. The limitations of our study also include the relatively low number of testing volunteers. These questions will be addressed in the upcoming projects of our research team.

In the near future, after the recommended modification of the model, and after obtaining the necessary ethical approval, cadaver studies are planned in order to prepare the devices for clinical trials. We believe that the findings could be implemented in everyday patient care as well as remote medicine applications.

5. Conclusions

Summarizing the results of the mechanical and structural analyses and the validation process, it is highlighted that CFR technology with carbon-reinforced composite materials can be suitable for the development and production of handheld medical devices. Furthermore, these devices can be reused for several sessions and procedures. The findings of this study also aim to inspire and support further research in the field of medical device design and fabrication based on 3D printing technologies.

Supplementary Materials: All of the necessary files to reproduce the experiments are available online under the following URL: https://data.mendeley.com/datasets/64sjwzcfz9/2. The files are in *.stl* and *.stp* format. The dataset contains the files to reproduce the tooth model, as well as the flexural test addons. The 3D model of the forceps is available upon request from the corresponding author.

Author Contributions: Conceptualization, R.T. and P.M.; methodology, R.T.; software, A.P.; validation, G.M., S.R., B.N., and F.J.M.; formal analysis, R.T.; investigation, R.T. and A.P.; resources, P.M.; data curation, F.J.M.; writing—original draft preparation, R.T., A.P., and P.M.; visualization, B.N.; supervision, P.M. and S.R.; project administration, P.M.; funding acquisition, P.M. All authors have read and agreed to the published version of the manuscript.

Funding: This research was supported by grants from National Research, Development, and Innovation Fund of Hungary tenders EFOP-3.6.1-16-2016-00004 and GINOP-2.3.2.-15-2016-00022. The work is related to the Thematic Excellence Program 2020—National Excellence Sub-program; Biomedical Engineering Project ("2020-4.1.1-TKP2020") of the University of Pecs, and provided support for P.M., A.P., R.T., B.N., and S.R. The project was also supported by NTP-NFTÖ-20-B-0071, granted to P.M.

Institutional Review Board Statement: The study was conducted according to the guidelines of the Declaration of Helsinki, and approved by the Institutional Ethics Committee University of Pecs. (Study based on the following approved protocol: 5825/2016).

Informed Consent Statement: Informed consent was obtained from all subjects involved in the study.

Data Availability Statement: Data are contained within the article or Supplementary Materials.

Acknowledgments: We would like to express our sincere thanks to the University of Pecs Medical School and the affiliated departments. We are especially thankful to our colleagues at the 3D Printing and Visualization Centre: Tamas Bulsz, Peter Hillebrand, Bence Manfai, and Andras Szoke. Additionally, we are grateful to our respected colleagues from the Department of Dentistry.

Conflicts of Interest: The authors declare no conflict of interest.

Appendix A

Table A1. Likert scale; summary of the results of the simulation of tooth extraction.

		Dentist 1	Dentist 2	Dentist 3	Dentist 4	Dentist 5	AVG
Metal Upper Right Molar Forceps	How firmly could you grab the 3D-printed tooth with the metal dental forceps?	5	4	4	4	5	4.4
	How similar was the torque you could use to what is necessary in a clinical environment?	5	4	4	4	4	4.2
	How similar was the axial extraction force you could use to what is necessary in a clinical environment?	5	5	4	4	5	4.6
Composite Upper Right Molar Forceps	How firmly could you place the forceps on the model?	3	4	4	4	3	3.6

Table A1. *Cont.*

	Dentist 1	Dentist 2	Dentist 3	Dentist 4	Dentist 5	AVG
How could you compare the similarity of the torque you used to what is necessary in a clinical environment?	3	3	4	4	3	3.4
How could you compare the similarity of the axial extraction force you used to what is necessary in a clinical environment?	3	4	3	4	4	3.6
To what degree could the flexibility of the forceps influence the proper positioning of the equipment during extraction?	4	4	5	3	4	4.0
How convenient did you find the composite forceps during the procedure?	4	5	4	5	4	4.4
Would you use such forceps during patient care?	No	Yes	No	Yes	No	40%
If you would not use our composite forceps in patient care, could it be made appropriate for you with some modifications?	Yes	Yes	Yes	Yes	Yes	100%

References

1. Nadagouda, M.N.; Rastogi, V.; Ginn, M. A Review on 3D Printing Techniques for Medical Applications. *Curr. Opin. Chem. Eng.* **2020**, *28*, 152–157. [CrossRef]
2. Sparks, D.; Kavanagh, K.R.; Vargas, J.A.; Valdez, T.A. 3D Printed Myringotomy and Tube Simulation as an Introduction to Otolaryngology for Medical Students. *Int. J. Pediatr. Otorhinolaryngol.* **2020**, *128*, 109730. [CrossRef]
3. Tejo-Otero, A.; Buj-Corral, I.; Fenollosa-Artés, F. 3D Printing in Medicine for Preoperative Surgical Planning: A Review. *Ann. Biomed. Eng.* **2020**, *48*, 536–555. [CrossRef]
4. Okolie, O.; Stachurek, I.; Kandasubramanian, B.; Njuguna, J. 3D Printing for Hip Implant Applications: A Review. *Polymers* **2020**, *12*, 2682. [CrossRef] [PubMed]
5. Toth, L.; Schiffer, A.; Nyitrai, M.; Pentek, A.; Told, R.; Maroti, P. Developing an Anti-Spastic Orthosis for Daily Home-Use of Stroke Patients Using Smart Memory Alloys and 3D Printing Technologies. *Mater. Des.* **2020**, *195*, 109029. [CrossRef]
6. Pentek, A.; Nyitrai, M.; Schiffer, A.; Abraham, H.; Bene, M.; Molnar, E.; Told, R.; Maroti, P. The Effect of Printing Parameters on Electrical Conductivity and Mechanical Properties of PLA and ABS Based Carbon Composites in Additive Manufacturing of Upper Limb Prosthetics. *Crystals* **2020**, *10*, 398. [CrossRef]
7. Haleem, A.; Javaid, M.; Khan, R.H.; Suman, R. 3D Printing Applications in Bone Tissue Engineering. *J. Clin. Orthop. Trauma* **2020**, *11*, S118–S124. [CrossRef]
8. Wake, N.; Rosenkrantz, A.B.; Huang, R.; Park, K.U.; Wysock, J.S.; Taneja, S.S.; Huang, W.C.; Sodickson, D.K.; Chandarana, H. Patient-Specific 3D Printed and Augmented Reality Kidney and Prostate Cancer Models: Impact on Patient Education. *3D Print. Med.* **2019**, *5*, 4. [CrossRef] [PubMed]
9. Javaid, M.; Haleem, A. Current Status and Applications of Additive Manufacturing in Dentistry: A Literature-Based Review. *J. Oral Biol. Craniofac. Res.* **2019**, *9*, 179–185. [CrossRef]
10. Lin, L.; Fang, Y.; Liao, Y.; Chen, G.; Gao, C.; Zhu, P. 3D Printing and Digital Processing Techniques in Dentistry: A Review of Literature. *Adv. Eng. Mater.* **2019**, *21*, 1801013. [CrossRef]
11. Lee, K.-Y.; Cho, J.-W.; Chang, N.-Y.; Chae, J.-M.; Kang, K.-H.; Kim, S.-C.; Cho, J.-H. Accuracy of Three-Dimensional Printing for Manufacturing Replica Teeth. *Korean J. Orthod.* **2015**, *45*, 217. [CrossRef]
12. Turkyilmaz, I.; Wilkins, G.N. 3D Printing in Dentistry—Exploring the New Horizons. *J. Dent. Sci.* **2021**, *16*, 1037–1038. [CrossRef]
13. Wong, J.Y. 3D Printing Applications for Space Missions. *Aerosp. Med. Hum. Perform.* **2016**, *87*, 580–582. [CrossRef]
14. Wong, J.Y.; Pfahnl, A.C. 3D Printing of Surgical Instruments for Long-Duration Space Missions. *Aviat. Space Environ. Med.* **2014**, *85*, 758–763. [CrossRef] [PubMed]
15. Wong, J.Y.; Pfahnl, A.C. 3D Printed Surgical Instruments Evaluated by a Simulated Crew of a Mars Mission. *Aerosp. Med. Hum. Perform.* **2016**, *87*, 806–810. [CrossRef]
16. Saharudin, M.S.; Hajnys, J.; Kozior, T.; Gogolewski, D.; Zmarzły, P. Quality of Surface Texture and Mechanical Properties of PLA and PA-Based Material Reinforced with Carbon Fibers Manufactured by FDM and CFF 3D Printing Technologies. *Polymers* **2021**, *13*, 1671. [CrossRef]
17. Saleh Alghamdi, S.; John, S.; Roy Choudhury, N.; Dutta, N.K. Additive Manufacturing of Polymer Materials: Progress, Promise and Challenges. *Polymers* **2021**, *13*, 753. [CrossRef] [PubMed]
18. Kabir, S.M.F.; Mathur, K.; Seyam, A.-F.M. A Critical Review on 3D Printed Continuous Fiber-Reinforced Composites: History, Mechanism, Materials and Properties. *Compos. Struct.* **2020**, *232*, 111476. [CrossRef]
19. Torres, J.; Cole, M.; Owji, A.; DeMastry, Z.; Gordon, A.P. An Approach for Mechanical Property Optimization of Fused Deposition Modeling with Polylactic Acid via Design of Experiments. *Rapid Prototyp. J.* **2016**, *22*, 387–404. [CrossRef]

20. Ayatollahi, M.R.; Nabavi-Kivi, A.; Bahrami, B.; Yazid Yahya, M.; Khosravani, M.R. The Influence of In-Plane Raster Angle on Tensile and Fracture Strengths of 3D-Printed PLA Specimens. *Eng. Fract. Mech.* **2020**, *237*, 107225. [CrossRef]
21. Gomez-Gras, G.; Jerez-Mesa, R.; Travieso-Rodriguez, J.A.; Lluma-Fuentes, J. Fatigue Performance of Fused Filament Fabrication PLA Specimens. *Mater. Des.* **2018**, *140*, 278–285. [CrossRef]
22. Landaeta, F.J.; Shiozawa, J.N.; Erdman, A.; Piazza, C. Low Cost 3D Printed Clamps for External Fixator for Developing Countries: A Biomechanical Study. *3D Print. Med.* **2020**, *6*, 31. [CrossRef]
23. Ramesh, M.; Panneerselvam, K. PLA-Based Material Design and Investigation of Its Properties by FDM. In *Advances in Additive Manufacturing and Joining*; Shunmugam, M.S., Kanthababu, M., Eds.; Springer: Singapore, 2020; pp. 229–241.
24. Yasa, E.; Ersoy, K. Dimensional Accuracy and Mechanical Properties of Chopped Carbon Reinforced Polymers Produced by Material Extrusion Additive Manufacturing. *Materials* **2019**, *12*, 3885. [CrossRef] [PubMed]
25. Dietrich, T.; Schmid, I.; Locher, M.; Addison, O. Extraction Force and Its Determinants for Minimally Invasive Vertical Tooth Extraction. *J. Mech. Behav. Biomed. Mater.* **2020**, *105*, 103711. [CrossRef] [PubMed]
26. Ahel, V.; Ćabov, T.; Špalj, S.; Perić, B.; Jelušić, D.; Dmitrašinović, M. Forces That Fracture Teeth during Extraction with Mandibular Premolar and Maxillary Incisor Forceps. *Br. J. Oral Maxillofac. Surg.* **2015**, *53*, 982–987. [CrossRef]
27. Loy, J.; Tatham, P.; Healey, R.; Cassie, L. Tapper 3D Printing Meets Humanitarian Design Research: Creative Technologies in Remote Regions. In *Creative Technologies for Multidisciplinary Applications*; Connor, A.M., Marks, S., Eds.; IGI Global: Hershey, PA, USA, 2016; pp. 54–75. ISBN 978-1-5225-0016-2.
28. Longhitano, G.A.; Nunes, G.B.; Candido, G.; da Silva, J.V.L. The Role of 3D Printing during COVID-19 Pandemic: A Review. *Prog. Addit. Manuf.* **2021**, *6*, 19–37. [CrossRef]
29. Oladapo, B.I.; Ismail, S.O.; Afolalu, T.D.; Olawade, D.B.; Zahedi, M. Review on 3D Printing: Fight against COVID-19. *Mater. Chem. Phys.* **2021**, *258*, 123943. [CrossRef]
30. Rendeki, S.; Nagy, B.; Bene, M.; Pentek, A.; Toth, L.; Szanto, Z.; Told, R.; Maroti, P. An Overview on Personal Protective Equipment (PPE) Fabricated with Additive Manufacturing Technologies in the Era of COVID-19 Pandemic. *Polymers* **2020**, *12*, 2073. [CrossRef] [PubMed]
31. Culmone, C.; Smit, G.; Breedveld, P. Additive Manufacturing of Medical Instruments: A State-of-the-Art Review. *Addit. Manuf.* **2019**, *27*, 461–473. [CrossRef]
32. Nikitakos, N.; Dagkinis, I.; Papachristos, D.; Georgantis, G.; Kostidi, E. Chapter 6—Economics in 3D printing. In *3D Printing: Applications in Medicine and Surgery*; Tsoulfas, G., Bangeas, P.I., Suri, J.S., Eds.; Elsevier: St. Louis, MO, USA, 2020; pp. 85–95. ISBN 978-0-323-66164-5.

MDPI

St. Alban-Anlage 66

4052 Basel

Switzerland

Tel. +41 61 683 77 34

Fax +41 61 302 89 18

www.mdpi.com

Polymers Editorial Office

E-mail: polymers@mdpi.com

www.mdpi.com/journal/polymers